UNDERSTANDING BIOSTATISTICS

Thomas H. Hassard, PhD

Director, Biostatistics Unit
Professor of Biostatistics
Department of Community Health Sciences
University of Manitoba
Winnipeg, Manitoba

With 50 illustrations

Mosby
Year Book

St. Louis   Baltimore   Boston   Chicago   London   Philadelphia   Sydney   Toronto

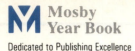
Mosby
Year Book
Dedicated to Publishing Excellence

Editor: Kimberly Kist
Assistant Editor: Penny Rudolph
Project Manager: John A. Rogers
Senior Production Editor: Celeste Clingan
Designer: Susan Lane

Printed in the United States of America

Mosby–Year Book, Inc.
11830 Westline Industrial Drive
St. Louis, Missouri 63146

Library of Congress Cataloging-in-Publication Data

Hassard, Thomas H.
    Understanding biostatistics / Thomas H. Hassard.
       p.   cm.
    Includes index.
    ISBN 0-8016-2078-3
    1. Biometry.   I. Title.
    [DNLM: 1. Biometry—methods.   WA 950 H353u]
    QH323.5.H36   1991
    574′.072—dc20
    DNLM/DLC   90-13708
    for Library of Congress   CIP

CL/CL/DC 9 8 7 6 5 4 3 2 1

To **Kay, Stephen** and **Juliet** with love

Preface

It's an open secret that biostatistics is feared and dreaded by many (perhaps most) health scientists. After all, the subject is complex, far removed from reality, and understandable only if you are a mathematics whiz, isn't it? Like so many of our modern myths, the truth could hardly be more different. The ideas that underlie biostatistics are almost embarrassingly simple, the only mathematical skill required is the ability to carry out simple arithmetic (with the assistance of a calculator or computer to eliminate the drudgery) and the issues that the subject addresses could hardly be more relevant. When called upon to make a decision, we all need to be able to weigh the available evidence in some fair, reasonable, and objective way. When those decisions have a direct impact on the health status and health care of individuals or entire communities, it is absolutely vital that the decision makers have a real understanding of how facts should be gathered and how one goes about evaluating evidence that is usually limited in extent and often apparently contradictory in nature.

The tragedy is that, all too often, health professionals never get the chance to acquire these vital and fascinating skills in any meaningful way. Learning how to mechanically massage numbers in accordance with some mysterious and apparently arbitrary formula is not learning about biostatistics. Being able to carry out a statistical test is, of course, a useful skill. What is vastly more important, however, is acquiring an understanding of the simple ideas that underlie the subject of biostatistics, of how these ideas address the very real problems that confront all investigators and of how they build together to provide tools that offer practical solutions to these problems. Only then can the tools be used in a meaningful and responsible way. My primary motivation in writing this book has been to try and de-mystify this fascinating and very relevant subject to help provide that basis of understanding. Biostatistics is accessible to anyone who has ever had to try and make sense of evidence from the intriguing, and often perplexing, world we live in. Try it and see.

Being able to gather that evidence in a fair and efficient way is, of course, just as important a part of the investigative process as being able to interpret it. Research design and data analysis are intimately related, and it is vital that investigators understand the issues that underlie research design and their practical implications. *Understanding Biostatistics* therefore includes discussions of the principles and the planning processes that are applicable to both experimental and observational research projects.

This book is intended to serve as a text for either a one-semester (Chapters 1 to 7) or two-semester course in biostatistics for graduate students, or senior undergraduate students, in the health sciences. It is also intended to serve as a stand-alone self-teaching resource for researchers working in the health sciences and for practicing health professionals who wish to understand the research process and keep up to date with the scientific literature. I have deliberately made no assumptions that my readers have any particular level of mathematical ability or algebraic skills because such a background is simply not necessary. The only essential prerequisite is an interest in finding out what biostatistics is all about and how it can help you solve some very real problems.

Although *Understanding Biostatistics* intentionally focuses on the health sciences environment, the problems it addresses are directly applicable to the life sciences in general. It is my hope that readers from this broader background will find this book helpful and informative and that they will find the ideas discussed in it just as relevant and useful in their own disciplines.

Acknowledgments

This book and the ideas underlying it have been developed over a number of years of teaching courses in biostatistics in universities in the United Kingdom, New Zealand, and Canada. Over the course of an academic career I have had the pleasure of meeting and working with many individuals who have enriched my appreciation of statistics and of life. Special thanks are due to Reg Parker, Eric Chesseman, and Desmond Merritt at The Queens University of Belfast; Brian Hayman, Dick Brook, Greg Arnold, and Doug Stirling at Massey University, New Zealand; and David Fish, Brian Postl, Bob Tate, and Mary Cheang at The University of Manitoba. I hope that the many other colleagues, too numerous to name, who have been such a pleasure to work with will accept my global thanks. Thanks are also due to Mrs. Ann James, who ably and enthusiastically helped me turn these ideas into readable form.

Finally, and most important of all, I wish to acknowledge my debt to my wife, Kay, and my children, Stephen and Juliet, whose love and unwavering support helped me survive many a lonely night in the basement and who ensured that I never lost track of what is truly important in life.

Thomas H. Hassard

Contents

CHAPTER 1

Describing Data

TYPES OF MEASUREMENT

The basic building blocks of any health research project, no matter how grandiose or humble, are the data—the measurements or values that characterize or describe some important aspect of the individuals under investigation. Appreciating the various forms that data can take (and what we can and cannot read into the data) is an important first step toward understanding how we can collect data in a fair and reasonable way, summarize it in a simple yet informative way, and use it to draw sensible, practical, and realistic conclusions about the problems we are investigating. Individuals can be described or measured in four distinctive ways, and the type of measurement being employed has a direct (and very sensible) effect on the way the data can and should be analyzed and interpreted.

The simplest way to describe someone is to place the individual in one of a number of mutually exclusive categories. For instance, a patient will have a particular blood type (with the possible types being O, A, B, or AB) or will have been born in a particular country (with the possible categories for this measurement being much more extensive). Data of this type is known as a **categorical** or nominal measurement.

The important feature of categorical data is that the categories involved are in no sense better or worse, or bigger or smaller, than one another. They are simply different from one another. Although categorical measurements can involve locating individuals in one of a wide variety of categories (four for blood type, many more for country of origin), one commonly encountered situation involves locating an individual in one of two possible categories. This is often referred to as a **binomial** measurement, because it involves nominating one of two possible categories. A patient's gender is a classic example of this, as is perhaps the ultimate binomial variable, whether an individual is alive or dead. A **variable** is simply any measurement that can take a range of possible values. Binomial measurements are discussed in more detail in Chapter 2.

Because the various categories in a categorical variable are simply different from one another, it makes no sense to try to combine them by, for instance, averaging. The idea of an average gender is of course quite nonsensical. Individuals are either male or female, and there simply is nothing in between the two. A categorical variable can be described very simply and effectively by stating either the number or percentage of individuals falling into each of its categories (for example, 34 of the 50 patients in the study, or 68%, were male). It is worth appreciating that, simply by commenting on the male category, we have also implicitly described the female category (16, or 32%, of the 50 patients have to be female). In general, description or knowledge of one less than the total number of categories will describe a categorical variable completely.

The realization that it is not necessary to know all the facts explicitly, and that any one of the facts—for example, the percentage of either males or females—implies another fact, is a theme we will encounter frequently in this book. In fact we refer to it again in the Degrees of Freedom section in this chapter. The analysis of categorical data is discussed in Chapter 6.

A slightly more sophisticated form of measurement involves placing an individual into one of a number of ordered categories. A patient's condition, for instance, might be described as seriously ill, moderately ill, mildly ill, or not ill. The important characteristic of this type of measurement is that the various categories now have an inherent logical order. (Therefore this type of measurement is referred to as an **ordinal** measurement.) It is clearly better to be moderately ill than to be seriously ill, and better still not to be ill at all. Thus ordinal data can reasonably be used to rank individuals in order. It cannot, however, be used to do more than that. There is no built-in guarantee that the magnitude of change between seriously ill and moderately ill is the same as the change between moderately ill and mildly ill. In other words, the steps up the ordinal scale very well may not be consistent. For that reason, calculating an average level of illness is not appropriate or sensible. Methods of analyzing ordinal data (based on ranking or ordering individuals) are discussed in Chapter 14.

The richest and most sophisticated form of measurement is a process in which individuals are placed on a (usually continuous) scale on which the intervals really do have a consistent interpretation. Measurements of this type are extremely common and occur when we measure an individual's weight, temperature, blood pressure, and so forth. Adding a pound to an individual's weight implies exactly the same amount of weight gain whether the change is from 100 to 101 lbs or from 240 to 241 lbs. Measuring individuals on this type of scale (called an **interval** scale) means that differences between individuals, or between an individual and a reference value, can be described in a consistent and meaningful way. (Indeed, this type of approach is used to help establish the extent of the interindividual differences, or variation, in a study group). This consistency also means that a set of interval measurements can be combined to calculate an average. It is as sensible to talk about an average weight or average age as it is nonsensical to talk about an average gender.

Because interval measurements can display great subtlety, they have attracted a rich body of analysis techniques; most of this book concerns ways of analyzing interval data. All the techniques we discuss are, however, built on a few very simple and very sensible ideas that are established in Chapters 2, 3, and 4. The remainder of this chapter addresses the question of how to reasonably summarize and describe the essence of a collection of interval measurements.

Before leaving the question of types of measurement, we should point out that some interval scales have a (usually unimportant) deficiency in that they lack a true zero point. Classical examples of this are the Celsius and Fahrenheit temperature scales. A temperature of 0° C does not mean a total absence of heat, and a temperature of 40° C is not considered to be twice as hot as a temperature of 20° C. By contrast, a weight of 0 lbs does mean a total absence of weight, and it is reasonable to say that an individual who weighs 240 lbs is twice as heavy as one weighing 120 lbs. A measurement scale with a constant interval and a true zero point is more accurately called a **ratio** scale because it can be legitimately used for "ratio" statements such as "twice as big." However, the distinction between interval and ratio measurements is usually unimportant because both exhibit the key characteristic of a consistent interpretation of measurement differences.

THE PROBLEM OF VARIATION

The principal problem encountered in working with living organisms is their inherent variability. This variability is especially marked in human beings. If, in an investigation of potential hypertension problems in young males who are heavy smokers, we measured the diastolic blood pressures of 56 subjects, we would almost certainly obtain 56 different blood pressure values, ranging perhaps from 60 to 120 mm Hg. This intrinsic variation, even within an apparently very similar group of individuals, makes clinical decisions difficult and makes even describing what we have observed far from straightforward. After a set of data has been collected, the major initial requirement is to obtain a general understanding of what we have found. To paraphrase an old adage, a picture is worth a thousand numbers, and a graphic description of a set of data is often an excellent way to get a feel for its principal characteristics.

One very useful technique is to display the data in a diagrammatic form known as a **histogram.** In a histogram the data is simplified by grouping it into intervals and then displaying it as a series of columns, each column proportional in height to the number of individuals falling into that interval.

With categorical or ordinal data, the groupings will naturally suggest themselves. (When there is no natural ordering to the groupings, such a diagram is usually referred to as a bar chart.) When the data is measured on a continuous scale, artificial groupings can be created by arbitrarily defining intervals on the measurement scale being used (for example, all individuals with a diastolic blood pressure between 60.0 and 64.9 mm Hg). The size of the interval should be chosen so that it gives between 8 and 14 groupings and is convenient to handle. The intervals should be mutually exclusive and ideally should be the same width.

Those 56 diastolic blood pressure results, when grouped into 5 mm intervals, result in the following classification:

INTERVAL	NUMBER OF INDIVIDUALS
60.0-64.9	1
65.0-69.9	1
70.0-74.9	2
75.0-79.9	5
80.0-84.9	9
85.0-89.9	10
90.0-94.9	11

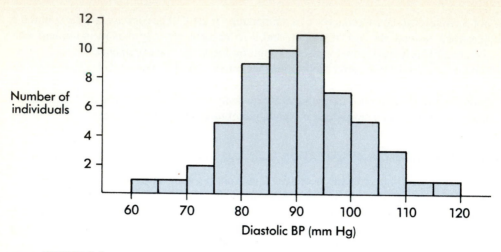

Histogram of diastolic blood pressures (56 young male smokers).

95.0-99.9	7
100.0-104.9	5
105.0-109.9	3
110.0-114.9	1
115.0-119.9	1

When this data is drawn, the resulting histogram (Figure 1-1) shows very clearly most of the major features of this data set. We immediately see, for instance, that most of the individuals have levels between 80 and 100 mm Hg, with the most common value being close to 90 mm Hg, and that relatively few have dramatically high or low levels (over 110 or under 70 mm Hg).

Histograms are very simple devices that convey a lot of relevant information very effectively. However, when making written reports of an investigation, we will usually need to summarize our data more succinctly than even a histogram allows. One or two simple facts can tell an audience, at a seminar or via a journal, the key characteristics exhibited by data sets resulting from the study of hundreds or even thousands of individuals.

MEASURING THE TYPICAL RESULT

The most obvious need when trying to summarize a set of results is to obtain some measure of the typical value or level encountered. The most obvious and most useful measure of this type is obtained by simply calculating the mean or average value. The **mean** is obtained by adding all the values together and dividing by the number of values:

$$\bar{x} = \Sigma x / n$$

A very simple shorthand notation is often used to define any statistical calculations we need to carry out and is well illustrated by the formula for the mean. The x denotes whatever we happen to be interested in (in this case, blood pressure levels); Σ, the summation sign, tells us to add all the values together; n denotes the total number of items that were added together; and $\bar{x}$ denotes the resulting mean value. These are the principal symbols encountered in statistical notation; practically all statistical formulas are built from this simple set of calculation instructions.

The mean is by far the most common measure of the typical value in normal use. However, it can be calculated meaningfully only for interval or ratio data. Even for interval measurements there are some fairly uncommon situations in which the mean will make a nonsensical statement about the typical result. These arise when a few individuals in a study group yield results that are dramatically higher (or less common, dramatically lower) than those of the rest of the study group. Calculating a mean that lumps these few extreme results in with the main body of data can result in a value that is not representative of any individual in the study.

For instance, 10 individuals who have tested HIV positive report the following number of sexual contacts in a 6-month period.

| 2 | 4 | 4 | 6 | 7 | 8 | 10 | 12 | 15 | 93 |

The mean value of 16.1 ($\Sigma x = 161$; $n = 10$) is higher than that reported by 9 of the 10 individuals and yet far below that reported by the tenth individual. It is not, in any real sense, typical or representative of anyone in the study group.

In a situation like this the **median** value may well be more informative. The median value is the value that has 50% of the recorded values above it and 50% below it. To obtain it we rank our observations in order from smallest to largest. If we had a total of 61 individuals, the median value would be the 31st (because 30 individuals would have smaller values and 30 larger). If we had a total of 60 individuals, the median value would be the value halfway between the 30th and 31st values (because, again, 30 values would lie below this and 30 above it).

For our hypothetical HIV results the median value is 7.5 contacts (halfway between the fifth and sixth results of 7 and 8 contacts respectively). This result certainly seems to be a fair summarization of typical sexual behavior in this group of individuals.

The median is much less influenced by extreme values than the mean is because the median discards much of the absolute meaning of a measurement and concentrates on its ranking. (Note that the median value of the HIV results would be quite unchanged whether the tenth individual reported 23, 93, or 193 contacts.) Normally we prefer to use the mean value (provided it is a fair and reasonable description of the typical result) precisely because it does retain all the original measurement information.

One other "typical" measure is sometimes used if the phenomenon we are studying is discrete, that is, it can take only a restricted range of values (such as family size). This is the **mode,** the modal value being the value that occurs most often. When, as is usually the case, the phenomenon can take a continuous range of values (such as blood pressure levels), and all the recorded values are slightly different, the concept of the mode is of little practical use.

MEASURING VARIATION

The results obtained from any group of individuals are inevitably going to show a spread of values. Although we certainly need to be able to make some statement about the typical individual, this, on its own, is clearly not enough. We must also have some indication of the extent of the variation or inconsistency that is present in the data set under study. Do the individuals show a high degree of consistency (and, by implication, all lie near the mean value), or are there very big differences between the individuals under study?

One obvious measure of variation is the difference between the largest and smallest values encountered in the study. This value is known as the **range.** It is very easy to calculate but is rarely used in practice. After all, its calculation uses information from only two individuals in the entire study (and the two most unusual at that). Surely a value that seeks to measure variation within a group of individuals should reasonably be expected to be based on information gathered from all the individuals under study, not a select and unrepresentative two.

A measure of variation that uses information from all the studied individuals can be calculated fairly easily. It is known as the **variance,** and it (or a very simple derivative of it known as the standard deviation) is by far the most widely used measure of variation. The variance is calculated in the following way.

Because the mean, $\bar{x}$, measures the typical value, it seems a sensible and logical choice to serve as a reference point or baseline for the entire data set. We can now easily calculate how much each individual varies from the mean by simply subtracting the mean from each individual value.

$$x - \bar{x}$$

If we did this for each individual value and added them together, we might well expect to have a measure of the overall variation present in the study group.

$$\Sigma(x - \bar{x})$$

We would expect this to be large if the variation is large, and small if most individuals are very similar to the mean value and hence show relatively little variation. Unfortunately, totalling the differences or deviations in this straightforward way tells us nothing at all. Those individuals who have values larger than the mean [$(x - \bar{x})$ positive] will simply cancel out those who have values below the mean [$(x - \bar{x})$ negative] because the mean, by definition, represents a typical or "middle" value. Any data set, highly consistent or highly variable, will therefore result in a zero value when all the deviations are added together.

Fortunately, we can get around the problem of positive and negative variation by simply squaring each difference so that both positive and negative values end up as positive and no longer cancel out one another. Our new measure of the variation in a group of individuals is therefore

$$\Sigma(x - \bar{x})^2$$

This measure often is referred to as the **sum of squares** or SS (Its full, and rather long-winded, title is the *sum of squared deviations from the mean*). The larger the amount of variation present in a set of data, the larger the value of the sum of squares. This method does, however, have some limitations as a universal measure of variation. A study based on 100 results is likely to result in a larger sum of squares than a study based on 10 results simply because it will result in the totalling of far more squared differences. To allow for fair comparisons between studies of different sizes, and to provide a truly universal measure of variation, it seems reasonable to take the study size into account by calculating an "average" variation. This measure of variation, the variance, is calculated as

$$\frac{\Sigma\,(x - \bar{x})^2}{n - 1}$$

You might very reasonably express some surprise that we have used the sample size less one ($n - 1$), rather than n, to "average out" the variation. The reason for this is an important and subtle one that will be discussed in the Degrees of Freedom section. I ask your indulgence until then.

The variance has one major drawback as a truly useful measure of variation. It is based on squared differences (to avoid the differences simply cancelling out one another, as we pointed out earlier), and hence it measures variation in squared units, (for example, in squared millimeters of mercury for a blood pressure study, or in squared pounds for a weight study). It is rather hard to relate to squared units of measurement, to put it mildly. Fortunately, the solution to this problem is extremely simple. Taking the square root of the variance will transform the variation measurements into sane and sensible units once again (lbs, mm Hg, and so forth). This most meaningful and most widely used measure of variability is known as the **standard deviation,** often referred to in statistical shorthand as s. Because the variance is the square of the standard deviation, we will not be at all surprised to see it referred to as s^2.

$$s = \sqrt{\Sigma\,(x - \bar{x})^2/(n - 1)}$$

It seems fairly clear from the way in which the standard deviation was worked out that it measures the "average" amount of variation (as defined by differences or deviations from the mean or typical value) present in a set of data. We will put more flesh on these bones in Chapter 2. However, thoughtful readers may believe that there is an easier way to achieve the same end. Indeed there is, although—sadly—it leads to a disappointing dead end.

The problem of the positive and negative differences that cancel out one another could also be solved, not by squaring them, but by simply ignoring their signs. This is a perfectly respectable mathematical strategy known as taking an absolute value; it is denoted by:

$$|(x - \bar{x})|$$

With no signs to worry about, the deviations could simply be added and averaged, with none of the problems of having to first square them and then find the square root of the final average:

$$\frac{\Sigma|(x - \bar{x})|}{n - 1}$$

Such a measure exists and is known as the **mean deviation.** As the name suggests, it too yields a measure (close to, but not the same as, that offered by the standard deviation) of the "average" variation in a set of results. However, it turns out (as we shall see in Chapter 2) that the standard deviation offers the potential to achieve a very detailed understanding of a data set and the lessons it might have to offer us. By contrast, the mean deviation makes a statement about "average" variation and nothing more. As a result the standard deviation has become (as its name implies) the standard measure of variability, and the simpler mean deviation has become nothing more than a biostatistical footnote.

Degrees of Freedom

If the variance, and its alter ego, the standard deviation, are measures of "average" variation, then why on earth do we divide them by $n - 1$ rather than n, the study size? The reason centers on the use we would like to make of this measure. If all we wanted to do was make a statement about the variability of the group of individuals under study, then dividing by n would suffice. However, we usually want to do something more than this. We might, for instance, want to study blood pressures in young males who are heavy smokers. To do that we will have to collect a small group, or sample, of such individuals and measure their blood pressures. The information that results will, we hope, tell us something about typical blood pressure levels and the variation in blood pressures in all young males who are heavy smokers, and not just the few we have been able to study. In other words, we almost invariably want to use the evidence from sample or study data to learn something about much wider groups or populations of individuals.

The sample average is a fair and reasonable indication of the probable average in the population as a whole. (It would, of course, be unreasonable to expect that the average of a few individuals would be absolutely identical to the average we would get if we could magically measure everyone in the population, a point we will discuss in more detail in Chapter 3.) However, the variance of a sample will tend to underestimate the amount of variation present in a whole population. Why should this be? If we want to learn about variation in the population (as we usually do), then the most logical baseline or reference point to use is the population average or mean. This is great in theory, but it is totally impractical. The only baseline available to us in practice will be that provided by the sample average. The snag is that, as we mentioned earlier, the sample mean is likely to be somewhat different from (although we hope very close to) the true population mean. The result is that the sample sum of squares will be somewhat smaller than it should be.

Why should it? So what if, by necessity, we have to use a sample mean diastolic blood pressure of, say, 90.0 mm Hg when the true population mean value might be 91.0? Some of the deviations $(x - \bar{x})$ on which the sum of squares is built will be 1.0 mm Hg smaller than they should be, but then others will be 1.0 bigger, and it will all balance out; or will it? Unfortunately, it will not. Once the deviations are squared, this balance disappears and the sum of squares based on the (impractical but more appropriate) population baseline will be different from, and always larger than, that based on the (very practical but not really appropriate) sample baseline. To confirm this, consider a sample of two diastolic blood pressures (a rather small sample, admittedly, but useful for illustrative purposes):

	CASE 1 80	MEAN 90	CASE 2 100
deviation	-10		$+10$
squared deviation	100		100

The sum of squares based on the sample mean of 90 is 200 (100 + 100). If we were to magically learn that the true population mean was 91, the sum of squares about this reference point would be 202 (121 + 81):

	CASE 1 80	MEAN 91	CASE 2 100
deviation	-11		$+9$
squared deviation	121		81

Any discrepancy between the sample mean and the true population mean will result in a sample referenced sum of squares that is smaller than it should be. (Try the preceding exercise with a value of 89, or indeed, any value other than 90.) When you allow a sample to define its own reference point (as, of course, you must) and in effect act as both judge and jury, then things look slightly more impressive (less variable) than they actually are.

What do we do about this problem? If the sum of squares is smaller than it should be, then the balance could be restored by dividing it by a value that is smaller than we might expect. It turns out that division by $n - 1$, rather than n, is exactly what is needed. This seems to suggest that one of the results is missing, or at least redundant, and that is, indeed, what has happened. Once you use a data set to create its own reference point (by calculating the sample mean) you can lose or destroy any one of the individual results without compromising the study at all. Again let us consider a hypothetical blood pressure study in which 10 diastolic blood pressures are measured. The sample average is calculated to be 90 mm Hg. Some time later one of the patient records is accidentally destroyed and is lost forever. Has the study been irrevocably damaged? Not in the slightest. Just list the remaining nine blood pressures.

84   112   71   94   85   82   99   102   92

These add up to 821. Because the original 10 results must have totalled 900 (the only way you can divide by 10 and get an average result of 90) the apparently lost result must be 79. Once the data is used to establish the sample mean, only $n - 1$ of the original observations are actually needed to describe the study situation in complete detail. This value, the number of truly independent items of information in a set of data, is usually referred to as the set's **degrees of freedom.**

The degrees of freedom are a reminder that there is a price to be paid for everything in this life. If we collect 10 items of data and use them once to establish the sample mean (as we are forced to), then there are only 9 genuinely meaningful facts left to use, not 10.

Calculating the Variance

We must admit that calculating the variance can seem a little daunting! Calculating it in the way we have discussed involves totalling all the values to find the mean, subtracting each individual value from the mean, squaring each of these deviations, totalling the squared values again, and so forth. Luckily, there is an alternative version of the variance formula that is especially easy to use with a calculator, involves many fewer calculations, and yields exactly the same value:

$$\text{Variance} = \frac{\Sigma x^2 - (\Sigma x)^2/n}{n - 1}$$

All that has happened is that the sum of squares is being calculated in a more convenient (but admittedly much less obvious) way. It is in fact very easy to show that the two sum-of-squares formulas are, indeed, one and the same.

$$\Sigma(x - \bar{x})^2 = \Sigma[x^2 - 2x\bar{x} + (\bar{x})^2]$$

$$= \Sigma x^2 - 2\Sigma x\bar{x} + \Sigma(\bar{x})^2$$

but

$$\Sigma(\bar{x})^2 = n(\bar{x})^2$$

because it is the same number added together *n* times.
Hence

$$\Sigma(x - \bar{x})^2 = \Sigma x^2 - 2 \Sigma x\bar{x} + n(\bar{x})^2$$

but

$$\bar{x} = \frac{\Sigma x}{n}$$

and so

$$\Sigma(x - \bar{x})^2 = \Sigma x^2 - \frac{2 \Sigma x \Sigma x}{n} + \frac{n (\Sigma x)^2}{n^2}$$

$$= \Sigma x^2 - \frac{2 (\Sigma x)^2}{n} + \frac{(\Sigma x)^2}{n}$$

$$= \Sigma x^2 - \frac{(\Sigma x)^2}{n}$$

However, if you feel ill at ease with algebra, do feel free to take my word for it. Better still, take a set of data, calculate the sum of squares using both approaches and confirm that they give the same answer. The experience will also confirm that the alternative version does indeed involve much less work; this is the approach that we will routinely use from now on.

A word of warning—please remember:

$(\Sigma x)^2$ means add all the numbers together (that is, Σx) and square the final answer. Σx^2 means square each individual number and then add these together.

These are two very different procedures that will give very different values. Make sure you know which is which.

To see the variance and standard deviation calculations in action, we will use the set of 10 diastolic blood pressures (complete with the resurrected lost value) that we last encountered in the previous section:

| 84 | 112 | 71 | 94 | 85 | 82 | 99 | 102 | 92 | 79 |

$$n = 10$$

$$\Sigma x = 84 + 112 + \ldots + 79$$

$$= 900$$

$$\bar{x} = 900/10$$

$$= 90.0 \text{ mm Hg}$$

(Incidentally, the median is 88.5, halfway between the fifth largest value of 85 and the sixth largest value of 92.)

$$\Sigma x^2 = 84^2 + 112^2 + \ldots + 79^2$$

$$= 7056 + 12{,}544 + \ldots + 6241$$

$$= 82{,}336$$

$$\text{Variance, } s^2 = \frac{82{,}336 - 900^2/10}{9}$$

$$= \frac{82{,}336 - 81{,}000}{9} = \frac{1336}{9}$$

$$= 148.44 \text{ mm Hg}^2$$

$$\text{Standard deviation, } s = \sqrt{148.44}$$

$$= 12.18 \text{ mm Hg}$$

(Remember that, because the variance and standard deviation are both based on a sum of squares, they must both be positive. Variation can never be less than zero. A negative variance is a sure sign of a calculation error.)

Interpreting a mean value from a common sense perspective is usually easy. Interpreting a standard deviation is slightly less obvious. As we mentioned earlier, it measures the "average" variation in a set of data, effectively a halfway point between those individuals who are very close to the mean value and those individuals who deviate very markedly from the mean. (Our calculated standard deviation of 12.18 is, in fact, almost exactly halfway between the smallest deviation of 2, resulting from the value of 92, and the largest deviation of 22, resulting from the value of 112). This perspective on the standard deviation will be developed in more detail in a section of Chapter 2, Describing a Normal Distribution.

Key Terms

Categorical	**Histogram**	**Standard deviation**
Binomial	**Mean**	**Mean deviation**
Variable	**Median**	**Degrees of freedom**
Ordinal	**Mode**	**Sum of squares**
Interval	**Range**	
Ratio	**Variance**	

PROBLEMS

1. The following results give plasma digoxin levels (in ng/ml) for 16 patients currently receiving a daily digoxin dose of 0.25 mg.

1.65	.60	.75	.95
.85	.55	.65	1.30
.70	.80	.90	.80
.95	1.10	.85	.60

Summarize the major features of this data set by
a. Calculating the mean and the median
b. Calculating the range, variance, and standard deviation

Try calculating the variance using both $\Sigma(x - \bar{x})^2$ and $\Sigma x^2 - \dfrac{(\Sigma x)^2}{n}$ approaches.

2. The serum cholesterol levels (in mg/dl) of 20 patients were measured, and the following results were obtained:

260	210	240	230	260
150	220	190	210	240
160	220	250	180	200
200	210	170	250	220

Summarize the major features of this data set by calculating
a. The mean and the median
b. The range, variance, and standard deviation

3. In a study of plasma digoxin concentrations in 58 patients on a continuous daily digoxin dose of 0.15 mg, the following results (in mg/ml) were obtained:

.43	.78	.88	.96	1.07	1.26
.52	.78	.89	.97	1.07	1.29
.59	.79	.90	.97	1.08	1.36
.61	.80	.91	.98	1.10	
.63	.82	.92	.99	1.12	
.66	.82	.92	1.01	1.13	
.69	.84	.93	1.02	1.15	
.72	.85	.93	1.03	1.17	
.73	.86	.94	1.05	1.17	
.75	.87	.94	1.05	1.19	
.76	.87	.95	1.06	1.22	

Find the mean and standard deviation of these observations and draw a histogram.

4. A national data base on breastfeeding among Indian and Inuit women was established in Canada in 1983. A sample of birth weights (in kg) of 60 children born to Indian mothers has been drawn from this data base and is reproduced below.

3.63	3.79	3.46	2.82	3.36	4.17
3.54	3.57	2.70	3.52	3.42	3.52
3.15	3.44	3.12	3.27	4.14	3.93
3.90	2.58	3.82	3.04	4.02	3.41
4.29	4.23	3.75	4.36	3.89	3.49
4.06	4.76	3.54	3.70	3.60	2.64
2.91	3.09	3.34	3.74	3.46	5.02
3.36	3.65	3.00	3.19	4.50	2.52
3.38	3.23	4.44	2.36	4.09	3.85
2.22	3.97	1.91	4.26	3.50	3.54

Calculate the mean, variance, standard deviation, and median of these results.

CHAPTER 2

Patterns in Data

PATTERNS OF VARIATION

The key dilemma facing every health sciences investigator is the fact that any set of data will inevitably show a great deal of variation among the individual results. In Chapter 1 we developed a reasonable and sensible way of measuring this variation. If this were all we could do, then making any real sense of our data would be very difficult. Luckily for us, most of the data that we encounter in practice demonstrates consistent, simple, and well-understood patterns of variation. This chapter will outline the key properties of some of the most commonly encountered patterns or distributions. Coupling our knowledge of these distributions with the basic descriptive measures introduced in Chapter 1 will take us an important step closer to being able to develop a real understanding of what the information we have collected is trying to tell us.

THE NORMAL DISTRIBUTION

The diastolic blood pressure levels that we looked at in Chapter 1 provide a good example of the way most sets of continuous data seem to behave in practice. Reexamining the histogram depicted in Figure 1-1, we would immediately deduce that the mean value is probably somewhere between 85 and 95 mm Hg, with a range of results from 60 to 120 approximately. We would also note, however, that the vast majority of individuals lie within a much narrower range than this (say, from 75 to 105), and that all these bands are centered fairly symmetrically around the mean value. The fact that this basic pattern recurs time and time again in all sorts of studies from blood pressure levels to body weights is not at all surprising. It is, of course, common sense to expect that most individuals will be reasonably similar to our mythical "typical" individual, and that extreme results will be encountered fairly infrequently. (It is almost a circular argument to reason that the more extreme or unusual a result, the less often we would expect to encounter it.) In addition, we usually have no reason to believe that unusually high results should be any more or less uncommon than unusually low results.

The basic pattern becomes even clearer as the number of results available for study increases and the categories that we use to construct our histograms can be made narrower. The histogram depicted in Figure 2-1 was obtained when diastolic blood pressures were measured for a total of 365 young male heavy smokers and the intervals reduced to 2.5 mm Hg in width.

The basic pattern of this set of results is now extremely clear. The main body of results is concentrated close to the mean value of 86, and as we move away from this typical value (that is, as the results become less and less typical or more and more extreme) the number of results observed falls off quite rapidly. This falloff seems to happen fairly symmetrically irrespective of whether we move to increasingly high levels (up to 115) or to increasingly low levels (down to 55). It is so common to find continuous results, from almost any source, distributed in this basic pattern that they are said, very unoriginally, to follow a **Normal distribution.**

If we could collect larger and larger sets of results and draw histograms with finer and finer divisions, we would find that we would end up eventually with a pattern (Figure 2-2) that follows the classical shape of the Normal distribution.

This is a symmetrical bell-shaped curve that shows all the characteristics that a sensible pattern of results should have, that is, with most individuals lying close to some typical value (which is the most common), and then with the number of individuals falling off quite markedly as we move to more extreme values. Again we would expect this falloff to be symmetrical (unless, for example, there were some inherent physiological reason why, say, extremely high values of a certain blood pressure parameter should be more common than extremely low values).

We cannot, of course, realistically expect the relatively small sets of data that are collected in most investigations to follow this idealized smooth curve exactly, if we were to graph them. However, most of them will almost certainly exhibit the major characteristics of a Normal distribution.

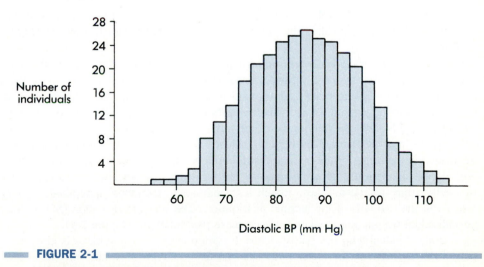

FIGURE 2-1

Histogram of diastolic blood pressure (365 young male smokers).

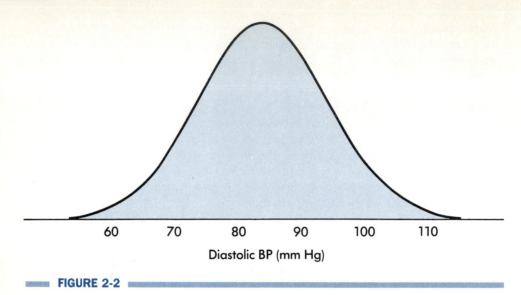

60   70   80   90   100   110

Diastolic BP (mm Hg)

FIGURE 2-2

Normal distribution of diastolic blood pressure in young male smokers.

Describing a Normal Distribution

In Chapter 1 we commented on the usefulness of having a few simple values that would summarize the major features of the data we are studying; first, some measure of the typical result and second, some measure of how variable our results are. In this section we have been developing the idea that most sets of continuous data tend to follow the same basic simple pattern. Logic tells us there must be some connection between the two ideas. What, then, do the mean and standard deviation actually tell us about the Normal distribution?

The mean, of course, is quite obvious. It measures the typical individual who, all other things being equal, is the type of individual we would expect to meet most often. The mean, therefore, measures the peak or center of our Normal distribution (Figure 2-3).

The standard deviation is less immediately obvious. (Remember we use the standard deviation, rather than the variance, because it measures variation in ordinary units and not in exotic squared units.) As we said in Chapter 1 the standard deviation measures the "average" amount of individual variation and hence is a measure of just how spread out the Normal distribution is. Some individuals exhibit values very close to the mean value (and are themselves very typical or unexceptional). These individuals will be located in the center of the Normal distribution. Some individuals, on the other hand, will exhibit dramatically above average or below average values and will deviate very markedly from the mean. These (much less common) exceptional individuals will be located in the two tails of the Normal distribution. The standard deviation (SD) defines an individual who is intermediate between these two situations (Figure 2-4).

An individual who lies one standard deviation away from the mean is almost exactly halfway between the mean value (that is, a "very typical" result) and those individuals who lie a long way from the mean and who are "very non-typical." Again, be-

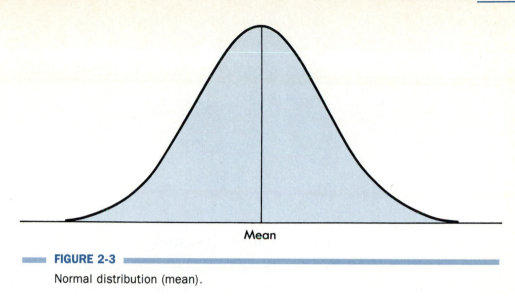

Mean

FIGURE 2-3

Normal distribution (mean).

cause of the symmetry of the Normal distribution, this is equally true for results lying one standard deviation above the mean and results lying one standard deviation below the mean.

The mean and standard deviation are the only two values we need to know to be able to describe a Normal curve completely. They are technically known as the defining parameters of the Normal distribution. The rest of our discussion of the Normal distribution will assume that we know the true values of the two parameters. This is almost

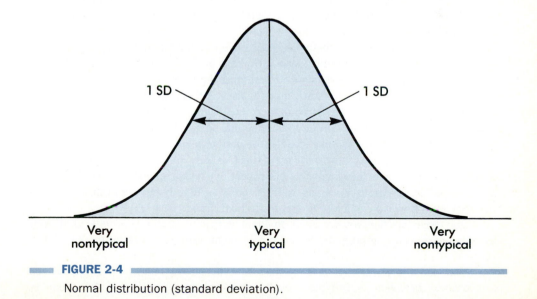

1 SD   1 SD

Very   Very   Very
nontypical   typical   nontypical

FIGURE 2-4

Normal distribution (standard deviation).

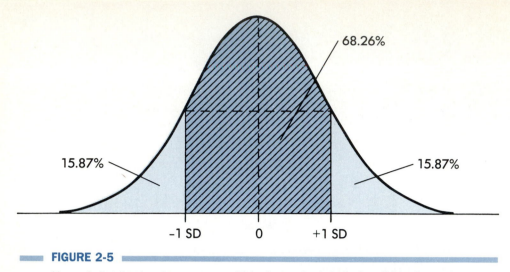

15.87%   68.26%   15.87%

−1 SD   0   +1 SD

FIGURE 2-5

Normal distribution (percentage within 1 standard deviation (SD) of mean).

always quite unrealistic. As we discussed in Chapter 1, we usually have access to only a relatively small set (or sample) of results that we must use to learn about the behavior of all the individuals (the population) with the characteristics we are interested in. This means that the mean and standard deviation of our sample, although we hope they are a very good indication, are not a perfect measure of the mean and standard deviation we would find if we could magically measure everyone in the population. Having to use an "educated guess," rather than the true but unknown values, does have some important practical implications that are discussed in Chapter 3. However, if we are prepared to put this problem on hold for a chapter, we can still learn a very great deal about how we can use our knowledge of the Normal distribution to start to answer questions about our research results.

Because the Normal distribution is such a well-defined and simple shape, a great deal is known about it. For example, we know that 68.26% (roughly two thirds) of all individuals will lie within one standard deviation of the mean value (Figure 2-5).

Most, although certainly not all, individuals will therefore lie within this "reasonably typical" band. Because the curve is, very considerably, symmetrical, we can also say that 15.87% of all results will lie more than one standard deviation above the mean, and 15.87% will lie more than one standard deviation below the mean.

In real life, of course, we will not expect exactly 68.26% of our results to lie within this band. This is the way we would expect our results to behave on average. For an individual set of results we might well get 65% or 72% lying within this band. We hope it would, however, be close to this theoretical figure.

One other fact that you are likely to encounter quite often is that 95% of all individuals (that is, nearly everybody) lies within 1.96 standard deviations of the mean (Figure 2-6).

Note that this ties in with our presentation of the standard deviation as measuring a point halfway between a very typical result and a very nontypical result. (If 1 standard

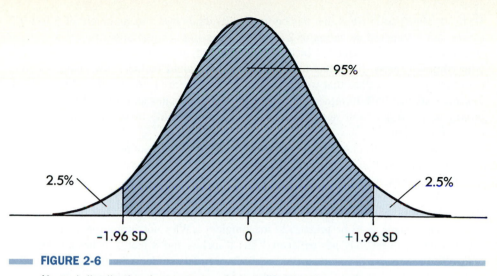

95%

2.5%

2.5%

-1.96 SD   0   +1.96 SD

FIGURE 2-6

Normal distribution (percentage within 1.96 SDs of mean).

deviation is halfway between the very typical and the very nontypical, then 2 standard deviations should reach out to encompass the very nontypical results. Rather than attempting to memorize 1.96, you may prefer to remember that nearly everybody lies within approximately 2 standard deviations from the mean.

We cannot, of course, realistically expect this band of ±2 standard deviations about the mean to include absolutely everybody. Even if it includes the very typical, reasonably typical and the very nontypical, there will always, by the nature of things, be a few very, very nontypical results lurking in any set of data. Regarding the most extreme 5% of our results as being very, very nontypical seems a fairly reasonable attitude.

The "vital statistics" of the Normal distribution are given in most statistical textbooks. For convenience, the major figures are reproduced in Table 2-1.

To check exactly how we use these figures, consider the following hypothetical study of the heights of male medical students. Suppose that a population survey (unre-

TABLE 2-1
The Normal Distribution

Number of standard deviations from mean	Results lying outside this (%)
1	31.74
1.64	10
1.96	5
2.58	1
3.29	0.1

alistic, I admit) established that this body of individuals had a mean height of 5 feet 9.2 inches and a standard deviation of 2.9 inches. This implies that about two thirds of the students lie between 5 feet 6.3 inches and 6 feet 0.1 inch, which seems in accordance with common sense. If we had to pick a band that would include most male students, we might pick the values that would include 95% of them ($1.96 \times 2.9 = 5.68$ inches); this suggests that 95% have heights between 5 feet 3.5 inches and 6 feet 2.9 inches (approximately). Again, this seems reasonable; heights outside these limits are certainly rare, but do occur.

If we wished to cover just about every possibility, we could use the fact that less than one individual in a thousand lies more than 3.29 standard deviations from the mean. In our present example, this means that practically everyone lies within a band of 9.54 inches (3.29×2.9) either side of the mean, that is, between 5 feet 0 inches and 6 feet 7 inches (approximately). This is certainly difficult to dispute, but statements like this are so general as to be practically meaningless. (Why not forget the statistics, just say that nearly everybody lies between 4 feet 0 inches and 8 feet 0 inches and be done with it?) The statement that most (that is, 95% of) male students lie between 5 feet 3.5 inches and 6 feet 3 inches (approximately) seems to be a reasonably happy balance between being comprehensive and still being meaningful.

Believable and Unbelievable Results

Trying to decide whether a certain statement or apparent result is true is one of the central problems we all face from time to time. The problem is that we rarely have total or irrefutable evidence one way or the other. We have to decide which conclusion is more believable or probable and make our decisions accordingly. (We will discuss this in the context of asking and answering scientific questions in Chapter 4.)

If someone accused of armed robbery states that he was at home alone at the time of the robbery, then it is very difficult (indeed, impossible) to prove conclusively that he is lying. If he has six previous convictions for armed robbery and was identified by two witnesses, then it would, however, seem reasonable to conclude that his alibi was very improbable and to reject it. Life is a process of looking at the evidence we are given and deciding what the most reasonable conclusion is. Luckily, our knowledge of the Normal distribution makes it very easy to decide whether our results are believable or unbelievable (certainly much easier than it is in the rest of our activities!).

If, in our height survey, we found a student with a recorded height of 7 feet 5 inches, we might consider this to be a little unlikely. It is, in fact, very easy to work out just how unlikely it is. The height of 7 feet 5 inches is 19.8 inches above the mean value of 5 feet 9.2 inches, and we know that the standard deviation for our students' heights is 2.9 inches. In other words, our tall friend lies 6.8 (19.8/2.9) standard deviations away from the mean. We already know that less than one person in a thousand lies more than 3.29 SDs from the mean; 6.8 SDs is, of course, far more extreme than even this. We could reasonably conclude that the chance of a student as tall as this being registered at our local university is so remote as to be practically unbelievable. It would seem sensible to decide that the result must have been a mistake and to consider some other possibility, such as that the recorder actually meant to write down 5 feet 7 inches.

Most real life situations are, of course, considerably less obvious. In these cases, being able to decide just how uncommon a result is becomes even more valuable. Our study group has a mean blood pressure level of 86 mm Hg with a standard deviation of 14 mm Hg. Does an individual with a blood pressure level of 70 have a particularly unusual level? To find out, we simply work out how many standard deviations away from the mean it lies:

Deviation from mean in mm Hg,

$$86 - 70 = 16.00$$

Deviation from mean in SD

$$16/14 = 1.14$$

Because it lies just over 1 SD away from the mean, we know that roughly one third of all individuals tested will have more unusual values; the result is, therefore, not at all remarkable.

Expressing our results in terms of standard deviations (known as "**standardizing**" the results and often denoted by the letter *z*) achieves two purposes. First, it enables us to check just how unusual a result is. Second, because the calculation is expressed as so many standard deviations, it removes the complications of differing units of measurement, such as inches, mg/100 ml, or kilograms. Irrespective of whether we are studying the use of Lidocaine in the prevention of primary ventricular fibrillation or the role of social factors in high-risk pregnancies, just how unusual a result is can be checked by doing one simple division and comparing the calculated number of standard deviations with the tabulated values.

To date we have encountered some key facts about the Normal distribution, for example, the fact that only 5% of individuals have results that lie more than 1.96 SDs from the mean, and only 1% have results that lie more than 2.58 SDs from the mean. What can we say when we carry out a calculation and get a value of, say, 2.12 SDs? We refer to the next lowest tabulated figure. Because 5% of individuals have results that lie more than 1.96 SDs from the mean, then less than 5% will have results that lie more than 2.12 SDs from the mean.

The notation used in scientific journals is slightly different from what we have used so far. We have expressed how unusual a result is by giving the percentage of individuals who we would expect to have results even less typical. In most research publications, this is expressed as a proportion of 1, that is, 5% is written as .05 and 1% as .01. The chance of something more unusual happening, when expressed as a proportion of 1, is known as a **probability.** The result we discussed in the last paragraph (that is, 2.12 SDs) would normally be written as $p < .05$. This means that the probability of seeing a more unusual result is less than .05, that is, less than 5 in 100 individuals, or 5%, would have a more unusual result.

Because the Normal distribution is so well known and so comprehensively documented, highly detailed tables are available to describe its properties, if required (see table T/1 in the appendix). Knowing only the mean and standard deviation of a particular population, we can use such a table to make very detailed statements about the way we would expect this population to behave (assuming, of course, that it does follow a Normal distribution).

Tables of the Normal distribution usually operate by telling us what proportion of the population would lie between the mean and a specified number of standard deviations above it. (Because the Normal distribution is symmetrical, this also, of course, gives us the same information about the corresponding number of standard deviations below the mean.)

The questions we might wish to answer would be of the form (using our male medical student example), "What proportion of students in general would we expect to be over 5 feet 10 inches?", "What proportion would we expect to be between 5 feet 10 inches and 6 feet 1 inches?", and so forth.

The first step, as you would expect, is to convert the question into its standardized form (that is, state it in terms of standard deviations):

A height of 5 ft 10 in is (5 ft 10 in − 5 ft 9.2 in)/2.9 in standard deviations above the mean:

$$= .8/2.9$$
$$= .28 \text{ standard deviations above the mean}$$

The Normal distribution table tells us that 11.3% of individuals lie between the mean and 0.28 SD above the mean (Figure 2-7).

Hence, because 50% of the population is above the mean, we know that 38.97% of students will have heights over 5 feet 10 inches (that is, over +.28 SD). Alternatively, we could say that the chance or probability of a student having a height over 5 feet 10 inches is .3897.

A height of 6 ft 1 in is (6 ft 1 in − 5 ft 9.2 in)/2.9 in standard deviations above the mean:

$$= 3.8/2.9$$
$$= 1.31 \text{ standard deviations above the mean.}$$

Our Normal table (Table T/1 in the appendix) tells us that 40.49% of all students will lie between the mean and +1.31 SD (Figure 2-8). Using this and our knowledge about the lower limit (that is, that 11.03% of all students lie between the mean and 5 feet 10 inches), we can readily deduce that 29.46% of all male students will lie between 5 feet 10 inches and 6 feet 1 inch (that is, the probability of a student being within these limits is .2946).

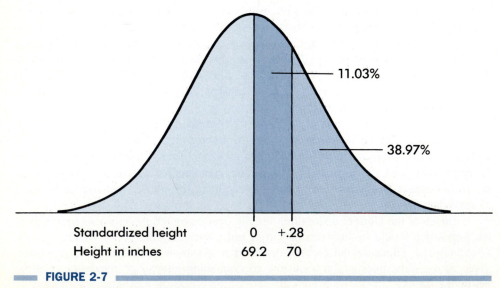

	0	+.28
Standardized height		
Height in inches	69.2	70

FIGURE 2-7

Normal distribution (heights of male students).

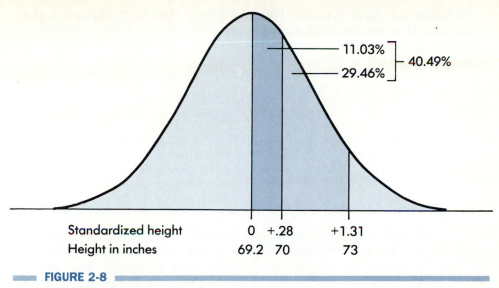

| Standardized height | 0 | +.28 | +1.31 |
| Height in inches | 69.2 | 70 | 73 |

FIGURE 2-8

Normal distribution (heights of male students).

Probability in Discrete Situations

In the previous section we explored the use of the Normal distribution to describe the pattern of results we might expect to find when investigating phenomena that are measured on a continuous scale, such as height or blood pressure. With measurements of this type we saw that it was a relatively simple job (provided we have access to a set of Normal tables) to discover how likely we would be to observe, say, an individual with a height greater than 5 feet 10 inches or a height between 5 feet 10 inches and 6 feet 2 inches. What we cannot do, of course, is ask how likely it is that an individual will have a height of 5 feet 7 inches. With a continuous measurement scale such a question is meaningless because the chance of anyone having a height of exactly 5 feet 7 inches (to the nearest millionth of an inch) is remote in the extreme.

The other major type of variable that occurs in the health sciences is the **discrete variable,** which can take only a limited range of specific and mutually exclusive values. Such variables very commonly take two alternative values, that is, a baby is male or female, a patient survives an operation or dies, and so forth. Many of the most crucial aspects of health investigation can be phrased in terms of alternative outcomes, most dramatically, life or death.

When the outcome of a particular situation is limited to a range of discrete possibilities, it becomes meaningful to ask how likely it is that one of these outcomes will be observed in practice. Sometimes common sense tells us what these probabilities should be. In the classic example of tossing a coin, we can say that the two outcomes of heads or tails are both equally likely (that is, probability of each = .5). Similarly, the principles of Mendelian genetics suggest to us that a parent carrying a recessive gene can either transmit it or not transmit it, with equal probability, to an offspring.

More frequently, however, the chance of something happening or not happening has to be estimated or assessed on the basis of whatever evidence is available to us. A patient suffering from a particular type of cancer will either survive for at least a 6-

month period after diagnosis or die during the 6 months. Here we would have to appeal to previous studies and other information to assess the chances of a patient surviving. This might indicate, for instance, that only one in every five such patients survives. We, therefore, say that

$$p \text{ (probability of survival)} = .2$$

$$q \text{ (probability of death)} = .8 = 1 - p$$

The letter p is usually used to denote the chances of the event of interest occurring, and q is used to denote the probability of the alternative event. (The total of p and q must, of course, be 1, because one outcome or the other must occur.)

PROBABILITY OF MULTIPLE EVENTS

We are rarely interested in single events, such as, "Will one particular patient survive?" We are much more frequently interested in what might happen to a group of patients we are studying. For example, "Of the 20 patients in a particular cancer program, what is the chance of four or more of them surviving at least 6 months past diagnosis?" To be able to answer questions like this we must be aware of the basic rules for determining just how likely multiple events are.

There are two basic rules to remember. The first is as follows:

1. The probability of independent events is multiplicative.

Independent events are simply events that do not influence one another. In most situations we might reasonably assume that patients will behave independently of one another and that the chance of one patient surviving will not be influenced by the fate of another patient.

Suppose there are two patients, Mr. Smith and Mr. Jones, in the cancer follow-up study. The chance of Mr. Smith and Mr. Jones both surviving is given by

$$.2 \times .2 = .04$$

Multiplying the probabilities effectively means that the overall probability will be substantially reduced, that is, that the chance of both events happening together will be considerably less than the chance of any one of the individual events happening. This multiplication of probabilities is very reasonable. Unfortunately, the outlook for an individual patient in this situation is very poor, and the prospect for both patients surviving the 6 months must, of course, be very, very poor indeed.

Note that the appearance of the word "and" in the description, "Mr. Smith and Mr. Jones survive," is a good indication that the probabilities should be multiplied.

The second basic rule is as follows:

2. The probability of alternative outcomes is additive.

This means that, in this situation, the overall probability will be increased. The wider the number of options we allow ourselves, the easier it must be to achieve some one of them. We can easily illustrate this property. The probability that Mr. Smith survives or Mr. Smith dies is given by

$$.2 + .8 = 1$$

as, of course, it must because this covers all the options open to him.

The use of the word *or* is a good indication that probabilities should be added.

These ideas have an important and not immediately obvious consequence. What, for instance, is the chance of one patient surviving out of our two-patient group?

$$\text{Because } p \text{ (survive)} = .2$$

$$\text{and } q \text{ (die)} = .8$$

we might be tempted to say that the chance of one surviving and one dying is:

$$.2 \times .8 = .16$$

This would, in fact, be incorrect. What we have forgotten is that there are two quite different ways in which one patient can survive. If the two patients involved are Mr. Smith and Mr. Jones, then the survivor could be either Mr. Smith or Mr. Jones. Do appreciate that these two outcomes are very different from one another, a fact that would be very apparent to you if you were either Mr. Smith or Mr. Jones. With this in mind, let's examine the chances of seeing one survivor, using our knowledge of the probability of multiple events.

Mr. Smith survives and Mr. Jones dies:

$$.2 \times .8 = .16$$

or   +

Mr. Smith dies and Mr. Jones survives:

$$.8 \times .2 = \underline{.16}$$

One of the two patients survives:

$$= .32$$

In other words, the probability of one patient out of the two surviving is given by

$$2\,pq$$

when p is probability of survival

$$= .2$$

and q is probability of death

$$= .8 = 1 - p$$

We can confirm that .32, rather than .16, is the correct description by examining the probabilities associated with all the possible outcomes of this particular two-patient study.

probability (both patients survive)

$$p^2 = .2 \times .2 = .04$$

probability (one patient survives)

$$2pq = 2 \times .2 \times .8 = .32$$

probability (no patient survives)

$$q^2 = .8 \times .8 = \underline{.64}$$
$$\overline{1.00}$$

The three possible outcomes do indeed total to a probability of 1.0, as they must.

BINOMIAL DISTRIBUTION

The situation we have just looked at is an example of a set of results that can be described by a pattern known as the **Binomial distribution.**

In a binomial situation we have a set of n independent events (possibly n patients in a study group), each of which can have two (hence the *bi* prefix) possible outcomes (for example, survival or death) with probabilities p (of survival) and q (of death, $1 - p$) respectively. The Binomial distribution describes the pattern of results we might expect to see in such a situation. It describes this pattern by showing us just how likely (that is, how probable) each possible overall outcome of our study is, that is:

What is the probability of no patient surviving?
What is the probability of one patient surviving?

. . .

What is the probability of all n patients surviving?

(The probability of these alternative outcomes must, of course, total to one, because they cover the range of all possibilities.)

In the previous section we saw how to calculate the probabilities of the three possible outcomes associated with a two-patient group (no survivors, one survivor, or two survivors). To enable us to deal easily with a study group of any size, we need to extend these simple ideas and develop a common strategy for working out the probability of any of these possible outcomes (for example, the probability, say, of two patients surviving and $n - 2$ dying).

We know that, because the patients are all separate independent entities, the desired probability is basically obtained by multiplying the individual patient outcome probabilities together:

$$\underbrace{p \times p}_{2} \times \underbrace{q \times q \times \ldots \times q}_{n-2} = p^2 \, q^{n-2}$$

However, we also recall that two patients can survive in a variety of alternative ways, that is:

Smith and Jones are the survivors.
Smith and Black are the survivors, and so forth.

To get the correct overall probability, we must add together the probabilities corresponding to each of these possible situations (that is, two patients survive, and $n - 2$ die). More conveniently, (and quite equivalently), we can simply multiply this probability by the number of ways in which two patients can survive, that is:

Smith and Jones survive or Smith and Black survive, and so forth

$$= p^2 \, q^{n-2} + p^2 \, q^{n-2} + \ldots + p^2 \, q^{n-2}$$

$$= \text{number of ways in which two patients can survive} \times p^2 \, q^{n-2}$$

It was obvious from our earlier, very small-scale example that there were two ways in which one patient could survive from a two-patient study group (Smith survives or Jones survives). However, once the scale of the problem becomes even a little bigger, the number of ways in which certain events can happen becomes far from obvious. Luckily, there is a convenient and easy-to-use formula that we can utilize to work out the number of possible ways in which certain events can occur.

The number of ways in which x individuals can survive out of a total of n individuals is given by

$$\frac{n!}{x!(n-x)!}$$

The symbol "!" in this context is read "factorial" and means "multiply together the number n and all numbers smaller than it in decreasing steps of 1."

That is,

$$n! = n \times (n-1) \times (n-2) \times \ldots \times 1$$

(Note that 0! takes the value 1.)

Luckily, this actually results in a much less complicated formula than you might fear, because the numbers on the top line and bottom line of the formula largely cancel out one another.

For example, two individuals can survive out of a total study group of six in:

$$\frac{6!}{2!4!} \text{ ways}$$

$$= \frac{6 \times 5 \times 4 \times 3 \times 2 \times 1}{2 \times 1 \times 4 \times 3 \times 2 \times 1}$$

$$= \frac{6 \times 5}{2} = 15 \text{ different ways}$$

To get a feeling for why this formula does, in fact, measure the number of ways in which x individuals can survive out of a total study size of n, think about trying to solve the "two patients out of six" problem by hand. Purely for convenience, let's call our patients A, B, C, D, E, and F. We could try writing down all the possible pairs on a systematic basis, for example:

PATIENTS

A and B	BA	CA	DA	EA	FA
AC	BC	CB	DB	EB	FB
AD	BD	CD	DC	EC	FC
AE	BE	CE	DE	ED	FD
AF	BF	CF	DF	EF	FE

This is, a total of 6×5 or 30 possible pairs. However, a moment's reflection will confirm that AB is exactly the same as BA. (It really doesn't matter if we introduce the survivors as "Mr. Smith and Mr. Jones" or " Mr. Jones and Mr. Smith.") In fact, each possible pair has been counted twice (AC, CA, and so forth), and the actual number of truly distinctive pairs is

$$\frac{6 \times 5}{2} = 15$$

The formula simply mimics the logical process that we would instinctively employ to answer this question ourselves.

The probability of any two patients surviving and four dying is, therefore:

$$15\,p^2\,q^4$$

If p is .2 and q is .8, then this gives us a final overall probability of

$$15 \times .2^2 \times .8^4$$

$$= 15 \times .04 \times .4096$$

$$= .246$$

The possible outcomes of a binomial data set (that is, the Binomial distribution) are in general given by the following relationship.

$$\text{Probability } (x \text{ individuals surviving out of a total of } n) = \frac{n!}{x!(n-x)!}\,p^x\,q^{n-x}$$

where

$$p = \text{Probability of survival}$$

$$q = \text{Probability of death} = 1 - p$$

For an example, of a group of six patients followed over 6 months with a probability of survival of each being .2, the probabilities of the various possible outcomes we might see are

$$\text{Probability that none survives, } p\,(0) = \frac{6!}{0!6!}\,(.2)^0\,(.8)^6$$

$$= .8^6 = .262$$

$$\text{Probability that one survives, } p\,(1) = \frac{6!}{1!5!}\,(.2)^1\,(.8)^5$$

$$= 6 \times .2 \times .3277 = .393$$

$$p\,(2) = \frac{6!}{2!4!}\,(.2)^2\,(.8)^4$$

$$= 15 \times .04 \times .4096 = .246$$

$$p\,(3) = \frac{6!}{3!3!}\,(.2)^3\,(.8)^3$$

$$= 20 \times .008 \times .512 = .082$$

$$p\,(4) = \frac{6!}{4!2!}\,(.2)^4\,(.8)^2$$

$$= 15 \times .0016 \times .64 = .015$$

$$p\,(5) = \frac{6!}{5!1!}\,(.2)^5\,(.8)^1$$

$$= 6 \times .00032 \times .8 = .002$$

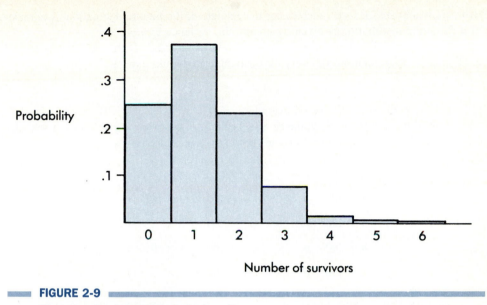

FIGURE 2-9

Binomial distribution; $n = 6$, $p = .2$.

$$p\,(6) = \frac{6!}{6!0!}\,(.2)^6\,(.8)^0$$

$$= 1 \times .000064 = .000 \text{ (approximately)}$$

Just as the Normal distribution illustrates how likely we are to see the range of results possible in a continuous situation, so the Binomial distribution illustrates the relative likelihood of the limited number of outcomes (seven in this case) that we can see in the binomial situation (Figure 2-9).

Although we can ask specific questions in the binomial situation (such as, what are the chances of two patients surviving?), we can also ask questions of a more general nature, similar to those we asked in the continuous situation (such as, what are the chances of three or more patients surviving?) by adding the relevant outcomes.

$$
\begin{aligned}
\text{Probability of 3 surviving} &= .082 \\
4 \quad &= .015 \\
5 \quad &= .002 \\
6 \quad &= \underline{.000} \\
&\ .099
\end{aligned}
$$

that is, less than a 1 in 10 possibility.

Just like the Normal distribution, the key characteristics of the Binomial distribution (what is the typical or average outcome we would expect to see if the study were repeated a number of times, and how wide a spread of outcomes might we expect to see?) are given by its mean and standard deviation (or equivalently, variance). These two parameters are, in fact, much easier to calculate in the binomial situation. The mean of a Binomial distribution is given by np and the variance by npq.

These are both quite logical when you think about them. If the chance of survival is .2 (one in five), then out of every 100 patients, we would expect:

$$100 \times .2 = 20$$

(one in five) patients to survive on average.

Similarly, note that the variance will be largest when $p = q = .5$, and will decrease as we move toward very low (or very high) survival rates.

$$n = 100 \quad p = .5 \quad npq = 100 \times .5 \times .5 = 25$$

$$n = 100 \quad p = .2 \quad npq = 100 \times .2 \times .8 = 16$$

$$n = 100 \quad p = .8 \quad npq = 100 \times .8 \times .2 = 16$$

$$n = 100 \quad p = .1 \quad npq = 100 \times .1 \times .9 = 9$$

$$n = 100 \quad p = .9 \quad npq = 100 \times .9 \times .1 = 9$$

At very low (or very high) survival rates, common sense tells us that practically all (or practically none) of the patients will die. The range of outcomes that will occur in practice will, therefore, be very limited, and the variance must, therefore, be correspondingly small. When survival is a 50/50 prospect, however, what will happen to each individual patient can be predicted with much less certainty. The range of outcomes that could occur in practice will be at its widest, and the variance will correspondingly be at its maximum.

The Binomial distribution looks and is quite different from the Normal distribution. It defines a discrete series of steps rather than a continuous smooth curve, and if p is not .5 it will not be symmetrical. However, there are situations in which it can look enough like the Normal distribution to allow us to draw on our knowledge of that distribution (for example, that 95% of our results should fall within approximately 2 standard deviations of the mean). To look like a Normal pattern, n, the number of individuals in the study group, must be fairly large. This means that the Binomial probability plot will have a large number of steps, with relatively small changes from step to step, and will start to approximate a smooth curve. In addition, the value of p should not be too far away from .5, in order to ensure that the Binomial pattern is reasonably symmetrical (this becomes less important as n becomes larger and larger). These two ideas come together in the following rule of thumb.

If both np and nq are greater than 5, then the Binomial distribution can be approximated by a Normal distribution with

$$\text{Mean} = np$$

and

$$\text{Standard deviation} = \sqrt{(npq)}$$

To see this idea in action, first reexamine Figure 2-9. This illustrates a Binomial distribution with $n = 6$ and $p = .2$. This does not satisfy our rule of thumb ($np = 1.2$)

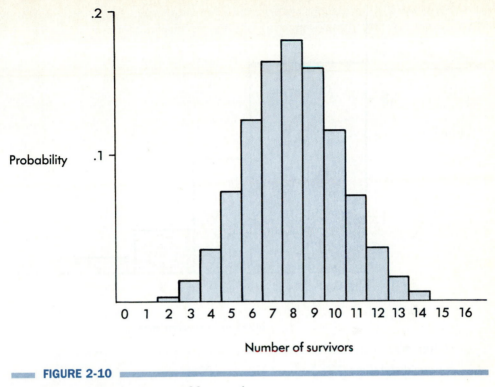

FIGURE 2-10

Binomial distribution; $n = 20$, $p = .4$.

and looks nothing like a Normal distribution. Now look at Figure 2-10, which illustrates a Binomial distribution with $n = 20$ and $p = .4$. This does satisfy our guidelines ($np = 8$, $nq = 12$), and it does indeed seem to approximate the classic Normal curve.

It is, of course, just an approximation, but it can provide a very quick and easy way to make some key statements about the anticipated behavior of a group of patients. For the $n = 20$, $p = .4$ situation depicted in Figure 2-10 we can say, for example, that

$$\text{Mean number of survivors} = 20 \times .4 = 8.0$$

$$\text{Standard deviation} = \sqrt{(20 \times .4 \times .6)} = 2.19$$

Because it behaves approximately like a normal distribution we can further say that, if a large number of such groups were studied, 95% of the time we would expect to see the number of survivors falling within 1.96 standard deviations of the mean, that is, somewhere between 3.71 [8.0 − (1.96 × 2.19)] and 12.29 [8.0 + (1.96 × 2.19)] survivors.

If we were running such a study (following 20 patients with an anticipated chance of survival of .4) as a one off proposition, as we usually would be, this is a reminder that, although we might expect to end up seeing 8 survivors, the actual number could be as low as 4 or as high as 12.

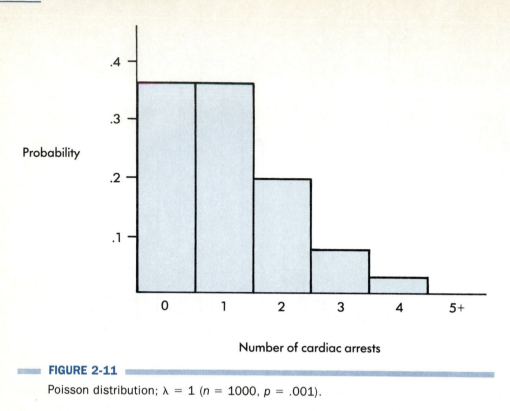

FIGURE 2-11

Poisson distribution; $\lambda = 1$ ($n = 1000$, $p = .001$).

POISSON DISTRIBUTION

Fortunately, most disease is relatively uncommon, at least in the sense that often only a small proportion of individuals in a given population will have a particular condition. The yearly death rate from heart attacks in males 30-44 years old will, for instance, be of the order of one per 2,000 such individuals. Hence, the probability of an individual in this age range dying from a heart attack in a given year is $p = .0005$.

We frequently wish to study phenomena of this type. This is, of course, still a binomial situation because it has only two distinct outcomes: An individual either dies of a heart attack or he does not. It has, however, two distinguishing features.

First, n, the number of individuals involved, is very large (usually of the order of thousands).

Second, p, the probability of the event of interest occurring is very, very small (for example, .0005).

We have already discovered from bitter experience that, even with very small numbers (that is, $n = 6$), it can be very tedious to describe the pattern of results that we might expect to see by calculating the various terms of the Binomial distribution. With a very large n and hence, a very large number of possible outcomes, most of which are totally implausible (for example, the possibility of 500 out of 1000 40-year-old males dying of heart attacks is simply unbelievable), use of the Binomial distribution as an investigative tool is both totally impractical and basically nonsensical.

Luckily, as numbers become very large and the event of interest becomes very rare, the pattern of results we would expect to see takes a form that is very easy to describe. This form is known as the **Poisson distribution** (named after G.S. Poisson who first described it.)

The Poisson distribution is a special case of the Binomial distribution that occurs when *n*, the study size, is very large and *p*, the chance of the outcome of interest, is very small.

Recall that the Binomial distribution has

$$\text{Mean} = np$$

$$\text{and variance} = npq$$

If *p* is very, very small, then $q = (1 - p)$ will, for all practical purposes $= 1$. Hence, in this situation:

$$\text{Variance} = npq = np = \text{mean}$$

The fact that the mean = the variance is uniquely characteristic of the Poisson distribution. Unlike the Normal distribution, it, therefore, needs only one parameter to define and describe it completely. This parameter is conventionally referred to as **lambda (λ).**

$$\lambda = np = \text{mean} = \text{variance}$$

We saw that the shape of the Binomial distribution (in other words, the chances of the various possible outcomes actually happening) was obtained by working out a series of terms, one corresponding to each possible outcome. Luckily, when *n* is very large and *p* very small (the Poisson situation) these Binomial terms (which would be a nightmare to work out) can be replaced by quite equivalent Poisson terms that are, in fact, extremely easy to work out.

OUTCOME	BINOMIAL SITUATION	POISSON SITUATION
		(*n* very large, *p* very small)
0 occurrences	q^n	$e^{-\lambda}$
1 occurrence	$n \; p \, q^{n-1}$	$\lambda e^{-\lambda}$
2 occurrences	$\dfrac{n!}{2!(n-2)!} \; p^2 \, q^{n-2}$	$\dfrac{\lambda^2}{2!} e^{-\lambda}$
3 occurrences	$\dfrac{n!}{3!(n-3)!} \; p^3 \, q^{n-3}$	$\dfrac{\lambda^3}{3!} e^{-\lambda}$
. .	. .	. .
. .	. .	. .
. .	. .	. .

The Poisson terms seem totally different from the Binomial terms. They are, however, absolutely identical (just a great deal simpler). For those of you who need convincing, the next few lines demonstrate the way G.S. Poisson went about changing those daunting Binomial terms, with a huge *n* and a tiny *p*, into their much simpler Poisson equivalents, which all depend on just one medium-size value, λ. For those of you who are allergic to algebra, please feel free to skip the next few paragraphs and take my word for it.

The chance of seeing no occurrences in the entire study group is q^n (where q is, say, the probability of not dying from a heart attack). This is equivalent to

$$(1 - p)^n$$

or

$$(1 - \lambda/n)^n$$

because

$$\lambda = np$$

Poisson recognized this expression as being a mathematical expansion that can be shown to be equivalent to e raised to the power of $-\lambda$, provided that n is very large. (The symbol, e, denotes the number 2.7183, which is the base of the natural logarithms).

The $e^{-\lambda}$ term, which lies at the heart of all the Poisson terms, can be calculated by using your calculator to take the antilogarithm of $-\lambda$, or using it to raise 2.7183 to the power of $-\lambda$.

Once this first term is established, the other terms fall into place very easily.

The chance of seeing one occurrence of the outcome is

$$n \, p \, q^{n-1}$$

which effectively equals

$$n \, p \, q^n$$

(because n is so very large that a change of one is virtually meaningless)

$$= \lambda \, e^{-\lambda}$$

since

$$q^n = e^{-\lambda}$$

and

$$np = \lambda$$

and so on.

(G.S. Poisson's arguments all depend on the fact that n, the study size, must be very large. If this is not true, all this crafty simplification falls flat on its face and the Poisson terms cannot be used. What constitutes very large is, of course, a matter of opinion but generally n should be at least several hundred for the Poisson approach to be justified. In addition p must be very small, ideally well below .01.)

To see the Poisson distribution in action, suppose you are following a group of 1000 individuals. If previous studies have indicated that the risk of cardiac arrest for this group is one per 1000, then it is a simple matter to calculate the likelihood of the various results you might obtain (Figure 2-11).

$$\text{Probability of cardiac arrest } (p) = .001$$

$$\text{Number of individuals in study } (n) = 1000$$

$$\lambda \text{ (mean)} = np = 1.0$$

Probability of no arrests,

$$p(0) = e^{-\lambda} \quad = e^{-1.0} \qquad\qquad = .368$$

Probability of 1 arrest,

$$p(1) = \lambda \, e^{-\lambda} = \lambda \, p(0) = 1.0 \times .368 = .368$$

Probability of 2 arrests,

$$p(2) = \frac{\lambda^2}{2!} \, e^{-\lambda} = \frac{\lambda}{2} p(1) = \frac{1.0}{2} \times .368 = .184$$

Probability of 3 arrests

$$p(3) = \frac{\lambda^3}{3!} \, e^{-\lambda} = \frac{\lambda}{3} p(2) = \frac{1.0}{3} \times .184 = .061$$

Probability of 4 arrests,

$$p(4) = \frac{\lambda^4}{4!} \, e^{-\lambda} = \frac{\lambda}{4} p(3) = \frac{1.0}{4} \times .061 = .015$$

Note how each probability can be systematically worked out from the previous probability, making the calculation of Poisson probabilities even easier than expected. Once the probabilities get very small (that is, it is clearly very unlikely that we would see 5 or 6 cardiac arrests, much less 15 or 16), then the most meaningful approach is to lump all other outcomes under "5 or more cardiac arrests" and obtain the probability for this by simply totalling the 0, 1, 2, 3, and 4 probabilities and subtracting from 1.

$$p \text{ (5 or more)} = 1 - [p(0) + p(1) + p(2) + p(3) + p(4)]$$

$$= 1 - (.368 + .368 + .184 + .061 + .015)$$

$$= 1 - .996$$

$$= .004$$

In the Poisson situation we are, essentially, involved in counting the occurrence of some phenomenon of interest (the number of cardiac arrests and so forth). In general, the Poisson distribution can be applied to most situations in which the event of interest can be counted (such as number of bacterial colonies on agar plates), provided that the events are relatively rare and occur independently of one another.

There might appear to be little connection (other than the basic act of counting) between counting the number of bacterial colonies on an agar plate and counting the number of heart attacks in 10,000 young men. The philosophical similarities run much closer than you might think. We can imagine the agar plate subdivided into a very large number of very small subdivisions. Each of these subdivisions may prove to be the site of a bacterial colony or it may not. Since the number of potential subdivisions is very large, and the number of colonies on the plate is very small, the chance of any one of the huge number of potential subdivisions actually ending up hosting a colony is actually very remote. The situation does indeed fulfill the basic requirement for a Poisson distribution. This argument can be developed for most counting situations, and it explains the role of the Poisson distribution as the pattern most likely to be followed by study outcomes that involve counting the occurrence of a particular phenomenon.

In the discussion of the Binomial distribution we commented that it can look like a Normal distribution under certain situations. Since the Poisson distribution is just a special case of the Binomial distribution, the same is true of it and the same rule of thumb holds.

If np is greater than 5, then the Poisson distribution can be approximated by a Normal distribution with

$$\text{Mean} = np$$

$$\text{Standard deviation} = \sqrt{(np)}$$

(Since q is effectively 1, there is no need to check nq because it is bound to be very large).

Suppose, for instance, we observe an agar plate with 6 bacterial colonies growing on it. We will never know what n or p are in this situation (since the plate subdivisions exist only as an intellectual exercise), but the available evidence suggests that the mean number of colonies is 6 (a fairly easy calculation!) and that the standard deviation is $\sqrt{6} = 2.45$. Using the Normal approximation to the Poisson, we can say that we would expect 95% of our future plates to yield between 1.2 $[6 - (1.96 \times 2.45)]$ and 10.8 $[6 + (1.96 \times 2.45)]$ colonies, an effective range of 1 to 11 colonies.

Key Terms

Normal distribution	**Binomial distribution**
Standardizing	**Poisson distribution**
Probability	**Lambda (λ)**
Discrete variable	

PROBLEMS

1. a. Reexamine the histogram for the digoxin data presented in problem 3 in Chapter 1. Does it demonstrate the key characteristics of a Normal distribution?
   b. How many observations lie more than 1 standard deviation from the mean? How does this compare with what you expect to find with a Normal distribution?
   c. Within what approximate limits would you expect most of your data to lie (assuming that "most of it" means 95% of it)? In practice, what percentage of your data actually falls outside these limits?

2. Draw a histogram of the birth-weight data presented in problem 4 in Chapter 1. What type of pattern do these results exhibit? What proportion of the birth-weights actually falls within 2 standard deviations of the mean? Is this what you might intuitively expect?

3. In an industrial city in northern England, birth weights were recorded for all children born over a 5-year period. For the 18,977 white infants born in this period, the mean birth weight was 3061 grams with a standard deviation of 381 grams. Assuming that these results follow a Normal distribution, what proportion of the birth weights would you expect to see
   a. Under 2500 grams?
   b. Between 2750 and 3500 grams?
   c. Between 3250 and 4000 grams?

4. A population of potentially hypertensive women has a mean diastolic blood pressure of 100 mm Hg with a standard deviation of 14. Assuming that diastolic blood pressure follows a Normal distribution, what proportion of these women would you expect to have diastolic blood pressures:
   a. Between 96 and 104?
   b. Below 80?
   c. Above 96?
   d. Above 125?
   e. Between 104 and 125?

5. With advances in neonatal intensive care, the prognosis for children born with exomphalos has improved. Current evidence suggests a survival rate (defined as the proportion of such children surviving for 10 years) of .69. Use this figure to describe the likelihood of the various survival outcomes (that is, no survivors, one survivor, and so forth) that face a group of eight exomphalos babies in 10 years' time. How likely is it that at least six of the children will survive?

6. Skin cancer has an incidence in northern North America of 24 cases per 100,000 individuals per year. Use this figure to describe the likelihood of the various possible numbers of skin cancer cases that might present themselves in a community of 12,200 over the course of a year. How likely is it that fewer than three cases will occur?

7. In a particular community 10% of the population is colorblind. You intend to screen a group of eight students for colorblindness. Assuming that the rate of colorblindness among students is the same as in the population at large:
   a. Work out the likelihood of the various possible results you might find (no students colorblind, one student colorblind, and so forth).
   b. What are the chances of finding three or more colorblind students?

8. The available evidence suggests that the risk of a child being born with spina bifida is .56/1000 births. You are involved in reviewing the 6200 births reported in a Midwestern city during a particular year, in order to determine the number of spina bifida cases reported. Determine the likelihood of the various possible results you might obtain.

CHAPTER 3

Estimation

THE NEED TO STUDY SAMPLES

Investigators carrying out a health research project usually have one of two objectives in mind. They may, for example, be interested in assessing typical systolic blood pressure levels in a certain group of individuals whom they believe may be potentially hypertensive. An investigation of this possibility would clearly call for individuals from the group of interest, normally called the **population,** to have their blood pressures measured to provide some indication of blood pressure levels within the population. Since it is highly unlikely that the investigator would have the facilities to screen every member of the population of interest, he or she would, by necessity, have to use the limited number of results that can be obtained, the **sample,** to measure or "estimate" the probable average systolic blood pressure levels in the total population. Using the results from a relatively small sample of individuals as a measure or indication of the levels (usually the mean level) in a much larger population of individuals is known as **estimation.**

The other classical research objective involves the comparison of the levels of a certain phenomenon in two or more populations. For instance a researcher might wonder whether the mean systolic blood pressure of elderly male smokers differed from that of elderly male nonsmokers. This comparative approach to research is known as **hypothesis testing,** and the ideas underlying it will be explained in Chapter 4. Again only a small number of individuals are likely to be available for study from each of the populations being compared, and the researcher will, by necessity, have to draw his or her conclusions from very limited evidence. An important part of that process involves using the sample evidence to indicate or estimate the most likely mean systolic blood pressures in the smoking and nonsmoking populations. Estimation and hypothesis testing are, in fact, very intimately related, and the ideas developed in this chapter are central to practically all of biostatistics.

Irrespective of the phenomenon in which we are interested—blood pressure levels in expectant mothers, malnutrition in socially deprived groups, or hearing loss in patients following inner ear surgery—one sad but unavoidable fact has to be faced. The total population of individuals that we are interested in studying will inevitably be far larger than the resources we will have available to study it. The population of interest may be of the order of hundreds of thousands, or even millions. Frequently, the population of interest is so general that it simply does not exist in a conventional sense, such as the population of all individuals who will require open heart surgery in the next decade. The practicalities of research life, therefore, dictate that almost every health research project be carried out on a sample (usually a relatively small one) drawn from the population of interest. (We will explore the dos and don'ts of selecting a sample in Chapter 15). This inevitably poses some major problems.

We have already seen in Chapters 1 and 2 that wide variations can be expected among the results from different individuals. We might therefore expect to find similar wide variations between different samples of individuals. Can we really expect to draw realistic conclusions from just one such sample, consisting of a relatively small number of individuals? If we collected a fresh sample, would we find a completely different result? Just how much reliability can we expect from a single set of results? Luckily the situation is not nearly as bad as we might fear. Sample means are much less variable, and correspondingly more reliable, than the individual results on which they are based. It is also very easy to work out just how variable they are. These two facts make sample means potentially very informative.

We noted in Chapter 2 that the pattern of results described by the Normal distribution is by far the most frequently encountered in health research. In fact, the discussion of Binomial and Poisson distributions points to the fact that even results that are technically quite dissimilar to the Normal pattern can frequently be quite acceptably approximated by the Normal distribution. As a result of this, we can use our basic knowledge of how the Normal distribution behaves and how it can be described, to tackle the problem of how much we can really read into a set of sample data.

We know that the Normal distribution is described by two parameters that measure or describe the most typical individual and the between-individuals variation. These parameters are often referred to as **mu (μ)** and **sigma squared (σ^2)** respectively. Strictly speaking, these refer to the parameter values that describe the properties of the population that interests us. These population parameters are, of course, unknown and, as we have just commented, in real life it is never practical to measure them since we would have to measure every individual in the population. What we do know are $\bar{x}$ and s^2, the mean and variance of our sample results. Technically these are known as sample estimators of the unknown population parameters. That is exactly what they are — readily available values that we can calculate from our sample results and that will give us a good indication (or estimate) of what those unknown population parameters (that we would ideally like to know) might be. We are usually interested in estimating the population mean. The key question that we still have to consider is, "Just how good an indication of this is the sample mean?"

HOW RELIABLE IS A SAMPLE MEAN?

We have already seen that any set of results will show quite a lot of internal variation. We can rarely do anything to control this variation directly, but we can at least measure and describe it, using a value like the standard deviation. When we measure a sample of subjects we are, however, usually not very interested in their individual values as such. We will use the individual values to calculate the mean value for that sam-

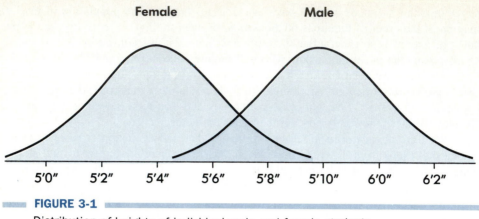

Female **Male**

5'0" 5'2" 5'4" 5'6" 5'8" 5'10" 6'0" 6'2"

FIGURE 3-1

Distribution of heights of individual male and female students.

ple, with the hope that it will tell us something about the typical value we might expect to find if only we could measure all the individuals who make up the particular population we are investigating.

The mean values of sets of results vary considerably less than the individual results on which they are based. In other words, if we collected a second sample of individuals from our potentially hypertensive population, we would expect the mean blood pressure level for this sample to be quite similar to the mean level given by the first sample (although we would not, of course, expect the two to be identical). We would certainly expect the two mean values to be much more similar than the values we would get if we just selected two individuals to be measured.

The reason that mean values are much less variable than individual values is extremely obvious once you think about it. A mean value is an average of a number of individual values. Even if one or two quite extreme individual values were to occur in a sample, the effect of these would be smoothed out to a large extent by the process of averaging them with the other values in the samples. Thus means show considerably less fluctuation from sample to sample than we would expect to find among individual values in a single sample. This makes drawing sensible conclusions about samples a much more practical proposition than we might have anticipated.

Supposing, to take a fairly trivial example, we wished to compare the heights of male and female medical students. If we simply selected one male and one female student and compared their heights, we would probably anticipate that the male student would be taller than the female student. However, we would not be all that surprised if the female student turned out to be the taller of the two. Both male and female heights show considerable variation, and the distributions for the two genders overlap substantially, even though the mean height for males is about 6 inches higher than that for females (see Figure 3-1).

Any attempt to draw general conclusions from samples of one is, of course, patently stupid because of their inherent variability and, in consequence, unreliability. If, however, we selected a sample of, say, 16 male students and 16 female students, we would be very surprised indeed if the mean female height turned out to be greater than

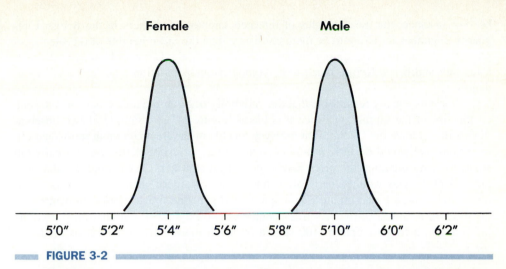

Female   **Male**

5'0"   5'2"   5'4"   5'6"   5'8"   5'10"   6'0"   6'2"

FIGURE 3-2

Distribution of mean heights of repeated samples of 16 male and 16 female students.

the mean male height. The presence of a few unusually tall females or unusually short males in the sample would have a minimal effect on the respective sample means since their influence would be balanced out by the much more typical results that would make up the vast majority of sample values. Although the individual students in the study would be every bit as variable as they always were, their sample means would be considerably more reliable (precisely because the dangers posed by variability would have been greatly reduced) and there would be a much greater chance of drawing sensible conclusions from the sample data.

If we were to repeat such a study a number of times, collecting fresh samples of male and female students, we would of course find that the average male and female heights would differ from one sample to another. However, because of the smoothing out effect of the averaging process that we have just discussed, this sample-to-sample variation would be greatly reduced, and the distributions of the male and female samples would scarcely overlap at all. (Compare Figure 3-2, which illustrates the pattern of variation we would expect to see if we collected many samples with 16 individuals, with Figure 3-1, which illustrates variation in individual heights.) In other words, virtually any such sample of male and female students would indicate that males are indeed taller than females.

THE STANDARD ERROR

We have just seen that mean values are much less variable than individual values, and that this reduction in variability is a direct consequence of averaging out all the individual values in the sample. The question is, "Just how much less variable is a mean value, and what does the variability depend on?" Obviously, the variability of a mean depends on the variability of the individual observations on which it is based—the more variable the individuals are, the more variable the means drawn from them are likely to be. For instance, the mean heights of repeated samples of soldiers in the British Life-guards Regiment are likely to be more consistent than the mean heights of repeated samples of medical students, simply because the soldiers in the Lifeguards must all be over 6 feet tall and hence form a much more homogeneous population than the medical students.

It seems equally reasonable that the variability of a sample mean must also depend on the size of the sample on which it is based (Figure 3-3). A sample of two observations will "average out" some, (but not very much!) of the fluctuations in individual observations and would do little to minimize the impact of encountering an unusually tall student when estimating the mean height of medical students. On the other hand a sample of 100 students would clearly offer much more protection against unusual observations, and a mean based on such a sample would generate much more confidence than the mean of a sample of two. (A fresh sample of two might differ dramatically from the first sample; a fresh sample of 100 should be much more consistent with its predecessor.)

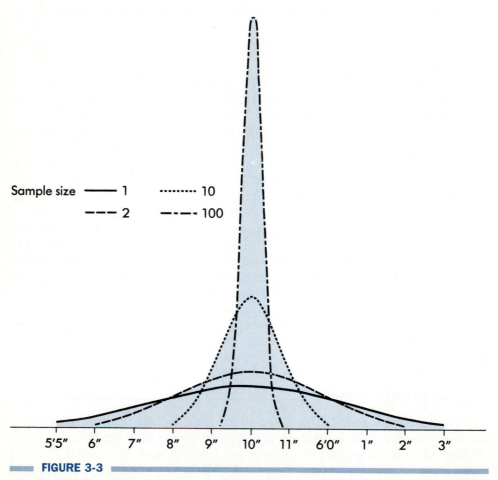

FIGURE 3-3

Distribution of mean heights of repeated samples (various sample sizes) of male students.

To summarize, the variability of sample means increases as the variability of the individual observations (which we can measure by their standard deviation) increases, and decreases as the sample size increases.

As we know, the term "standard deviation" is used to describe the variability of a group of individual observations. The term **standard error** is used to describe the variability of a sample mean calculated from these observations. The relationship between the two is given by

$$\text{Standard error (SE)} = \frac{\text{Standard deviation (SD)}}{\sqrt{(\text{Sample size, } n)}}$$

This relationship is very much in line with our conclusions of two paragraphs ago. The only item that might cause some surprise is the fact that it is the square root of the sample size that is involved and not the sample size itself. If, however, we remember that the standard deviation (the most practical measure of variation) is itself the square root of the variance (the basic measure of variation), then it is perhaps not surprising that a square root is involved on the bottom line as well. Like most formulae, this relationship makes some very obvious points in a concise, but possibly not very obvious, way. All other things being equal, it is more informative to study large samples than small ones, and if the individuals you are studying show wide variations, then you will need to study a pretty large sample of them if you are to have a chance of reaching some meaningful conclusions (an issue that will be addressed again in Chapter 9).

A RANGE OF CONFIDENCE

When we collect a sample of, say, potentially hypertensive individuals and calculate its mean, we are using this mean as an indication of the typical blood pressure level in the population from which the sample was drawn. The mean of the sample should be a very good indication of this typical value, but it would be unrealistic to claim that it must be exactly right. After all, if we were to collect a fresh sample the following day, we would, of course, find a different mean value. We would certainly expect it to be very similar to the first sample mean value (and by implication, very similar to the unknown mean value for the population), but we could not reasonably expect it to be identical.

It follows that when we are trying to estimate the mean level for a particular population, it is not sufficient or, indeed, honest merely to give the mean value of our sample. After all, other samples would have yielded other values, all equally plausible. What we really need to do is to give a range of values within which we would expect to find all the possible sample means (or more realistically, nearly all of them). The implication is that if most of the possible sample means lie within these values, then the true but unknown group or population mean should, in all probability, lie between them as well. This is a much more honest approach than pretending that our sample mean must have given the true mean for the whole population.

The problem is very similar to one we encountered before in Chapter 2 in our discussion of the Normal distribution. We saw that most (that is, 95% of) individuals lie within approximately 2 standard deviations of the mean, provided that the sample does indeed follow a Normal distribution. Since we are now studying means rather than individuals, it seems equally reasonable to say that most (that is, 95% of) sample means

44 *Understanding Biostatistics*

will lie within approximately 2 standard errors of the mean value. (Note the change from standard deviation to standard error, since we are now considering means rather than individual observations.)

Once we collect a set of results, we can calculate its standard deviation in the way outlined in Chapter 1. The standard error can then be very easily calculated from this simply by dividing the standard deviation by the square root of the number of observations involved. This, in effect, gives a measure of how variable future sample means would be if we had the time and money to collect more samples. When we know the standard error, it becomes a simple matter to work out the values that lie 2 standard errors above and 2 standard errors below our sample mean. We are now fairly certain that most (that is, 95% of) all possible samples would give mean values within this range and we are, therefore, fairly certain that the true but unknown value that really interests us lies within these limits as well. (If repeated sampling kept producing mean heights between 5 feet 9 inches and 5 feet 11 inches, surely we could hardly believe that the population mean height could be, say, 5 feet 4 inches. It seems far more reasonable that it should be somewhere between 5 feet 9 inches and 5 feet 11 inches.)

"Fairly certain" is, of course, a rather vague comment. We can be a little more precise. Since we know that 95% of all possible samples would probably give values within these limits, it seems reasonable to say that we are 95% certain (or 95% confident) that the true mean lies between them. We cannot, of course, be completely (that is, 100%) confident that this is true. We might, after all, have collected a very strange and nontypical sample, but this is very unlikely. It is possible to achieve a higher degree of confidence by calculating the range within which we would expect, say, 99% of all future sample means to fall (approximately 2⅔ standard errors above and below the sample mean). The advantage is that we are now more confident (99% confident) that the population mean is within these limits. The disadvantage is that inevitably, the limits are wider and more vague. Limits based on 95% confidence (known as 95% **confidence limits**) offer a widely accepted and reasonable compromise between confidence and precision.

The proposition that 95% of future sample means would be expected to fall within a band of (approximately) 2 standard errors either side of the existing sample mean only makes sense if the sample means would, in fact, follow the familiar Normal distribution. If the individual observations follow a Normal distribution it seems (and is) quite reasonable to assume that the means of samples involving these observations will also follow a Normal distribution. It turns out that the "smoothing out" effect of averaging not only reduces the variability of the mean, it also quite literally "smooths out" the pattern or distribution that would be followed by the means resulting from repeated sampling. (This phenomenon is a consequence of a statistical property known as the Central Limit Theorem.) In other words, even if the individual observations do not follow a Normal distribution, the means of fresh samples will effectively follow a Normal distribution anyway, provided that the original distribution is not grossly abnormal and the sample size is reasonably large (as always, the bigger the sample size, the more it will smooth things out).

This means that the concept of confidence limits, and the closely related ideas of hypothesis testing that occupy much of this book, can be used if your data follows a Normal distribution and also if it does not, but comes reasonably close to it.

INTRODUCING THE t DISTRIBUTION

The statement that a band of plus or minus two standard errors should encompass 95% of all future sample means is, as always, an approximation to the truth. Finding the correct value is not quite as obvious as it might seem. (Shouldn't he really have said 1.96 standard errors?) To calculate confidence limits we call upon the sample mean and the sample standard deviation (which we use to calculate the standard error). The sample mean provides a fair and reasonable (more formally, an "unbiased") indication of the mean of the whole population but, since it is only a sample, it is not the definitive word on the subject. This, of course, is exactly why we calculate confidence limits. Similarly, the sample standard deviation provides a fair and reasonable indication of the true population standard deviation. Like the sample mean, however, it is not the definitive word on the subject since a new sample would inevitably result in a slightly different standard deviation and consequently a slightly different standard error.

This presents a real problem. To calculate confidence limits about a sample mean, because we believe, quite correctly, we should take it with a pinch of salt, we are forced by sheer necessity to use a sample standard deviation that we should also take with a pinch of salt. This classic vicious circle posed a major dilemma for early researchers. Many tried, very unconvincingly, simply to ignore the problem and pretend that the sample standard deviation did indeed measure the population standard deviation perfectly (a reasonable argument if you can measure everyone in the population or, at the very least, study a huge sample; a very flimsy argument if, like nearly all of us, you have to work with relatively small samples). Research took a gigantic step forward when William Gosset came up with a solution to this problem in 1908.

Gosset, a statistician with the Guinness Brewery in Dublin (surely a statistician's dream job!) had a brilliantly simple idea. Since the standard deviation (and therefore the standard error) we are forced to use is only an estimate and not to be taken totally seriously, why not make some allowance for this and be extra cautious when making statements such as, "95% of all future sample means would lie within 1.96 standard errors"? Gosset calculated the extent of allowance that one should make in order to cope with this problem and published his results (essentially amendments to the familiar Normal distribution) as the *t* **distribution** (sometimes called the Student's *t* distribution). To its eternal shame, Guinness would not let him publish his results under his own name, and Gosset, a very modest genius, published them under the pen name "Student." Why call it the *t* distribution? No one knows. Perhaps it was a tip of the hat to Britain's favorite nonalcoholic beverage by someone slightly peeved by attitudes in the brewing industry.

The size of the required allowance depends on the size of the sample involved. The larger the sample, the more faith we can put in its standard deviation and the less cautious we need to be. For instance, for a sample size of 15, Gosset's research (summarized in tables known as *t* tables) says that 95% of future sample means would probably lie within 2.14 (rather than 1.96) standard errors. If the sample size is 30, the *t* table indicates that 95% of future means would lie within 2.04 standard errors, and, if the sample size is 60, 95% of future means would be within 2.00 standard errors (very close to the Normal distribution figure of 1.96 and some justification for our use of the phrase, "approximately 2" in previous chapters). In other words, the allowance that must be made for not really knowing the population standard deviation is usually small and only really noticeable if the sample size is very small.

████ **TABLE 3-1** ██

The *t* distribution

	df	0.10	0.05	0.01	0.001
	10	1.81	2.23	3.17	4.59
	20	1.72	2.09	2.84	3.85
	40	1.68	2.02	2.71	3.55
	60	1.67	2.00	2.66	3.46
(Normal distribution)	∞	1.65	1.96	2.58	3.29

Since practically all research involves the study of samples and an acceptance that the evidence we have is less than perfect, *t* tables are extensively used by researchers in the health sciences, and it is important that you feel at ease with them. To illustrate their use (practically all statistical tables are laid out in the same basic style) an abbreviated section of the table is given in Table 3-1 (a fuller version of the table is given in table T/2 of the Appendix).

The horizontal axis denotes the proportion of values that lie more than a specified number of standard errors away from the mean. The vertical axis serves as an index of the sample size. More accurately, it measures the degrees of freedom (df) present in the study. Where fitting confidence limits to a sample mean, this is the sample size − 1 (that is, $n − 1$) because, as we discussed in Chapter 1, we have already been forced to use the sample mean once just to calculate the sample standard deviation. If we were, for instance, fitting 95% confidence limits to the mean of a sample of 21 results, we would look along the 20 degrees of freedom row and note that we would expect only .05 (expressed as a proportion of 1) or 5% (expressed as a percentage) of the means of any future samples we might collect to fall more than 2.09 standard errors away from the mean of the sample we have collected. We are therefore 95% confident that the true population mean should fall within a band of 2.09 standard errors either side of the sample mean.

Inevitably *t* tables are less detailed than tables describing the Normal distribution. After all, they have to describe a whole host of situations ("just how much allowance should we make if the sample size is 10, 25, and so forth?"). The result is that even very complete *t* tables tabulate only the most frequently used values (.05 outside, .95 or 95% inside; .01 outside, .99 or 99% inside, and so forth) and some selected degrees of freedom. Once the sample size gets sufficiently large, small changes to the degrees of freedom make very little difference to the extent of the allowance that must be made. If therefore the *t* tables do not have a degrees of freedom row that corresponds exactly to the sample size under study, simply use the row that comes closest.

The final row in a *t* table details the values that would be appropriate if the sample were infinite (or, more realistically, very large indeed). In this (usually totally implausible) situation, the population standard deviation would be known precisely, and there would be no need for any extra caution. These values are, of course, simply the values from the Normal distribution.

___ CALCULATING CONFIDENCE LIMITS ___

The following short example should clarify the few simple calculations involved in finding confidence limits. Suppose that blood pressure readings have been taken from 16 subjects and their diastolic blood pressure levels recorded. From the 16 individual results, it has been calculated that the mean value is 90.0 mm Hg, and the standard deviation is 14 mm Hg.

$$\text{Standard error} = \frac{\text{Standard deviation}}{\sqrt{n}}$$

$$= \frac{14}{\sqrt{16}} = \frac{14}{4}$$

$$= 3.50$$

An examination of a table of the t distribution shows that 95% of all future sample means should lie within 2.13 standard errors of the mean value.

$$\text{One standard error} = 3.50 \text{ mm Hg}$$

$$2.13 \text{ SEs} = 2.13 \times 3.50$$

$$= 7.46 \text{ mm Hg}$$

Therefore we are 95% certain that the true mean lies between

$$90.0 + 7.46 = 97.46$$

and

$$90.0 - 7.46 = 82.54$$

The 95% confidence limits are therefore 82.545 to 97.455 mm Hg.
This is often written as

$$90.0 \pm 7.46 \text{ mm Hg}$$

Other confidence limits can easily be calculated if required. From the tables of the t distribution for a sample size of 16 (15 df) we see that 99% of all possible means fall within 2.95 SEs.

$$2.95 \times 3.50 = 10.32 \text{ mm Hg}$$

$$99\% \text{ confidence limits} = 90.0 \pm 10.32$$

$$= 79.68 \text{ to } 100.32 \text{ mm Hg}$$

Note, as we insist on greater and greater confidence, our limits become wider and inevitably, more vague.

As we saw in Chapter 2, in certain situations the Binomial and Poisson distributions behave very much like the Normal distribution. When these conditions are met (when np and nq are both greater than 5), we can use the ideas we have discussed in this chapter to fit confidence limits to both Binomial and Poisson parameters.

In a typical Binomial situation, for instance, the individuals exhibit one of two possible characteristics or give one of two possible responses to a question or test. The proportion of the sample responding in a certain way is an estimate of the true population proportion and really needs confidence limits fitted to it to make it an honest and truly meaningful comment on the wider population.

From Chapter 2 we know that, if n individuals are sampled and the chance of a positive response is p, then the mean number of positive responses will be np with a standard deviation of $\sqrt{npq}$.

Expressing these results as a proportion of the sample size, n, we have that

$$\text{Mean proportion of positive responses} = np/n = p$$

$$\text{With a standard deviation} = \frac{\sqrt{npq}}{n} = \sqrt{\frac{pq}{n}}$$

The standard deviation of a proportion is, in fact, also its standard error. Recall that a standard error is a statement about the variability of a group measure such as a mean. A proportion is inherently a group measure (it simply does not exist as an individual concept), and hence the measure of its variability qualifies as a standard error.

Confidence limits of 95% for a proportion are given by

$$p \pm 1.96 \sqrt{\frac{pq}{n}}$$

Note that we use 1.96 rather than the corresponding t value. Using these ideas in the Binomial (and Poisson) context can only be justified if we feel satisfied that the data is effectively behaving "normally". If this is really true, we can simply go ahead and use the .05 Normal value of 1.96. If it isn't true, we really shouldn't be doing this at all.

Let's illustrate this with a brief example. As a result of a survey of 340 school children, 41 of them were diagnosed as being asthmatic. The estimated proportion of asthmatics in the school age population is therefore

$$\frac{41}{340} = 0.121$$

with 95% confidence limits of

$$0.121 \pm 1.96 \sqrt{\frac{.121 \times .879}{340}} = .086 \text{ to } .156$$

COMPARING A SAMPLE WITH A KNOWN STANDARD

The principles of confidence limits can be easily utilized if we are required to compare the mean of a set of sample data with some prespecified or standard value. In an investigation of this kind we are seeking to answer the question, "Does the evidence provided by the sample results suggest that the true mean of the population from which the sample is drawn differs from the standard value with which it is being compared?".

We have seen that we cannot reasonably claim that the true mean of the population must be exactly the same as the mean of a sample drawn from the population. However, we can state limits that we are confident (usually 95% confident) will enclose the true population mean. If the standard value lies outside these limits, then it seems reasonable to conclude that the population mean does, in fact, differ from the standard value. More precisely, we could say that we are 95% confident that the population mean differs from the standard value. (If the standard value also fell outside the 99% confidence limits, then we could further say that we were 99% confident that a genuine difference existed between the test material and the standard.)

If, on the other hand, the standard value fell within the 95% confidence limits, then it is quite possible that the true mean of the population and the standard value are indeed the same. There is certainly no strong evidence to suggest that they differ. In this situation we would, therefore, have to conclude that no significant or substantial differences existed between the population mean and the standard value.

In a study of obstetrical problems in an urban inner-city area we might wish, for example, to explore the possibility that young, socially deprived mothers tend to produce children with nonstandard birth weights, when we accept a figure of 3.20 kg, as obtained from a national birthweight census, as a birthweight standard. The following results represent the birth weights of 12 children of mothers fitting into this category. Does this sample indicate that there is a tendency for such mothers to have children with birth weights that differ from the standard value?

BIRTH WEIGHT (KG)

x	x^2
2.6	6.76
3.1	9.61
3.5	12.25
2.8	7.84
2.9	8.41
3.0	9.00
3.1	9.61
2.9	8.41
3.0	9.00
3.7	13.69
2.9	8.41
3.1	9.61
$\Sigma x = 36.6$	$\Sigma x^2 = 112.60$

$$\text{Mean birth weight, } \bar{x} = \Sigma x/n$$

$$= 36.6/12$$

$$= 3.05 \text{ kg}$$

$$\text{Variance, } s^2 = \frac{\Sigma x^2 - (\Sigma x)^2 /n}{n - 1}$$

$$= \frac{112.60 - (36.6)^2/12}{11}$$

$$= .0882 \text{ kg}^2$$

$$\text{Standard deviation} = .297 \text{ kg}$$

$$\text{Standard error} = \frac{.297}{\sqrt{12}}$$

$$= .086 \text{ kg}$$

Consulting t tables for 11 DF, we see that 95% confidence limits are given by $\pm$ 2.201 standard errors:

$$= \pm 2.201 \times .086$$

$$= \pm .189 \text{ kg}$$

We are therefore 95% confident that the true population mean lies between 2.861 and 3.239 kg.

The standard value of 3.20 kg lies within these limits. It is therefore possible that the true mean birth weight of the population of young disadvantaged mothers could indeed be the standard value. There is certainly no solid evidence that it is not, and it would be incorrect to infer from the available evidence that these mothers, in general, had underweight babies.

An alternative and slightly more convenient way of carrying out the same calculations would be to determine how many standard errors the standard value lies from the sample mean. By consulting t tables we can then tell if the standard value would lie inside or outside the 95% confidence limits and hence, what conclusions should be drawn:

$$\text{Sample mean} = 3.05 \text{ kg}$$

$$\text{Standard value} = 3.20 \text{ kg}$$

$$\text{Difference} = 3.20 - 3.05$$

$$= .15 \text{ kg}$$

$$\text{Standard error} = .086 \text{ kg}$$

$$\frac{\text{Difference}}{\text{SE}} = \frac{.15}{.086}$$

$$= 1.744$$

The standard value therefore lies 1.744 standard errors from the sample mean. Consulting a set of t tables for 11 df shows that 95% confidence limits would enclose all values lying within 2.201 standard errors of the sample mean. The standard value clearly lies within this range, and it is therefore quite possible that the standard value and the true population mean are indeed identical. This technique of assessing how substantial a difference in two values is by expressing it in terms of standard errors and then determining if it could reasonably be explained by chance variation is at the cornerstone of most of the applications of statistics in the health sciences.

To be honest, it is fairly unusual to have access to a standard or baseline value that is truly fixed. Even census means are, in practice, simply means of very large samples and are subject to some (but very little) variability. Minimum (or maximum) acceptable levels, as enshrined in quality control legislation, however, would provide a fixed baseline with which a sample could be compared. In the next chapter we explore the much more common situation in which the two values being compared are both sample means (rather than one sample mean and one fixed standard) and hence are both inherently unreliable.

▭ Key Terms

Population	**Sigma squared (σ^2)**
Sample	**Standard error**
Estimation	**Confidence limits**
Hypothesis testing	*t* **Distribution**
Mu (μ)	

▭ PROBLEMS

1. Problem 3 in Chapter 1 presented the digoxin levels of 58 individuals who were receiving a daily digoxin dose of 0.15 mg. Using this sample evidence, what comments can you reasonably make about the mean digoxin level in the population of all patients on this particular treatment regime? Is this evidence consistent with the possibility that the mean digoxin level in the population of such patients is actually 0.90 mg/100 ml? Is it consistent with a population mean of 0.80 mg/100 ml?

2. Problem 2 in Chapter 1 presented serum cholesterol values for 20 patients. Use the results obtained in that problem to fit 95%, 99% and 99.9%, confidence limits to the sample mean cholesterol value.

3. Fit 95% confidence limits to the mean of the plasma digoxin data presented in problem 1 in Chapter 1. Why are these limits much wider than the corresponding limits for the plasma digoxin data set presented in problem 3 in Chapter 1? (See problem 1 above.)

4. Use the birthweight data of problem 4 in Chapter 1 to make a statement about the probable mean birth weight of all children born to Indian mothers in Canada in 1983.

5. A total of 120 Inuit children in the Canadian Western Arctic were screened for myopia, and 20 of the children were determined to be myopic. Calculate 95% confidence limits for the true prevalence of myopia in this population.

CHAPTER 4

Comparing Two Groups

POSING SCIENTIFIC QUESTIONS

From childhood on, we expand our understanding of the world around us by asking questions. The most meaningful and most answerable questions are often posed in the form of comparisons. Do men and women differ in their ability to cope with a particular stressor? Does a newly developed drug produce an outcome superior to the drug currently used to treat the condition under study? Do two different laboratories produce different results when they are asked to analyze the same set of blood samples? A question such as, "Is the new drug any good?" is vague and almost impossible to answer. (What exactly do we mean by "any good"?) A question that involves a comparison ("Is the new drug any better than the one we are using right now?") is meaningful and answerable precisely because it includes within it a reference point or baseline against which the available evidence can be evaluated.

If questions are to be based on comparisons, then the comparisons involved must be fair, reasonable, and as informative as possible. The very important area of the planning of comparative studies (often referred to as Experimental Design) will be discussed in Chapter 8. In this chapter we will explore just how one can answer the comparative question itself ("Do the two drugs differ?") in a reasonable and realistic way when the available evidence is very often both scanty and inconsistent. Should we, for example, conclude that "drug A is definitely superior to drug B," or "I think that drug A is superior to drug B but I'm not certain," or "No, there's no difference," or "Gee, I just can't tell"?

The principles we will discuss in this chapter apply to all comparisons, but the technique we develop will enable us to cope with one of the most frequently encountered comparison problems, "Is there any evidence that two groups of results (such as two drugs or two genders) really differ in their average levels of performance?" In other words, the techniques presented in this chapter are generally applicable whenever two groups are being compared in terms of some value that is measured on a continuous (interval or ratio) scale. (For other important reservations about the use of this technique, see the section on Assumptions Involved in t Testing later in this chapter). Techniques for comparing more than two groups in terms of a continuous measurement are presented in Chapter 5, techniques for comparing groups in terms of categorical measurements are presented in Chapter 6, and techniques for comparisons involving ordinal measurements are presented in Chapter 14.

── MEASURING THE EVIDENCE ──────────────

In Chapter 3 we discussed the problems that arise when we have to rely on sample evidence to learn about the true average level of some particular phenomenon. The principal problem is, of course, that a fresh sample would inevitably result in a somewhat different mean value, and hence we should be cautious about taking our sample mean totally seriously. (The solution we produced was to come up with a range of values that we are reasonably confident make a realistic, if necessarily somewhat vague, comment about the true mean.)

These problems are compounded when we are faced with two groups or sets of results, both of which are simply samples (often very small samples), and asked to comment on the differences between them. If we carry out a comparison of two muscle relaxants, and the group of individuals on drug A gives better results on average than those on drug B, what have we actually proved, if anything? Do our results really mean that drug A is indeed a better muscle relaxant than drug B? Not necessarily. Perhaps if someone else repeated the study they might come to an entirely opposite conclusion.

The likely consistency of fresh data is a key issue in deciding just how impressive the evidence really is. If the study were replicated the results obtained would certainly be slightly different. They might even show major differences with, in our example, drug B now apparently giving results superior to drug A. If we are confident that our study, when repeated a number of times, would consistently give the same results, or at least very similar results, then we can confidently draw some hard and fast conclusions from it. If, however, we feared that if the study were repeated, the results would show a great deal of variation and inconsistency, then it would be very foolish to claim that the study has proved anything at all.

To get an intuitive feel for what is important when comparing mean performance in two groups, let us look at some hypothetical results that might arise from a study of this type. In the set of results illustrated in Figure 4-1, we see that the individuals treated with drug B do have a larger mean value than those on drug A. However, the improvement achieved by using drug B is fairly small. There are also large variations in the performances of the individuals receiving each treatment. Indeed, a number of the individuals on drug A do better than some of those on drug B. (There is a large overlap between the two distributions.) With results like these we would not be very confident about repeating the experiment. It is quite conceivable that, repeated with new subjects, the study might show that drug A apparently gives results superior to drug B. We would be foolish to claim that we had proven that drug B was inherently a more suitable treatment than drug A.

Suppose, on the other hand, we got results similar to those illustrated in Figure 4-2. Here the drug B individuals have a substantially higher mean value than those on drug A, and the improvement achieved by using drug B is very marked. Furthermore, the variation in the performance of the individuals is very much less. There is no over-

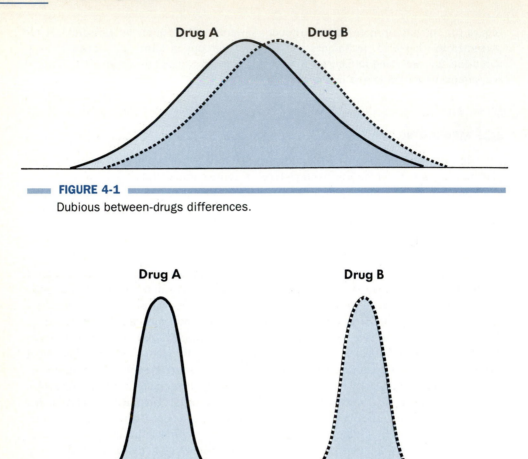

Drug A   **Drug B**

FIGURE 4-1

Dubious between-drugs differences.

Drug A   **Drug B**

FIGURE 4-2

Clear between-drugs differences.

lap between the two sets of results; even the worst of the drug B individuals do better than the best of the drug A group.

With results like these we would be very confident that, if our study were repeated, we would still find a superior performance from the individuals treated with drug B. We could, therefore, claim with a great deal of confidence that our experiment has proved that drug B is markedly superior to drug A and should be used in future routine practice.

From considering these two simple examples, we can see that the conclusions we reach about our experiments depend on two basic aspects of our results. First, how different are the mean values? The greater the difference, the more convincing the evidence. Second, how variable are our results? The more variable they are, the more cautious we have to be about interpreting them. The final conclusion we reach about our study must take both aspects of our results into account and must strike a balance between them.

Unfortunately, real-life results are rarely as clear-cut and obvious as the two hypothetical examples we have just discussed. Very often it is difficult to know exactly what, if anything, a study has proved. Let's explore how we can use the principles we have discussed to reach some sensible conclusions about the comparison of sets of sample evidence.

In a comparative situation, the means of the two groups of individuals under test are probably the values that most interest us. After all, these are the best indication we are likely to have of the average performance levels we can expect from our treatments in routine practice. It follows, therefore, that any difference in the mean results of our two groups is going to be of major interest to us. How should we measure such a difference? The answer seems so obvious as to be scarcely worth defining.

$$\text{Difference} = \bar{x}_1 - \bar{x}_2$$

The difference in mean performance is exactly that—the mean of one group of results minus the mean of the other group. (Which of our treatment groups we decide to call group 1 is totally irrelevant.)

The difference in the group means is, of course, very important. However, the variability in performance of the individuals in a study is a cautionary reminder that things would not necessarily be the same if we duplicated the study with fresh groups of individuals. It is clearly very important that we establish some measure of the variability of our results. We have already seen that, for a single set of results, variability is measured by the variance (that is, the sum of squares divided by the degrees of freedom):

$$\frac{\Sigma(x - \bar{x})^2}{n - 1}$$

or

$$\frac{\Sigma x^2 - (\Sigma x)^2/n}{n - 1}$$

In a two-group comparative situation there are, however, not just one but instead two groups of individuals to be considered, and the evidence from both groups must be taken into account. The overall measure of individual variability, therefore, simply combines or "pools" the information (that is, sums of squares and degrees of freedom) from the two sets of results.

$$\text{Pooled variance, } s_p^2 = \frac{\Sigma(x_1 - \bar{x}_1)^2 + \Sigma(x_2 - \bar{x}_2)^2}{n_1 - 1 + n_2 - 1}$$

$$= \frac{\Sigma(x_1 - \bar{x}_1)^2 + \Sigma(x_2 - \bar{x}_2)^2}{n_1 + n_2 - 2}$$

in which

$$\Sigma(x_1 - \bar{x}_1)^2$$

is simply the sum of squares for the first set of results (calculated in the usual way), that is, as

$$\Sigma x_1^2 - (\Sigma x_1)^2/n_1$$

and n_1 is the number of individuals receiving treatment 1, and so forth.

The pooled variance is, in fact, simply the weighted average of the two group variances. If there are equal numbers of individuals in each of the two groups, then the pooled variance is the straightforward average of the two group variances. However, if the groups are of unequal size, then pooling the sums of squares and degrees of freedom allows us to attach somewhat more importance to the variance of the larger group. (Since we have more information about this group it seems only fair that we should pay proportionately more attention to it when calculating an overall measure of variability.) The result is a single measure that fairly uses all the facts at hand to characterize the problems posed by variability in a two-group study. (The pooled standard deviation, s_p, is simply the square root of the pooled variance.)

A variance (pooled or otherwise) measures the variability between the individuals in a study. Assessing the results of a two-group comparison involves comparing the means of the two sets of results and attempting to determine if the difference between the means is, in some sense, impressive. We are, therefore, ultimately concerned with the variability of the means rather than of the individuals. We know from Chapter 3 that the variability of the mean of a single sample is measured by its standard error, which is determined by the sample variance, s^2, and sample size, n.

$$SE = \sqrt{(s^2/n)}$$

With two groups and two group means we are concerned with the variability of both means. More accurately, we are concerned with how they might vary relative to one another. Would the differences we see stay fairly constant in a fresh study, or might they fluctuate a lot and even change direction? In other words, what is really important is finding a way to measure the variability of the difference between the two means.

Since this involves both means, it is obtained by combining the variance and sample size characteristics of both sets of results. We have already combined the variability of both groups into our overall or pooled variance. Therefore we simply use this as our measure of individual variability in both groups.

The standard error of the difference between the two means

$$= \sqrt{\frac{s_p^2}{n_1} + \frac{s_p^2}{n_2}}$$

$$= s_p \sqrt{\frac{1}{n_1} + \frac{1}{n_2}}$$

where s_p is the pooled standard deviation.

We now have measures of the two aspects of our results that most interest us—the difference in the mean performance of the two groups and the overall variability of this difference. To try to determine if our study has actually proved anything, we must try to strike a balance between these two conflicting aspects of our results.

The most obvious way of striking a balance between these two values is to do quite literally that by working out the ratio of the two values.

$$\frac{\text{Difference between group means}}{\text{Variability of this difference}}$$

that is,

$$\frac{\bar{x}_1 - \bar{x}_2}{s_p \sqrt{\dfrac{1}{n_1} + \dfrac{1}{n_2}}}$$

Think for a moment about what this ratio will tell us about our problem.

In a situation where the difference between the means is large and the variability of the results is small (as shown in Figure 4-2), the ratio will take a large value. Results of this type are a very strong indication that the differences between our groups are real and substantial and would almost certainly be confirmed in future investigations.

On the other hand, if the difference between the means of the two groups is small, and the variability is large (as shown in Figure 4-1), then the value of the ratio will be small. Results of this type suggest very strongly that any apparent differences our study has detected are probably accidental byproducts of the variability of our data and might well disappear in a future replication.

Using the ratio of difference to variability (as measured by the standard error of the difference) should therefore give us a sensitive indication of just how convincing the results of our study are. Furthermore, we see that the larger the value of our ratio, the more convinced we are that we have detected really genuine differences between the groups being compared. The problem we are left with, however, is: How big a value would we need to get before we would be convinced that there was a real difference between our two groups of subjects?

——— EVALUATING THE EVIDENCE ———

Let us consider for a moment what our ratio actually measures.

$$\frac{\bar{x}_1 - \bar{x}_2}{s_p \sqrt{\dfrac{1}{n_1} + \dfrac{1}{n_2}}}$$

The top line measures the difference between the mean values. The bottom line measures the standard error of the difference between the means. Our ratio therefore measures the difference between our two sample means in terms of standard errors. For example, suppose that in a trial we found a difference of 0.6 mg/100 ml between the means of our treatment groups with a standard error of 0.2 mg/100 ml. What our ratio is actually telling us is that the difference between our two mean results is 3 standard errors (0.6/0.2).

All we have done, in fact, is convert the difference in the two sets of results from the original units (mg/100 ml or kilograms or whatever we happen to be studying) to standard error units. That is, we have standardized the difference, just as we discussed in Chapter 2 in the section on Believable and Unbelievable Results. This has two great advantages. First, we can use exactly the same criteria to judge all our research results irrespective of what we are studying. Second, it enables us to decide just how big our difference has to be to be really convincing evidence that the two groups being compared are genuinely different.

Think about it this way. If the two groups being compared are effectively identical, then there should be no difference in the mean performance of the two sets of results:

$$\bar{x}_1 - \bar{x}_2 = 0$$

Of course, because the study subjects are intrinsically variable, we could not reasonably expect the difference to work out to be exactly zero. However, we can at least say that most of the time (that is, 95% of the time) we would not expect the actual differences to be more than approximately 2 standard errors away from zero, assuming, of course, that the results follow a Normal distribution.

It follows that if the two groups really are genuinely different, we would expect the difference between the means of our results to be substantially different from zero, and almost certainly more than 2 standard errors away from zero. It seems that 2 standard errors (a ratio value of 2) is probably a reasonably good, if somewhat approximate, check value for deciding if our experimental results indicate real differences in our treatments or merely "accidental" differences, values greater than 2 suggesting that the differences are genuine, values less than 2 suggesting that any differences we see are quite possibly "accidental."

The ideas we have just developed are important, and it is worth taking a little time to think through them again. Much health research involves comparing the performance of two groups of subjects, often on alternative treatments for a specific condition, often a newly developed treatment being compared with some standard, current treatment. It is always possible that there is no real difference in the effectiveness of the two treatments (that is, the new treatment is no real improvement on the existing one). If that were the case, we might expect to find no difference in our two sets of results.

In the real world, however, we only have the time and facilities to study small groups of relatively variable individuals, so we could not reasonably expect the two groups to behave exactly alike even when no genuine differences exist. We would, however, expect any differences we might find to be small and almost certainly smaller than 2 standard errors. If the treatment difference we find from our experimental results is less than approximately 2 SE, it is quite possible that this difference could have happened when no genuine difference existed. In these circumstances, we would conclude that our new treatment was, in fact, no advance on existing treatments. (After all, we have no strong evidence to the contrary.)

Suppose, on the other hand, that the difference we found from our experimental results was greater than 2 standard errors. This is a large difference, and it seems very unlikely that a difference this big would arise unless the two treatments were genuinely different. In these circumstances we would feel that we had very strong evidence to suggest that real differences did exist, and we would conclude that one treatment (the new one, we hope) was indeed superior to the other.

We have used the value of 2 standard errors as a cutoff point to decide between genuine and non-genuine differences because 95% of all non-genuine differences will lie between ±2 standard errors. This, as we will realize, is an approximation. As we saw in Chapter 3, we can use a set of *t* tables to tell us the exact number of standard errors that would enclose 95% of such "accidental" differences. (Remember that we use *t* tables because the standard error we have calculated is itself simply a sample estimate.)

To look up the tables we need to know the sample sizes, or more precisely, the degrees of freedom. If one group has n_1 individuals ($n_1 - 1$ df), and the other group has n_2 individuals ($n_2 - 1$ df), then the total number of degrees of freedom is

$$(n_1 - 1) + (n_2 - 1) = n_1 + n_2 - 2$$

In other words, it is the total number of individuals in our experiment less 2. Based on past experience, this is fairly reasonable, since we are using 2 sample mean values in our calculations. (Note that this is the degrees of freedom we used to calculate the pooled variance.)

Suppose we had 12 individuals in one test group and 18 in the other. The total number of degrees of freedom (df) involved are

$$12 + 18 - 2 = 28$$

Consulting a set of *t* tables:

		10%	5%	1%	0.1%
DEGREES OF FREEDOM	**28**	1.70	2.05	2.76	3.67

we see that less than 5% of all non-genuine differences would give a ratio value of greater than 2.05 standard errors. (Again, very close to our approximation of 2.) We would, therefore, use the value of 2.05 as our exact cutoff point. If our ratio gives us a value of greater than 2.05, we would conclude that the differences between our two sets of treatment results are large and almost certainly indicate that one treatment is genuinely better than the other. On the other hand, if our ratio turns out to be less than 2.05, we would feel that this could quite possibly be an "accidental" difference. We would, therefore, conclude that our evidence does not support the proposition that the two treatments under test do genuinely differ from one another.

As we have just seen, the value of our test ratio has to be checked against a set of *t* tables before we can decide what our conclusions will be. For this reason the ratio value is normally referred to as a *t* value in the research literature, and the test for comparing two sets of results is usually called a **t test** or, sometimes, a Student's *t* test.

To see the *t* test in action, consider the following fairly typical research situation. A total of 24 hypertensive individuals were split into two groups of 12. Group 1 received a diuretic therapy while group 2 received a diuretic therapy in combination with other antihypertensive agents. After one month of treatment the patients' diastolic blood pressures were measured, and the results obtained are summarized below. Does the available evidence suggest that the mean diastolic blood pressures of the two groups genuinely differ from one another?

	SAMPLE SIZE	MEAN	SUM OF SQUARES	VARIANCE
Group 1	12	117.0	5170	470.00
Group 2	12	93.0	4502	409.27

The difference between the group means is, fairly obviously, 24.0 mm Hg (117.0 − 93.0).

The pooled variance is

$$\frac{5170 + 4502}{11 + 11} = \frac{9672}{22}$$

$$= 439.64 \text{ mm Hg}^2$$

This is, as we pointed out, simply the average of the two separate group variances. (If the two groups had not been of equal size it would be a weighted average with the variance of the larger group having more influence.)

The pooled standard deviation is

$$s_p = \sqrt{439.64}$$

$$= 20.97 \text{ mm Hg}$$

The standard error of the difference between the two means is given by

$$s_p \sqrt{\frac{1}{n_1} + \frac{1}{n_2}}$$

$$= 20.97 \sqrt{\frac{1}{12} + \frac{1}{12}}$$

$$= 8.56 \text{ mm Hg}$$

The t value (test ratio) is given by

$$\frac{\text{Difference between the means}}{\text{Standard error of this difference}} = \frac{24.0}{8.56}$$

$$= 2.80$$

Since there are 24 individuals in the study, and the study data has been used to calculate two sample means, the degrees of freedom involved are $24 - 2$ or 22. Consulting a set of t tables (Appendix T/2) shows that the 5% value is 2.07. The calculated t value comfortably exceeds this, and we therefore conclude that a genuine difference in mean diastolic blood pressure does indeed exist between the two treatment approaches.

It is worth recalling that the standard error of the difference is exactly what it says, a standard error, and can be used to fit confidence limits to the sample difference, as discussed in Chapter 3. For 22 degrees of freedom, 95% confidence limits will be given by ± 2.07 standard errors, and hence 95% confidence limits for the mean difference in diastolic blood pressure between the two treatment groups is given by

$$24.0 \pm (2.07 \times 8.56) = 6.28 \text{ to } 41.72 \text{ mm Hg}$$

In other words, even if the study were replicated many times, the diuretic treatment group would be unlikely to have a mean result less than 6.28 or greater than 41.72 mm Hg above that of the combined therapy group.

Since the confidence limits do not include zero, that is, it seems very unlikely that the two treatment groups could, in fact, be identical, they also point clearly to the conclusion that a genuine difference does exist between the two treatments, with the diuretic treatment resulting in a mean diastolic blood pressure somewhere between 6.28 and 41.72 mm Hg higher than the combined treatment. The t test and confidence limits approaches are, of course, equivalent, and the calculation of confidence limits about the sample difference is a very useful reminder that the situation might well change dramatically if the study in question were repeated on a fresh sample.

PRINCIPLES OF COMPARISON—HYPOTHESES

It seems self evident that a new treatment should never be put into clinical practice without thorough prior testing to determine if it is superior to the product it is designed to replace. This implies a very obvious but very important point. The onus is on the manufacturer of the new drug to demonstrate that the new product is indeed superior. The users or manufacturers of the standard or current treatment are not under any obligation to prove that their treatment is no worse than the new treatment with which it is being compared. The situation is comparable to that in a court of law; one is assumed innocent until proven guilty. Similarly, in a scientific comparison, the two groups being compared are assumed the same until proven different. This basic starting point for all comparisons, the proposition that the two groups are equivalent, is known as the **null hypothesis.**

Is there an alternative to the null hypothesis? It seems obvious that, if the null hypothesis or initial belief is "the groups being compared are the same," then the **alternative hypothesis** or belief to which we might switch if the evidence were very convincing (just as a juror might switch from "innocent" to "guilty" in similar circumstances) must be "the groups are different." Indeed this is usually (but not always) the alternative belief to which we are open. Please appreciate that the burden of proof is on the alternative hypothesis, and we will not sway from our belief in the null hypothesis unless the evidence of a difference is highly persuasive ("beyond a reasonable doubt," as the lawyers would say).

How does this philosophical framework help us assess the evidence at hand (as measured by the *t* value we developed in the section, Measuring the Evidence)?

Since we initially believe there is no real difference, we would expect to see a *t* value of zero. However, we realize that it is probably unrealistic to expect a value of exactly zero, and we would find a small *t* value quite consistent with the null hypothesis and certainly not compelling evidence in favor of the alternative hypothesis. On the other hand a very large *t* value (in either direction, since our alternative belief is simply that the groups are different, for example, that men might have faster reaction times than women or vice versa) would be convincing evidence that the study differences were too large to be dismissed as mere accidents. The only reasonable response to the facts at hand would be to reject the null hypothesis and embrace the alternative of a genuine difference.

From our knowledge of the *t* distribution we know that 95% of "accidental" differences would be no more than 2.07 standard errors (for 22 degrees of freedom) away from zero. It is possible to obtain differences larger than this even if the two groups under comparison are truly identical. They will occur, in fact, in 5% of such situations. This, however, is a fairly uncommon or unlikely possibility ("beyond a reasonable doubt"), and it seems much more reasonable to conclude that a *t* value in excess of 2.07 supports the existence of a genuine difference and calls for a switch to the alternative hypothesis. (That is, the argument that a large sample difference suggests a real difference is inherently more appealing than the remote (5%) possibility that the large sample difference is an unfortunate fluke.)

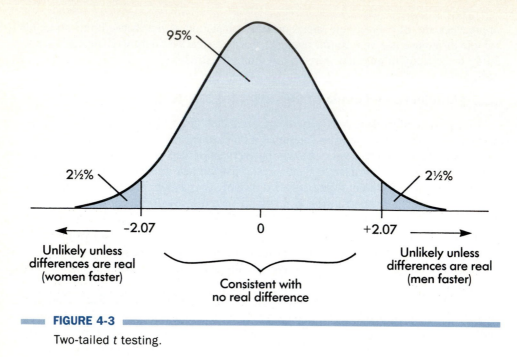

95%

2½%

2½%

−2.07   0   +2.07

Unlikely unless
differences are real
(women faster)

Consistent with
no real difference

Unlikely unless
differences are real
(men faster)

FIGURE 4-3

Two-tailed *t* testing.

Since we are interested in the existence of genuine differences in either direction (that is, that men react faster than women, or women react faster than men) we should be quite happy with the fact that the 5% zone in Figure 4-3 is, in fact, comprised of two 2.5% regions located symmetrically in the two tails of the *t* distribution.

For a hypothetical 24-individual (that is, 22 degrees of freedom) study comparing male and female reaction times, our decision game plan would therefore be

1. Measure the strength of the sample evidence by calculating the *t* value.
2. If *t* is greater than +2.07, switch to the alternative belief of a true difference (specifically that men react faster than women, if we arbitrarily subtract the female mean from the male mean), recognizing that there is a 2.5% chance that the apparently impressive evidence is in fact simply a fluke.

    If *t* lies between −2.07 and +2.07, stick with our initial belief that there is no difference in reaction times, since the sample evidence is unimpressive and could easily arise by accident, even if the two genders had identical reaction times.

    If *t* is less than −2.07, switch to the alternative belief of a true difference (specifically that women have faster reaction times than men), while recognizing that there is a 2.5% chance that the apparently impressive evidence is simply a fluke.

Overall then, there is a 5% chance that we may switch to the alternative hypothesis when the apparently dramatic sample differences are in fact simply the result of some unfortunate quirk in sample behavior, and the two genders really do not differ at all. Testing the evidence in this way, with the 5% risk of being misled being split into two symmetrical 2.5% regions in the two tails of the *t* distribution, is known as **two-tailed**

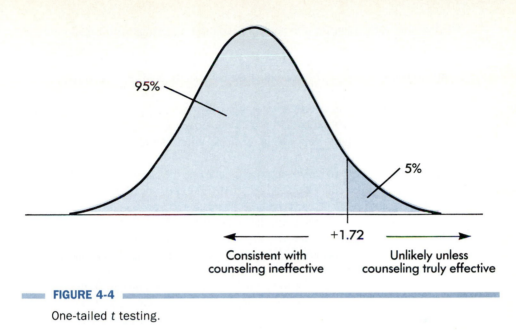

95%

5%

+1.72

Consistent with
counseling ineffective

Unlikely unless
counseling truly effective

FIGURE 4-4

One-tailed *t* testing.

testing. Two-tailed testing, that is, being open to the possibility of differences in either direction, is the standard way we approach a comparison.

There are, however, situations in which we are not interested in the general alternative hypothesis (the groups are different) but rather in a very specific alternative hypothesis (group A is better than group B). This focused interest arises either from the logic of the situation (when evaluating a new therapy our objective is to determine whether it is superior to the existing practice, rather than simply different from it) or from our prior knowledge of the situation (when evaluating counseling support for cancer patients we might dismiss the prospect of counseling being harmful as being untenable and feel that the real decision lies between "counseling is superior to no counseling," that is, a specific alternative, and "counseling is a waste of time" that is, the initial or null belief).

The decision to explore a specific alternative is taken before beginning the analysis; indeed it should be taken before the data is even collected, and it must be firmly rooted in the logic of the comparison. Exploring a specific alternative also changes the way we reach decisions. Let us assume that we feel we can live with a 5% risk of switching to the alternative belief by mistake. Since we are interested in only one specific alternative, the 5% does not have to be split between the two tails of the *t* distribution and can be concentrated in just one tail. (Testing a specific alternative is known as doing a **one-tailed test,** as depicted in Figure 4-4.)

This means that we have to be very careful about how we consult a set of *t* tables when carrying out a one-tailed test. Sets of *t* tables are always tabulated on the assumption that we will want to do a two-tailed test (certainly by far the most common situation). To find the appropriate 5% one-tailed value, we must determine the tabulated 10% value (the 10% is, after all, made up of two tails, each of 5%). For 22 df, the appropriate value is 1.72.

The decision game plan for a hypothetical counseling comparison based on 24 individuals is therefore

1. Measure the strength of the sample evidence by calculating a *t* value
2. If *t* is greater than +1.72 (assuming we have subtracted the uncounseled mean from the counseled mean), switch to the specific alternative belief that counseling is superior to no counseling, recognizing that there is a 5% risk that this change of belief may be based on faulty evidence.

   If *t* is less than +1.72, stick with the initial belief that counseling is ineffective.

Again, just like the two-tailed approach, there is the same overall 5% risk of switching to the alternative belief in error. The risk is simply sliced up differently.

One-tailed testing is attractive because it increases the prospects of switching to the alternative hypothesis while retaining the same overall 5% risk that the switch is a mistake. Please remember, however, that the decision to use a one-tailed test must be made before analysis, must be justified by the logic of the comparison, and must be clearly identified as such in any report.

Do also appreciate that sticking to the null hypothesis does not mean that we have proven that the null hypothesis is true. It simply means that we cannot reject it as a reasonable explanation for the facts. An innocent verdict does not necessarily mean that a jury is convinced of a defendant's innocence. It may mean that the facts were not strong enough to justify a guilty verdict, although guilt may be suspected.

—— PROBLEMS OF COMPARISON—ERRORS ———————————

In any situations in which human beings are asked to make decisions based on limited evidence, mistakes occur. Innocent people do get convicted; the guilty do go free. The same is true for scientific comparisons. The only difference is that we are able, at least, to determine just how often the mistakes are likely to happen, although not, of course, precisely when they will happen.

We have already met one of these mistakes or errors quite explicitly. The use of a two-tailed value of 2.07 or one-tailed value of 1.72 (when there are 22 df) means that we acknowledge the possibility that there is a 5% (or .05, expressed as a probability) chance that the apparently impressive evidence that caused us to switch to the alternative hypothesis was simply an unfortunate fluke, and the groups under comparison are, in fact, identical. This mistake, embracing the alternative hypothesis when it is actually not true at all, is known as a **type I (or α) error.** The famous research phrase

$$p < .05$$

is simply a statement about the size of the type I error and really means "I think you should join me in switching to the alternative belief, that is, that the two treatments really do differ, because the evidence I have is so strong that there is less than a 5% risk that it is simply a fluke, and that is a very small risk."

Is a 5% or .05 risk really very small? It would certainly be nice to have to live with a smaller risk than this of being fooled by apparently impressive evidence.

Should we not give serious consideration to toughening the criteria we use? After all, less than one "accidental" difference in 1000 will give a *t* value of greater than 3.79

(for 22 degrees of freedom). Why not use this as a decision value? We can be practically certain that a *t* value of greater than 3.79 represents genuine differences in the treatments under test, and we will have reduced our risk of a type I error to a trivial level (p < .001). Unfortunately, if we do this, we greatly increase the risk of a second, less obvious, error. The greater the level of proof we require before we are impressed, the more difficult it becomes for any study we carry out in the real-life, and less than ideal, world to meet these exacting criteria, even when there are quite genuine differences in the groups being compared. The more we guard against the risk of detecting non-genuine differences, the more we run the risk of failing to detect genuine differences.

This second type of error (sticking with the initial belief when, in fact, the groups are genuinely different) is known as a **type II** (or β) **error.** Determining the risk of a type II error in any given situation is more complex than simply looking up a set of tables and depends on several factors, such as how dramatic the genuine differences actually are and the size of the study. This will be considered in detail in Chapter 9. If the type II error measures the probability of not detecting a genuine difference, then 1 − the type II error must, of course, measure the probability of actually succeeding in detecting a genuine difference, which is exactly what we would like all our investigations to do.

This measure (1 − type II error) is known as the **power** of a study, and a well-planned study will have as large a power as possible. This usually means making sure that the study collects enough evidence to ensure that a genuine difference will be at little risk of being obscured by the inevitable variation in the results. The important issue of sample size and study power is discussed in Chapter 9.

We can say that the use of a .05 type I error seems, in most circumstances, to offer a reasonable degree of protection against being misled while ensuring that the type II error will not be unacceptably high. It is, in effect, a happy compromise between too little proof and too much proof.

A final word on the notation used to report *t* tests in scientific journals. If the value of *t* is less than the decision or critical value, the groups are said to be "not significantly different". If *t* does exceed the critical value, then the groups are said to be "significantly different at the *p* < .05 level." This reflects the fact that the chance of having made a type I error is less than 5 in 100 and is a measure of the author's confidence that the observed treatment differences are genuine.

If the calculated *t* value is very much larger than the critical value and, in fact, exceeds some of the other tabulated values, this, of course, should also be reported. Suppose we obtained a *t* value of 2.84 with 25 degrees of freedom. Consulting a set of *t* tables yields:

		10%	5%	1%	0.1%
df	**25**	1.71	2.06	2.79	3.73

This *t* value clearly exceeds the 5% critical value. However, it also exceeds the 1% value and would be reported in the following way:

$$t = 2.84, \text{ with 25 df}$$

The treatments are significantly different with $p < .01$.

PAIRED COMPARISONS

Genuine but relatively small differences in mean performance can easily be obscured by the wide variations in individual performance that are often observed in any group of variable individuals, resulting in what we now know as a type II error. The problem is further compounded when one is working with two such groups of diverse individuals. An alternative to comparing two groups might be to take one single group of individuals, give each individual in the group both treatments (usually in some random order), and determine, on an individual-to-individual basis, which treatment was superior by comparing each individual's performance on treatment A to the individual's performance on treatment B (in terms of the simple difference between them). Since there is a highly consistent basis for comparison in this approach, even relatively small differences in treatment effectiveness can often be detected.

There will, of course, be many occasions when treating individuals twice is simply not practical, such as in the comparison of surgical procedures. In these situations, pairs of individuals who are inherently very alike can be treated, with one member of the pair receiving treatment A and the other treatment B. Here the pair (not the individual) is the key unit in the study, and again the measure of real interest is "how differently did these two, very similar, individuals react to the treatments under comparison?"

Pairs can be formed on a natural basis (twins) or on a more arbitrary artificial basis (pairing off individuals of the same gender and age, and so forth). In all cases, for this paired comparison approach to be truly effective, the members of each pair must be as similar as possible. (In Chapter 8 we will discuss in more detail the subject of organizing your research material to make your study more sensitive.)

It is important to note at this stage that the test we developed in the section, Measuring the Evidence (often called the independent t test) should not be used when there is some inherent connection between two sets of results. When the results form logical pairs, it is the differences between the pairs of results in which we are interested, and any analysis carried out must be based directly on these differences. The version of the t test that we use to analyze results of this type is known as the **paired t test**.

The value of major importance in a paired study is d, the difference between the two values resulting from each pair of individuals. If one treatment is genuinely more effective than the other, we would expect to find large differences consistently favoring the superior drug and a correspondingly large mean difference $\bar{d}$. On the other hand, if no genuine difference exists between the treatments, the differences we observe will result solely from the small amount of natural variation between the pairs. The differences will, therefore, be small, some apparently favoring one treatment and some the other, and with an overall mean difference, $\bar{d}$, close to zero.

The appropriate null hypothesis is, of course, that $\bar{d} = 0$. The proposition we wish to test is, "Is the value of $\bar{d}$ we have observed so large that it is very unlikely that it could have occurred by sheer chance when the true difference is zero?" If it is very unlikely (say, less than one chance in 20 or .05), then it seems more sensible to reject the null hypothesis and conclude that the two treatments are genuinely different.

The variability of the differences can be measured by calculating their standard deviation and then their standard error (since we are concerned with the size of the mean difference). If the true difference is zero, then we would expect 95% of all the differences we might observe, if we replicated our experiment many times, to fall within approximately 2 standard errors of zero. If the difference we observe in practice lies more than approximately 2 standard errors away from zero, then we would conclude that it is large and almost certainly represents a genuine difference. Conversely, if it is less than approximately 2 SEs away from zero, it is quite possible that it could have arisen accidentally, and we would not be justified in rejecting our null hypothesis.

$$\text{Mean difference, } \bar{d} = \frac{\Sigma d}{n}$$

where

$$n = \text{number of pairs or number of differences}$$

$$\text{Variance of } d = \frac{\Sigma d^2 - (\Sigma d)^2/n}{n-1}$$

$$\text{Standard deviation, SD, of } d = \sqrt{\text{Variance}}$$

$$\text{Standard error, SE, of } \bar{d} = \frac{\text{SD}}{\sqrt{n}}$$

Number of standard errors that d lies from 0 (that is, t)

$$= \frac{\bar{d} - 0}{\text{SE of } \bar{d}}$$

$$t = \frac{\bar{d}}{\text{SE of } \bar{d}}$$

(Note that this corresponds in principle to the t test we developed earlier, in which we used the ratio of difference to variability (that is, standard error) as a test criterion. Here, however, we can work directly with the differences between pairs of results, something we cannot do when the two groups of individuals are quite independent.)

Since the fundamental units of study are the differences and only one mean is involved, the degrees of freedom involved are given by:

$$\text{df} = \text{Number of differences} - 1$$

and the precise critical value against which we test is given by consulting t tables for the appropriate number of degrees of freedom.

Although the paired t test is similar in principle to the t test for separate (unpaired) groups of individuals (the independent t test), it is important to appreciate the vital difference between them. A very common mistake is to attempt to analyze the results of a paired experiment using an independent t test, thus totally ignoring the paired nature of the results and losing all the advantages of this very sensitive form of investigation.

The very simple calculations involved in a paired t test can be illustrated by the following weight gain study in which seven pairs of twins were allocated at random to two alternative diets and their weight gain measured after a fixed time. Our initial belief is that the diets are identical in effectiveness, and our (two-tailed) alternative hypothesis is that they genuinely differ.

WEIGHT GAIN (KG)

	DIET A	DIET B	DIFFERENCE	
			d	d²
Pair 1	10	16	+6	36
Pair 2	17	20	+3	9
Pair 3	8	14	+6	36
Pair 4	15	15	0	0
Pair 5	17	16	−1	1
Pair 6	12	16	+4	16
Pair 7	14	17	+3	9
			$\Sigma d = 21$	$\Sigma d^2 = 107$

$$\text{Mean difference, } \bar{d} = 21/7$$
$$= 3.00 \text{ kg}$$

$$\text{Variance of } d = \frac{107 - (21)^2/7}{6}$$
$$= 7.33 \text{ kg}^2$$

$$\text{Standard deviation of } d = \sqrt{7.33}$$
$$= 2.708 \text{ kg}$$

$$\text{Standard error of } \bar{d} = \frac{2.708}{\sqrt{7}}$$
$$= 1.024 \text{ kg}$$

$$t = \frac{3.00}{1.024}$$
$$= 2.93$$

$$\text{Degrees of freedom} = 7 - 1$$
$$= 6$$

$$\text{5\% critical value of } t \text{ for 6 df} = 2.45$$

$$t = 2.93, \text{ 6 df, } p < .05$$

Our conclusion therefore would be that the diets do indeed differ and that diet B results in a higher weight gain than diet A.

A COMPARISON CHECKLIST

Before setting out to analyze a set of results from a study that seeks to compare two groups or treatments, it is worthwhile to ask yourself the following questions.

Can I state the null hypothesis being tested?

Can I state the most appropriate alternative hypothesis? (Are you interested in detecting any differences between the two treatments, or are you interested only in determining if one particular treatment is superior to the other?) This will determine if the test should be a two-tailed or a one-tailed test.

Are the two sets of results unrelated to one another, or can they be formed into logical pairs? This will determine if the test should be an independent *t* test or a paired *t* test.

Can I state the appropriate degrees of freedom? If the test is an independent *t* test, this is the total number of individuals in both groups less 2 (for the two means that are being compared). If it is a paired *t* test, this is the number of pairs less 1 (for the mean difference that is being compared with zero).

Can I state the appropriate critical value? This involves looking up a student's *t* table for the correct degrees of freedom (row) and the correct type I error (column). This will normally be 5% if you are carrying out a two-tailed test, or 10% if you are carrying out a one-tailed test (since you are only interested in one tail or half of the 10%, that is, 5%).

Does the calculated value of *t* exceed the critical value? If so, you can reject the null hypothesis and accept your alternative hypothesis. In other words, the observed differences are so large that it is reasonable to conclude that they are genuine and very unlikely to be the result of an accident. (The chance of this having happened, that is, the risk of a type I error, is less than 5%, or one in 20.)

If the calculated value of *t* does not exceed the critical value, then there is a real risk that any apparent treatment differences are purely accidental. Therefore you cannot, on the basis of this weak evidence, reject the null hypothesis.

Does the calculated value of *t* exceed any of the other tabulated critical values (such as the 2% or 1% values) for the appropriate degrees of freedom? If it does, say so. After all, this means that the evidence in favor of a genuine difference is even more impressive, and the risk of a type I error (that is, getting an apparently impressive difference accidentally) is even smaller (such as less than 2 in a 100). Always quote the highest critical value that was exceeded (such as $p < .02$ or $p < .01$).

Having carried out an analysis, you will want to report your conclusions.

Have you described what you did and your conclusions in clear English that a colleague can read and understand?

"The results were analyzed using an independent *t* test, and no significant differences between the treatments were observed."

"The results were analyzed using a paired *t* test, and oral contraceptive usage was observed to result in significant weight gains ($p < .05$)."

Have you also stated the principal results of your analysis in the standard form so that someone else can check them if they wish? These should consist of:

1. Your calculated value of *t*

2. The degrees of freedom involved

3. The type I error involved

For an analysis in which the calculated value of *t* was less than the basic 5% critical value, (and that would lead us to conclude that the treatments under test were not significantly different), the principal results might look like

$$t = 1.27, \text{df} = 34, \text{N.S.}$$

For an analysis in which the calculated value of t exceeded the basic 5% critical value and also the higher 1% critical value, (giving very impressive evidence of a real difference with less than a 1% risk of this being accidental, that is, a type I error), the principal results might look like

$$t = 2.83, \text{df} = 27, p < .01$$

Have you, somewhere in your report, given some basic facts about your analysis? People will want to know. Useful facts to give are

1. The number of individuals in each test group
2. The mean value of each group
3. The standard deviation of each group
4. The range of results (highest and lowest) you observed in each group.

Although this checklist of things to do and think about when carrying out a statistical analysis has been presented in terms of the t tests that we have encountered in this chapter, this checklist can also be applied to the other statistical tests we will encounter in the rest of this book.

ASSUMPTIONS INVOLVED IN t TESTING

All statistical methods require us to make certain assumptions about the way our data behaves, and the t test is no exception. The basic assumption is that the results follow a Normal distribution. We have used our knowledge of the properties of the Normal and t distributions to solve the problem of comparing sets of results, and the whole argument that a difference of greater than approximately 2 standard errors is not likely to be accidental, falls to pieces if the data does not follow a Normal distribution.

In addition, use of the independent t test assumes that the variability of the two groups is the same (or in practice, reasonably similar). If this is not the case, calculating a common (pooled) variance would obviously be a meaningless thing to do. (This assumption does not arise in a paired t test, since there is effectively only one set of results, that is, the differences, to consider.)

Luckily, the t test is known to give reliable results even when these ideal assumptions do not hold exactly. If, however, either of the assumptions is markedly incorrect, then the t test should not be used. In this case, alternative, but less desirable (because they are more prone to miss genuine differences), methods of analysis are available (the "nonparametric" methods). These will be discussed in Chapter 14.

Key Terms

t **test**	**Type I error**
Null hypothesis	**Type II error**
Alternative hypothesis	**Power**
Two-tailed test	**Paired t test**
One-tailed test	

PROBLEMS

1. In an investigation of the pharmacodynamics of digoxin, serum digoxin levels were determined in 12 males and 14 females (all healthy and aged between 20 and 45) 4 hours after rapid intravenous injection of the drug. Is there any evidence of gender differences? (Results given in ng/ml.)

MALES		FEMALES	
1.19	1.13	1.14	0.98
1.14	1.22	1.17	1.06
1.24	1.27	1.09	1.18
1.21	1.15	1.11	1.08
1.09	1.20	1.25	1.05
1.11	1.09	1.29	1.16
		1.13	1.13

2. As part of a study to determine the effects of a certain oral contraceptive on weight gain, nine healthy females were weighed at the beginning of a course of oral contraceptive usage. They were reweighed after 3 months. Do the results suggest evidence of weight gain?

SUBJECT	INITIAL WEIGHT (LBS)	3-MONTH WEIGHT (LBS)
1	120	123
2	141	143
3	130	140
4	150	145
5	135	140
6	140	143
7	120	118
8	140	141
9	130	132

3. In a study of the effects of storage conditions on the methemoglobin content of blood, blood samples were collected from 12 females ages 20 to 39 years. Each sample was then split into two subsamples, one being stored in darkness and the other stored under exposure to light. (All other storage conditions, including temperature and humidity, were held constant for all samples.) After 8 days the methemoglobin levels of the 24 blood samples were assayed.

   Do the results obtained suggest that exposure to light during storage affects the methemoglobin content?

METHEMOGLOBIN CONTENT (%)

SAMPLE	DARK	LIGHT
1	5.4	8.9
2	2.7	6.3
3	7.4	14.2
4	6.2	7.4
5	8.8	6.4
6	7.9	11.3
7	9.9	6.8
8	5.3	9.4
9	6.8	10.5
10	10.1	8.9
11	5.2	7.1
12	6.5	9.4

4. In an attempt to reduce blood loss following dilation and evacuation abortions, the use of a local vasoconstrictor (a paracervical injection of vasoperin) was compared with a saline solution injection. Total blood loss during the evacuation was measured (in ml) for a total of 24 patients (12 in each treatment group), and the results are presented below. Is there any evidence that the use of a vasoconstrictor reduces blood loss?

VASOCONSTRICTOR	SALINE
180	340
320	200
240	420
440	550
60	280
140	260
220	190
300	360
120	250
80	310
290	140
140	290

5. In a study of hematological changes in trained athletes following prolonged exercise, hemoglobin levels (in g/dl) were measured in 10 athletes before and after their participation in a continuous 60-kilometer march. Do the results presented below offer any evidence of hematological change following such a sustained period of exercise?

BEFORE	AFTER
14.6	13.8
17.3	15.4
10.9	11.3
12.8	11.6
16.6	16.4
12.2	12.6
11.2	11.8
15.4	15.0
14.8	14.4
16.2	15.0

6. As part of an investigation of the development of infant sleep patterns, the sleep of 20 infants (10 male and 10 female) was monitored on several occasions between 1 week and 6 months of age. The quiet sleep results (in minutes) at 1 week of age for the 20 study infants follow. Is there evidence of a difference in quiet sleep behavior between the two genders?

QUIET SLEEP (MALE)	QUIET SLEEP (FEMALE)
85	140
129	155
215	33
143	209
44	166
173	72
230	116
198	131
105	97
127	124

Can you fit 95% confidence limits to the mean difference in quiet sleep time? What do these limits suggest about the true difference between the genders? Is this consistent with the conclusions indicated by the appropriate statistical test?

7. The cohort of 20 children mentioned in problem 6 was monitored again at 1 month of age. The duration of quiet sleep observed at 1 month for each child is given below, together with the corresponding 1 week result for that child. Test for any evidence of a difference in mean quiet sleep times between these two ages.

QUIET SLEEP AT 1 WEEK	QUIET SLEEP AT 1 MONTH
105	122
131	147
230	238
72	70
166	176
209	201
44	48
97	91
173	178
140	136
155	164
198	212
33	54
85	96
129	138
116	122
124	130
127	125
215	215
143	148

Again fit 95% confidence limits to the mean difference in quiet sleep times and determine if these limits are consistent with the results of the statistical test.

8. Use the data presented in Chapter 1, problems 1 and 3, to test the proposition that a higher daily digoxin dose should result in a higher plasma digoxin level.

CHAPTER 5

Analysis of Variance

■ COMPARING SEVERAL GROUPS ■

In Chapter 4 we discussed the problems we face when trying to compare two sets of data, and we developed a solution to those problems, in the form of the *t* test. Although this is a very common situation, there is no reason a comparison has to involve only two groups. Many investigative situations involve collecting data on three or more groups of individuals, with the objective of determining whether any true differences in mean performance exist among the conditions under study. This often happens in the experimental situation where several different treatments (for example, various therapeutic approaches to a specific problem or, various dosage levels of a particular drug) may be under comparison and, for convenience, we will use the terminology of experiments and treatments in this chapter. The problem, and its solution, is of course relevant to any situation in which we want to compare the mean performance of several groups.

The very generality of the problem ("several" could, in theory, be anything from 2 to 200 or beyond) suggests that, if we can find a solution, it could prove to be very useful indeed. It also suggests that we may need to come up with a slightly more flexible way of looking at our data. Assessing the size of the difference between two means (by standardizing it) is fine if only one difference concerns us. With several means and a host of possible differences, this approach could very well be a major headache. In this chapter we will develop a very general, yet very simple and sensible, approach to the general problem of comparing any number of groups. In the sections, Exploring Experimental Results and Asking Specific Questions, we will face up to the very real problem of trying to decide exactly what we mean by "real differences" when there are a variety of differences involved in the study. (In other words, which differences are real and which are not?)

Could we use the *t* test techniques we developed in Chapter 4 to analyze the results of a multiple-treatment experiment? Perhaps. Suppose we were carrying out an investigation of the effectiveness of five different sterilization regimes in killing bacteria. One way to analyze the results of such a study would be to view it as a series of two-treatment experiments and to compare all the possible treatment pairs using the familiar *t* test. Unfortunately, this approach has a couple of very serious drawbacks. First, a rather large number of two-treatment pairs can be formed from such a multi-treatment experiment. In the five treatment situations there are a total of 10 (5 $\times$ 4 / 2; see the discussion of the number of ways of forming pairs presented in the Chapter 2 discussion of Probability of Multiple Events) possible treatment pairs. Carrying out 10 separate analyses of a single experiment is tedious, time-consuming, and quite possibly confusing. Common sense suggests that there has to be a more elegant way to tackle this situation.

There is, however, a far more important reason we should not analyze a multi-group experiment this way. When we analyze a conventional two-treatment experiment, we are prepared to run a 1 in 20 risk of an apparently significant result arising purely by accident (that is, a 5% chance of a type I error). We regard such a risk as being fairly unlikely and feel justified in accepting with confidence any significant results we obtain. Analyzing a single experiment as a series of 10 treatment pairs is a very different proposition. The chance of an apparently significant result arising purely by chance somewhere in the 10 analyses increases dramatically. We can show (using the ideas of probability discussed in Chapter 2) that the chances of a misleading result rise from .05 for a single analysis to .40 for a set of 10 analyses, that is, a very real risk indeed. (The chance of not making a type I error on a single test is .95. The chance of not making a type I error somewhere in a series of 10 tests is .95 multiplied together 10 times, which is approximately .60. Hence the risk of a type I error occurring somewhere along the line is .40.) This approach is the statistical equivalent of playing Russian roulette not once, but 10 times! Clearly a course of action not to be recommended! (What if we did find a significant difference between a pair of treatments? How could we take it seriously knowing that the chance of it being simply a fluke was very substantial?)

Using a series of *t* tests to analyze a multi-group experiment is therefore tedious and just plain hard work. Worse still, it will yield results that are potentially misleading. We need, therefore, to develop a method of analyzing general (that is, multi-group) experimental results that will give us a single, reliable answer to the question, "Is there any evidence of real differences in the effectiveness of the various treatments under test?"

SOURCES OF VARIATION

Apparently very similar individuals can differ quite markedly in the way they respond to a particular treatment. This inherent variability is what complicates the interpretation of any multi-group experiment. It can, however, be used to solve some of the problems it creates if we understand what is causing it and how to measure it.

Variation between experimental results has, at least potentially, two quite separate causes. Much of the variation will be due to the natural inherent variation that exists among all individuals. This variation is referred to as natural variation, random variation, or (rather unfairly) error variation. However, some of the variation may well have been imposed by the experimenter's use of different treatments. In a hypertension investigation, a highly effective treatment is likely to result in some of the patients (those receiving the best treatment) having markedly lower blood pressures than others. This "imposed" variation is sometimes referred to as the **treatment effect.**

Individuals who receive the same treatment will experience identical experimental conditions. The variation within each of the treatment groups (**within-treatments variation**) must therefore be a consequence of solely random variation.

Between-treatments variation (that is, differences between the group means) is, however, another matter. If some treatments are genuinely more effective than others (if there really is a treatment effect), then we would expect to see relatively large differences between the treatment means and a relatively large between-treatments variation. On the other hand, if there is no real difference in treatment effectiveness (no true treatment effect), we would still have to expect to see some between-treatments variation. It would be quite unrealistic to expect different groups of patients, even receiving identical treatments, to give exactly the same mean values. The small differences in the means would be an inevitable consequence of the familiar natural or random variation that we have already mentioned. Between-treatments variation will, therefore, always reflect random variation and, if (and only if) a treatment effect really exists, some additional "imposed" variation.

To summarize:

Between-treatments variation = Random variation (always) and imposed variation (maybe)

Within-treatments variation = Random variation (always)

This suggests a way of assessing whether the treatments in an experiment (irrespective of their number) really do differ in their effectiveness. We first break up the total variation in a set of experimental results into two parts:

1. Variation between treatments
2. Variation within treatments.

Since these two situations cover all the available options in a simple experiment, we can say that

$$\text{Total variation} = \text{Between-treatments variation}$$
$$+ \text{ within-treatments variation}$$

We are now in a position to compare the between-treatments variation and within-treatments variation in a very direct and informative way.

If there is no real differences between the treatments, that is,

$$\frac{\text{Between-treatments}}{\text{Within-treatments}} = \frac{\text{Random variation}}{\text{Random variation}} = 1$$

then the between-treatments and within-treatments variations should be very similar, yielding a ratio value of 1, or close to it.

If, however, there are real differences between the treatments, that is,

$$\frac{\text{Between-treatments}}{\text{Within-treatments}} = \frac{\text{Random variation} + \text{imposed variation}}{\text{Random variation}} > 1$$

then the between-treatments variation should be substantially larger than the within-treatments variation and should yield a ratio value much greater than 1.

MEASURING VARIATION

Having developed a strategy for exploring the question, "Do the groups in the study differ from one another in terms of their mean performance?" we now have to translate these ideas into practice. To do this, we need to be able to measure the amount of between-treatment variation and within-treatment variation that is present in a set of experimental results.

Purely for convenience, let's explore how to do this in the context of a hypothetical three-treatment experiment with n_1, n_2, and n_3 patients receiving the respective treatments. The total study size is N, where

$$N = n_1 + n_2 + n_3$$

The basic building blocks of variation are, as we saw in Chapter 1, the deviations of the individual results from the reference point provided by the mean result, and a very obvious and familiar measure of the amount of variation present in any given situation is the sum of these squared deviations (in other words, the sum of squares).

Considered in a global, very superficial sense, an experiment simply provides us with N individual results. Totally ignoring the distinction between the various treatment groups, we could therefore regard the experimental data as simply a single set of N results and measure the total variation present in the data set by calculating its overall sum of squares using the familiar formula:

$$\Sigma x^2 - \frac{(\Sigma x)^2}{N}$$

(Because we have a number of treatment groups, we must be more careful than usual with our notation. Where no subscript is specified, as above, summation is over the entire set of N results. When a subscript is specified, summation is only over the results from the specified treatment group.)

Ignoring the distinction between the various treatment groups in an experiment is, of course, missing the whole point of the study. We now have to face up to the need to measure the between-treatments and within-treatments variation. Let's first consider how we might go about measuring within-treatments variation. For any particular treatment, the within-treatment variation can be very directly measured by restricting our attention to the results from that treatment group and calculating their sum of squares.

For example, the variation within the first treatment group is given by

$$\Sigma x_1^2 - \frac{(\Sigma x_1)^2}{n_1}$$

Similarly, variation within the second group is given by

$$\Sigma x_2^2 - \frac{(\Sigma x_2)^2}{n_2}$$

and for the third group by

$$\Sigma x_3^2 - \frac{(\Sigma x_3)^2}{n_3}$$

Over the whole experiment, therefore, the within-treatments variation is given by

$$\Sigma x_1^2 - \frac{(\Sigma x_1)^2}{n_1}$$

$$+ \ \Sigma x_2{}^2 - \frac{(\Sigma x_2)^2}{n_2}$$

$$+ \ \Sigma x_3{}^2 - \frac{(\Sigma x_3)^2}{n_3}$$

$$= \Sigma x_1{}^2 + \Sigma x_2{}^2 + \Sigma x_3{}^2 - \left[\frac{(\Sigma x_1)^2}{n_1} + \frac{(\Sigma x_2)^2}{n_2} + \frac{(\Sigma x_3)^2}{n_3} \right]$$

$$= \Sigma x^2 \qquad\qquad - \left[\frac{(\Sigma x_1)^2}{n_1} + \frac{(\Sigma x_2)^2}{n_2} + \frac{(\Sigma x_3)^2}{n_3} \right]$$

We now have measures of the total variation and the within-treatments variation present in our study. However, we still need to calculate the amount of between-treatments variation present. Since all the variation in the study has to be either between-treatments variation or within-treatments variation, we could easily obtain the between-treatments variation by simply subtracting the within-treatments variation from the total variation:

$$\text{Between} = \text{Total} - \text{Within}$$

The idea of calculating two of the sums of squares and then finding the third by subtraction is an excellent labor-saving device. It turns out, however, that in practice it is easier and quicker to calculate the total and between-treatments variation and obtain the within-treatments variation by subtraction.

$$\text{Between} = \qquad \text{Total} \qquad - \text{Within}$$

$$= \Sigma x^2 - \frac{(\Sigma x)^2}{N} - \Sigma x^2 + \left[\frac{(\Sigma x_1)^2}{n_1} + \frac{(\Sigma x_2)^2}{n_2} + \frac{(\Sigma x_3)^2}{n_3} \right]$$

$$= \left[\frac{(\Sigma x_1)^2}{n_1} + \frac{(\Sigma x_2)^2}{n_2} + \frac{(\Sigma x_3)^2}{n_3} \right] - \frac{(\Sigma x)^2}{N}$$

or, as it is often presented:

$$\frac{T_1{}^2}{n_1} + \frac{T_2{}^2}{n_2} + \frac{T_3{}^2}{n_3} - \frac{(\Sigma x)^2}{N}$$

where T_1 is the total of the treatment 1 results (that is, Σx_1), and so forth.

(It may not seem much quicker to calculate the between-treatment variation and get the within-treatment variation by subtraction, rather than doing the opposite. It is, in fact, less work, but the real advantage of taking this approach comes when we learn how to analyze more complex experiments in Chapter 11. There we will see that very straightforward extensions of this approach enable us to analyze even very sophisticated experiments very easily and painlessly).

To date we have been talking specifically about a hypothetical three-treatment experiment. This approach can be applied very naturally to the comparison of any number of groups. If there were k treatment groups, the between-treatments variation would be given simply by

$$\frac{T_1{}^2}{n_1} + \frac{T_2{}^2}{n_2} + \frac{T_3{}^2}{n_3} + \ . \ . \ . + \frac{T_k{}^2}{n_k} - \frac{(\Sigma x)^2}{N}$$

Note that the term $(\Sigma x)^2/N$ appears in the calculation of both the total variation and between-treatments variation. It is known as the "correction factor" and reflects the fact that both these types of variation are measured relative to the overall mean. The term, Σx, means, of course, sum the entire set of experimental results and is sometimes referred to as the overall total or the grand total (GT), that is,

$$\frac{(\Sigma x)^2}{N} = \frac{(GT)^2}{N}$$

To see these ideas in action, let's consider an experiment to compare the effects of three different food regimes on lipoprotein levels in human infants. A group of 10 full-term newborn infants was fed human milk (HM), another group of 10 received milk formula (MF), and a third group received nucleotide supplemented milk formula (NMF). After four weeks their high density lipoprotein (HDL) levels were measured (in mg/dl) and are presented below.

HM	56	63	45	41	71	60	78	50	68	62
MF	40	48	60	38	28	44	66	22	45	54
NMF	71	57	64	44	73	50	79	67	84	61

For the HM group

$$\Sigma x_1 = 56 + 63 + \ldots + 62 = 594$$

$$\Sigma x_1^2 = 56^2 + 63^2 + \ldots + 62^2 = 36,504$$

and

$$n_1 = 10$$

The HM mean is therefore 59.4 mg/dl.
For the MF group:

$$\Sigma x_2 = 445 \qquad \Sigma x_2^2 = 21,449 \qquad \text{and } n_2 = 10$$

with a mean of 44.5 mg/dl.
For the NMF group:

$$\Sigma x_3 = 650 \qquad \Sigma x_3^2 = 43,658 \qquad \text{and } n_3 = 10$$

with a mean of 65.0 mg/dl.
The overall data set has

$$\Sigma x = 594 + 445 + 650 = 1689$$

$$\Sigma x^2 = 36,504 + 21,449 + 43,658 = 101,611$$

and $N = 30$

The total variation is given by

$$101,611 - 1689^2/30 = 6520.30$$

The between-treatments variation is given by

$$\frac{594^2}{10} + \frac{445^2}{10} + \frac{650^2}{10} - \frac{1689^2}{30} = 2245.40$$

Now that we know the total variation and the between-treatments variation, we can easily obtain the within-treatments variation by subtraction.

The within-treatments variation is given by

$$\text{Total} - \text{Between} = 6520.30 - 2245.40$$

$$= 4274.90$$

COMPARING VARIATION

We now have the ability to split up the total variability in a set of experimental data into its two constituent parts: variation between treatment groups (resulting from random variation and possibly a treatment effect) and variation within treatment groups (resulting solely from random variation). As we saw in the previous section, we normally do this by calculating the total variation and the between-treatments variation and obtaining the within-treatments variation by subtracting the between-treatments variation from the total variation.

Before we compare the between-treatments variation and within-treatments variation to check for the existence of a treatment effect we must recall a lesson we learned back in Chapter 1. Remember that the size of a sum of squares tells us as much about the number of observations on which it is based as it does about the inherent variability of the sample. To get a truly reasonable measure of variability we divided the sum of squares by its degrees of freedom and arrived at the basic measure of variability, the variance. To make the comparison between the between-treatments and within-treatments variation fair, we must do exactly the same thing and convert them both to variances by dividing them by their respective degrees of freedom.

The total variation is based on all N observations and is measured relative to the overall mean. As you would expect, the total degrees of freedom are, therefore, $N - 1$.

The between-treatments variation is based on the k treatment means (if there are k treatment groups) and, again, is measured relative to the overall mean. The between-treatments degrees of freedom are, therefore, $k - 1$.

The within-treatments degrees of freedom can be obtained directly by subtracting the between-treatments degrees of freedom from the total.

$$\text{Within} = \text{Total} - \text{Between}$$

$$= N - 1 - (k - 1)$$

$$= N - k$$

This makes extremely good sense. The variation within a particular treatment group is based on the n observations for that treatment and measured relative to the treatment mean. It will, therefore, have $n - 1$ degrees of freedom. The within-treatments variation for the overall experiment is obtained by adding together the individual within-treatment variations. Similarly, the within-treatments degrees of freedom for the complete experiment are obtained by summing the individual within-treatments degrees of freedom:

Within treatment 1 $\mathrm{df} = n_1 - 1$

Within treatment 2 $\mathrm{df} = n_2 - 1$

. . .

. . .

Within treatment k $\mathrm{df} = n_k - 1$

Within all treatments $\mathrm{df} = N - k$

Dividing the between-treatments and within-treatments variation (sums of squares or SS) by their respective degrees of freedom converts them into variances (sometimes also referred to as mean squares or MS). These variances can now be fairly compared with one another. This comparison is carried out by simply dividing the between-treatments variance by the within-treatments variance. If this ratio (the variance ratio) is much larger than 1, then it indicates the existence of real differences between the treatments under study. If the variance ratio is close to 1, it indicates that both variances simply reflect random variation and that no true difference exists between the treatment means.

The major results for an analysis of this type (known as the **analysis of variance,** since it is based on measuring and comparing variances) are usually presented in a table (called, very sensibly, an analysis of variance table) which reflects how the various sums of squares and degrees of freedom relate to the totals for the overall experimental results.

SOURCE OF VARIATION	SUM OF SQUARES (Variability)	DEGREES OF FREEDOM	MEAN SQUARE (Variance)	VARIANCE RATIO (F)
Between treatments	2245.4	2	1122.70	7.09
Within treatments	4274.9	27	158.33	
Total	6520.3	29		

The preceding analysis of variance table summarizes the analysis of the food regimes experiment introduced in the section on Measuring Variation.

We have already commented that a variance ratio of around 1 indicates no true treatment differences and a variance ratio of much greater than 1 signifies that there is a real treatment effect. In the absence of any real differences, we could not, of course, realistically expect to find a variance ratio of exactly 1. It might be somewhat smaller or somewhat larger. We have to decide, therefore, whether our calculated variance ratio is so close to 1 that it might easily have arisen in the absence of any real treatment effect, or so much greater than 1 that it is very unlikely to have arisen accidently and almost certainly indicates a real treatment effect.

This is exactly the same problem we face when interpreting the results of a t test (or indeed any statistical test). In that situation we consulted a set of t tables that tabulated values of t (decision or critical values) that had certain (.05, .01, and so forth) very small chances of being exceeded simply by accident. Tables that specify critical values of a variance ratio are available (see table T/3 in the Appendix). They are known as **F tables,** in recognition of Sir Ronald Fisher, the British statistician and geneticist who developed the concept of the analysis of variance. They are slightly less convenient to use than t tables since they involve not 1 but 2 degrees of freedom; between-treatments (the numerator or "rows" df) and within-treatments (the denominator or "columns" df) respectively. However, the principle of their use is exactly the same.

As with the t test (and indeed all statistical tests) the .05 value has come to be used as the basic critical value. Variance ratios (F ratios) larger than this are accepted as evidence of the existence of real treatment effects, since there is only a remote ($p < .05$) chance of such a large value arising accidentally (that is, a type I error). We would, therefore, reject the null hypothesis of no treatment differences in favor of the alternative hypothesis of real differences in treatment effectiveness. If the variance ratio is less than the critical F value, we view it as quite possibly simply a fluctuation from a true value of 1, and hence, quite consistent with the null hypothesis of no true difference, which we would therefore retain.

The variance ratio of 7.05 obtained for the food regimes experiment exceeds the .05 critical value for F, with 2 and 27 degrees of freedom, of 3.25. (and also the .005 critical value of 6.45). This clearly leads us to the conclusion that the differences in high density lipoprotein levels between the food regimes are real and much too large to be attributable to merely chance variation, and that the chance of this conclusion being wrong (a type I error) is less than .005.

Statistics (and, indeed, all investigative science) is concerned with measuring the amount of variation we encounter in real life research, trying to explain it by attributing it to various possible causes, and finally checking how believable these explanations are by testing them in some appropriate way. The analysis of variance does all of these. Little wonder, therefore, that it is by far the most important of all the techniques of bio-statistics. It can be used to analyze highly sophisticated experiments in which there are a variety of possible explanations for the variation (see Chapter 11), relationships in which one variable influences another one (see Chapter 7), and combinations of both these situations (see Chapter 12).

No matter how the analysis of variance is used, however, its basic principles remain exactly as we have outlined them in this chapter. The total variation in a set of data is split ("partitioned") and attributed to a number of potential causes or effects, and the reality of each effect is tested by comparing the appropriate variation against a "benchmark" that is known to represent purely random variation.

ASSUMPTIONS IN THE ANALYSIS OF VARIANCE

As with the t test, the analysis of variance is built on two basic assumptions that must be valid if the test is to give fully reliable results. The assumptions are very reminiscent of those underlying the t test. This is not at all coincidental, since the two tests are very closely related. The analysis of variance can be used to compare any number of treatments and could certainly be used to compare two groups, the situation for which we developed the t test in Chapter 4. Is there a potential conflict here? Not at all. In this situation the two tests give identical results, the only trivial difference being that the F value is the square of the t value, since the analysis of variance is based on squared values. (Compare F tables for 1 between-groups degree of freedom with a set of t tables, and you will see that the F values are the square of the corresponding t values.)

The first assumption is that the data should follow a Normal distribution. This, of course, applies to all statistical tests that use the principles of the Normal distribution to measure the rarity of a certain result.

The second assumption is that all the treatment groups are equally variable. If their variances differed widely, then it would be meaningless to pool their sums of squares in order to arrive at a single within-treatments variance.

Luckily, like the *t* test, the analysis of variance is known to be fairly robust. In other words, it will continue to give reliable results even if the assumptions are not totally borne out in practice. If the assumptions are seriously wrong, however, then the analysis of variance should not be used. In this case, alternative (but less powerful) tests are available and will be discussed in Chapter 14.

EXPLORING EXPERIMENTAL RESULTS

Analysis of variance is a very elegant way of answering the general question, "Do any of the treatments differ from the others?" (This is certainly the most common question that presents itself when we consider a study involving multiple treatments. There is another, quite different, situation that arises when we have some very specific hypotheses about our experimental results, which we will discuss in the section on Asking Specific Questions. In this section we concentrate on the more common situation, that is, we have a general interest in all or any treatment differences). If the answer to that question is "No, there is no convincing evidence that any real differences exist," then there is nothing more to be said, and no further analysis of the data is needed or would be appropriate. If, however, the answer is "Yes, the treatments do genuinely differ from one another," then the general nature of both the question and the answer leaves a still unresolved question, "What exactly does that mean? Are all the treatments different from one another or are only some of them truly different?"

In the case of the food regimes experiment discussed earlier in this chapter, all three regimes might differ from one another; the human milk and milk formula regimes might be effectively identical but both different from the nucleotide milk formula regime; and so on. The situation becomes even more complex when more treatment groups are involved. In the remainder of this section we will consider some ways of resolving this dilemma and obtaining a detailed understanding of exactly where the differences in a multiple-treatment experiment actually are.

At the beginning of this chapter we discussed one possible approach to the comparison of several treatments, that is, the testing of all pairs of treatment means using a series of *t* tests. This will certainly provide a detailed picture of exactly which treatments differ from one another. Unfortunately, as we noted, repeated testing in this way very greatly increases the risk of obtaining an apparently significant difference purely by accident (that is, a type I error) somewhere in our comparisons, and hence, undermines the credibility of all the conclusions.

There is a very simple way out of this dilemma, although, like many solutions, it has a price to be paid. If we were prepared to set more stringent standards for each individual comparison (that is, use a more severe significance level as the basic critical value for our *t* tests), then we could ensure that the overall risk of making a type I error somewhere in our comparisons was held at an acceptable level (preferably at the .05 level we have used so often).

How much we have to tighten our standards depends on the number of comparisons involved. The more comparisons, the greater the risk of a type I error occurring somewhere, and consequently the more cautious we must be about interpreting each individual comparison. (Recall that we saw in the section on Comparing Several Groups that the overall risk of a type I error is $(1 - .95^c)$ if we test c pairs of means using a .05 critical value for each comparison). To maintain the overall risk of a type I error at .05, we should use

$$\frac{.05}{c}$$

where c is the number of comparisons to be made, as the basic significance level for testing each individual comparison. (Please appreciate that c is the number of comparisons or pairs, not the number of groups. If there are k groups, then c is given by $[k \times (k - 1)/2)]$.

This modification is sometimes referred to as the Bonferroni correction. For instance, if we wished to test five pairs of treatment means, each individual difference would be viewed as significant only if the t value exceeded the .01 critical value. To check that this tightening of standards for each individual test will ensure that the overall risk of a type I error is held at .05, try calculating $(1 - .99^5)$, which is the risk of committing a type I error somewhere in the testing process, if each of the five individual tests is based on a .01 decision value (that is, a .99 chance of not committing a type I error). The chance of not committing a type I error anywhere in the five tests is $.99^5$, and the chance of a type I error happening somewhere along the line is, of course, 1 minus this. It will work out very close to .05.

There is one advantage to t testing after an analysis of variance has been carried out, namely that the pooled within-treatments variance has already been calculated. Since we assume that all the treatment groups have the same variability, there is no need to restrict the pooled variance calculation to the two particular groups we happen to be comparing. Using the pooled within-treatments variance for the overall experiment has the considerable advantage that it is based on the widest possible evidence and hence, is inherently more reliable than a pooled variance calculated from only two groups. (This increased reliability is reflected in the fact that we correspondingly use the within-treatments degrees of freedom of $N - k$ for such a t test). In addition, there is the secondary advantage that the pooled variance does not have to be recalculated for each comparison.

The t test for treatment means following analysis of variance is therefore given by

$$t = \frac{\bar{x}_i - \bar{x}_j}{\sqrt{s_w^2 \left(\frac{1}{n_i} + \frac{1}{n_j} \right)}}$$

where

$$s_w^2 = \text{the within-treatments variance}$$

$$\bar{x}_i = \text{the mean of treatment group number } i$$

$$n_i = \text{the number of individuals in treatment group number } i$$

and so forth, and the df involved are given by the within-treatments df.

In life there is rarely any free lunch, and exploring group differences via a series of extra-stringent t tests does have a major disadvantage. Because we have been forced to adopt a more severe criterion for assessing each individual comparison (to protect us against committing a type I error somewhere along the line), it now becomes much more difficult to obtain significant results. The risk of failing to pick up quite genuine treatment differences (that is, the risk of a type II error) is inevitably substantially increased. A technique that is overcautious in this way is said to be highly conservative or to lack power. This is clearly not a very satisfactory situation, although it is unavoidable. In addition, having to recalculate t values for each treatment pair is rather tedious.

When talking about analysis of variance, we have assumed that there may be differing numbers of patients in the various treatment groups. There are, however, major advantages in having the same number of patients in each group. (For the more complex applications of analysis of variance discussed in Chapter 11, it is important that the patient numbers be equal.) When the group sizes are equal, we can use several methods for comparing treatment means that overcome some of the drawbacks we have just noted. The material in the rest of this chapter assumes that all treatment groups will have the same number of patients. This is, quite simply, good experimental practice and enables you to use a variety of ways for further exploring your results.

Least Significant Difference

As you may recall from Chapter 4, a t test involves expressing the difference between two means (the top line of the t test formula) in terms of the standard error of the difference (the bottom line). The standard error depends on the within-treatments variability and the size of the treatment groups. The within-treatments variability is most reliably measured by the within-treatments variance and is constant for all comparisons (remember we assume that all the groups in the study have the same variability and that the pooled within-groups treatments variance therefore applies in all situations). If, in addition, the treatment groups are all the same size, then the standard error of the difference (the key element in testing differences between means) will be constant, irrespective of which particular treatments we are comparing. This fact can be used to simplify considerably the way we compare treatment means.

When the treatment sizes are equal (and we can simply use n to denote the sample size in each treatment), we can slightly rearrange the t test formula in a way that proves to be very useful.

$$t = \frac{\bar{x}_i - \bar{x}_j}{\sqrt{s_w^2\left(\dfrac{1}{n} + \dfrac{1}{n}\right)}}$$

$$= \frac{\bar{x}_i - \bar{x}_j}{\sqrt{(2s_w^2)/n}}$$

that is,

$$\bar{x}_i - \bar{x}_j = t\sqrt{2\,s_w^2/n} \quad \text{or} \quad \sqrt{2}\,t\sqrt{s_w^2/n}$$

In other words, the larger the difference between any pair of means, the larger the t value that would result from a t test (since the pooled variance and group size are constant for the experiment).

In conventional t testing we use the appropriate .05 value of t as our basic critical value. Using this approach, two treatments are viewed as significantly different if the resultant t value is greater than this critical value. Looking at the rearrangement above we can see that, if the difference between the two means

$$\bar{x}_i - \bar{x}_j \text{ is greater than } t_{.05}\sqrt{2s_w^2/n} \quad \text{or} \quad \sqrt{2}\ t_{.05}\sqrt{s_w^2/n}$$

when $t_{.05}$ is the appropriate .05 value of t, then the treatments must be significantly different since the t value calculated from them will exceed the .05 critical value.

The value

$$\sqrt{2}\ t_{.05}\sqrt{s_w^2/n}$$

when n is the number of individuals per group,

s_w^2 is the within-treatments variance,

and $t_{.05}$ is the .05 critical value of t based on

the within-treatments degrees of freedom,

is known as the **Least Significant Difference** (LSD).

Mean pairs that are farther apart than this are significantly different from one another. Mean pairs whose difference is less than the LSD are not significantly different. (The LSD is the smallest difference you can observe between two group means and still find them to be significantly different at the conventional .05 level).

Since the Least Significant Difference is a single constant value for the whole experiment, it is possible to compare all the treatment means (in effect, carry out the equivalent of a series of t tests) easily and painlessly by arranging them in order of size and simply comparing their difference with the LSD. Any pair of means exceeding the LSD can be viewed as significantly different.

The LSD for the food regimes experiment is

$$\sqrt{2} \times 2.052 \times \sqrt{(158.33/10)}$$

where   2.052 is the .05 critical value of t for 27 df

       158.33   is the within treatments variance

                (see the section on Comparing Variation)

and 10   is the sample size in each treatment group

$$= 11.547$$

In other words, treatments that differ by more than 11.547 are significantly different because their comparison via a conventional t test would result in a significant t value.

This suggests that the human milk and nucleotide milk formula regimes (with means of 59.4 and 65.0 respectively) do not effectively differ from one another and that the milk formula regime (with a mean of 44.5) results in high density lipoprotein levels that are significantly lower than either of the other regimes. Before taking these conclusions too seriously we should be aware, however, that the LSD has a major, and very obvious, flaw, which means that it should not be used in practice. The basic idea, however, is very appealing, and in the next section we will see how it can be very easily modified to overcome this problem.

Tukey's Multiple Comparison Test

The Least Significant Difference is appealing and very convenient to use. However, it suffers from one damning drawback. Since the LSD (which is simply a very convenient way of doing lots of t tests) is based on a .05 critical value of t, the risk of incurring a type I error somewhere in all the comparisons will be unacceptably high. We have also already met the solution to this problem in the section Exploring Experimental Results, namely, to use the $.05/c$ critical value of t (where c is the number of comparisons to be made), thus ensuring that over all c comparisons we run only a .05 chance of getting an apparently significant result by accident.

An acceptable way of comparing the means of several groups is known as **Tukey's multiple comparison test** and is used in exactly the same way as the LSD, except that it is based on a more severe critical value for t:

$$\sqrt{2}\, t_{.05/c}\, \sqrt{s_w^2/n}$$

Obtaining the appropriate value of t is, of course, not all that easy since t tables are usually tabulated at a very limited number of points (.05, .01, and so forth). Four groups, for instance, require the comparison of six different pairs of means, and very few t tables will tabulate the .00833 (that is, .05/6) value of t. Tables (known as "studentized range" tables) have been prepared that give the appropriate precalculated values of

$$\sqrt{2}\, t_{.05/c}$$

(These tables also incorporate a minor modification to ensure that the test is not excessively conservative.)

These values, known as q values, (see table T/4 in the Appendix) are tabulated by the appropriate (that is, within-treatments) degrees of freedom and by k, the number of treatment groups (k, of course, determines c, the number of comparisons, and is more convenient to use).

Tukey's multiple comparison test is, therefore, given by

$$q\sqrt{s_w^2/n}$$

Again the treatments are simply arranged in order of their means. Any pair of means whose difference exceeds the Tukey value can be regarded as being significantly different from one another.

For the food regime experiment, the appropriate Tukey's value is given by

$$3.51\,\sqrt{158.33/10}$$

where 3.51 is the q value for 3 groups and 27 within-groups degrees of freedom

158.33 is the within groups variance

and 10   is the sample size in each group

$$= 13.967$$

Note that the Tukey's value is noticeably bigger than the LSD value. As we said, we are setting more severe standards and demanding more impressive evidence before we will accept that any two groups do indeed differ from one another. By doing this, we have protected ourselves against the risks posed by multiple testing. (The down side to this is that tougher standards will inevitably mean that more genuine differences will be overlooked, that is, that the type II error will rise).

There are a number of multiple-comparison tests available; Duncan's test, the Newman-Keuls test, and others; all based on essentially the same philosophy as Tukey's test. They differ primarily in the extent to which they allow for the inevitable conservatism of multiple testing, and all yield broadly comparable results. One common and very useful way of displaying the results of a multiple-comparison test is to arrange the group means in order of size and denote those means that differ by less than the Tukey value (that is, are not significantly different) with a common underline.

The food regime experiment would be summarized by

MF	**HM**	**NMF**
44.5	59.4	65.0

indicating that a more detailed appraisal of our earlier conclusion (that the food regimes differ in terms of their high density lipoprotein levels) would be that the milk formula regime produces substantially lower high density lipoprotein levels than either the human milk or nucleotide milk formula regimes, which do not effectively differ from one another.

Our discussion has assumed that the treatment groups are all the same size. This is ideal but may not always be the case. Unequal group sizes present no problems as far as an analysis of variance goes (at least for the simple situations we have discussed in this chapter), and Tukey's multiple-comparison test can be used, on an approximate basis, in this situation by calculating a single "composite" sample size for the two unequal size groups being compared. The most appropriate size to use is not the mean of the two sample sizes but the slightly smaller, and hence more conservative, harmonic mean, which is given by

$$\frac{2 n_1 n_2}{n_1 + n_2}$$

To compare groups of size 10 and 16, we should therefore calculate a Tukey's value based on an *"n"* of 12.31, rather than 13.

ASKING SPECIFIC QUESTIONS

The previous section concerned itself with the very commonly encountered situation in which we have a general interest in the existence of differences between a number of treatment groups. The strategy we have developed to cope with this general interest involves carrying out a global test for the existence of differences and if, and only if, the differences are significant, following it up with an examination of the various treatment pairs.

There is, however, another situation, somewhat less common, in which an investigator embarking on a multiple-treatment experiment will have some very specific questions in mind, that he or she hopes the data will illuminate. An experiment might, for instance, involve comparing a control or untreated patient group, a group actively treated with medication at a relatively low dosage, and a group actively treated with the same medication at a higher dosage. The researcher might be interested in answering the specific question "Does actively treating this condition, irrespective of the dosage used, have any effect at all on the patient?" In other words, if we looked at the evidence afforded by the two actively treated groups and compared (or "contrasted") it with the untreated group, would we see any difference? Another experiment might involve comparing groups that all receive a particular drug, but in a range of equally spaced dosage levels. Here asking the specific question "Is there a trend in patient mean response that reflects the trend in dosage levels?" makes far more sense than asking the general question "Are there any differences?"

If specific questions are appropriate (these questions will flow from the nature and logic of the experiment and its underlying research objectives), then you should address these questions directly through the technique we will discuss in this section. It is not necessary, and not appropriate, to test the general question of "Any differences?" in the way we outlined in the section in this chapter that discussed Comparing Variation, since more direct and relevant questions are crying out to be addressed. Asking and answering specific questions involves the same general principles we have already established in this chapter, but applied in a more focused way. (We will, for example, still have to calculate the between-treatments variation, but this is used simply to yield the within-treatments variation, rather than as a key aspect of the testing process).

To answer a specific question involves three basic steps:

1. Describing the question in a simple, concise way that details exactly how we wish to compare (or "contrast") the treatment means.

2. Calculating the variation between treatments that is due to this specific difference (or "contrast").

3. Testing this specific between-treatments variation against the within-treatments variation in the usual way.

The initial and key step in testing specific research questions is to define the question. Remember that what you want to do is to compare or contrast the differences between the treatment means in a specific way that is of particular interest to you. (In the jargon of statistics, this is called "forming a **linear contrast**.") This is achieved by using numerical values, or "weights," to define the question. Their use is best explained by a simple example.

Suppose, for instance, you have carried out an experiment involving three treatment groups:

1. A control group receiving a dummy treatment and that gave the following results:

$$\bar{x}_1 = 10.3 \qquad n_1 = 6 \qquad \text{(that is, } T_1 = 61.8\text{)}$$

2. A group receiving an active treatment with a certain drug at a relatively low dosage level:

$$\bar{x}_2 = 14.7 \qquad n_2 = 6 \qquad \text{(that is, } T_2 = 88.2\text{)}$$

3. A group receiving an active treatment with the same drug at a relatively high dosage level:

$$\bar{x}_3 = 15.6 \qquad n_3 = 6 \qquad \text{(that is, } T_3 = 93.6\text{)}$$

Using the approach developed in the section on Measuring Variation, the between-treatments variation was calculated to be 96.52, and the within-treatments variation was calculated to be 167.96.

You wish to pose the specific question "Does the active treatment (irrespective of dosage) actually differ at all from the dummy treatment?"

The linear contrast that defines this specific question is:

$$C = 2\bar{x}_1 - \bar{x}_2 - \bar{x}_3$$

Note that the dummy treatment mean, $\bar{x}_1$, and the active treatment means, $\bar{x}_2$ and $\bar{x}_3$, have contrasting signs, denoting that we are interested in the contrast or difference between them.

In addition, the two active treatments have the same weight (both -1), denoting that they are to be considered as totally equivalent, that is, we wish to contrast the dummy treatment with the active treatments, without distinguishing between high or low dosage levels.

Also very important is that the weights total to zero. Between-treatments variation is measured about the overall mean. It is, therefore, vital that we retain exact balance about the overall mean and not "tip" the data toward either the higher ($\bar{x}_2$ or $\bar{x}_3$) or lower ($\bar{x}_1$) treatment means.

Surely a variety of other weights, such as $[-2, +1, +1]$, $[+1, -\frac{1}{2}, -\frac{1}{2}]$ would meet all these criteria? Indeed they would, and rest assured that all such weights would define this particular question equally well and give exactly the same answer. Any set of weights that reflect the question you wish to ask and that total to 0 are quite acceptable.

The variation between the treatment means that reflects the particular effect defined by the contrast is given by

$$\frac{nC^2}{a_1^2 + a_2^2 + a_3^2}$$

or more generally

$$\frac{nC^2}{\Sigma a_i^2}$$

where a_1 is the weight associated with treatment mean 1, and so forth.

Since variation is measured as a sum of squares, the contrast itself must be squared. The presence of n, the number of patients in each treatment group, is a consequence of the fact that a treatment mean reflects the impact of that treatment on all the individuals who receive it. The squares of the weights on the bottom line are included to ensure that any valid set of weights (which, of course, were squared when the contrast was squared) will give the same results. Doubling the weights simply quadruples the values top and bottom and leaves the final answer unaltered.

For our hypothetical experiment, the sum of squares due to the difference between active and dummy treatments is, therefore:

$$\frac{6\,[(2 \times 10.3) - 14.7 - 15.6]^2}{4 + 1 + 1} = 94.09$$

A linear contrast, since it represents one very specific and absolutely fixed question, has one degree of freedom associated with it.

The dummy/active treatment effect described by this linear contrast can now be tested in the usual way by comparing it with the within-treatments ("error") variation. (Note that some of the general between-treatments variation of 96.52 remains unexplained. This represents variation resulting from treatment differences not included in our specific question. Since we have contrasted the dummy treatment with the active treatments, irrespective of dose, the difference we have not considered is the difference between the two active dosage levels.) A complete analysis of variance based on the dummy/active comparison is, therefore, summarized in the following table.

SOURCE OF VARIATION	SS	DF	MS	F
Between treatments				
Due to dummy/active differences	94.09	1	94.09	8.40
Due to other differences	2.43	1	2.43	0.22
Within treatments	167.96	15	11.20	
Total	264.48	17		

The results (each tested against F with 1 and 15 df) confirm that very significant differences do exist between the dummy treatment and the active treatments and that no other significant differences exist in the data set (that is, that differences in the active dosage levels do not produce any corresponding differences in treatment effectiveness).

In a complex experiment with many treatments, you may wish to pose several specific questions. You can do this by asking them in the way we have just outlined. However, it is very important to ensure that the questions posed are completely independent of one another. If this is not the case, the between-treatments sums of squares cannot be subdivided since some of the variation will "overlap" more than one question. Independence of contrasts can easily be verified by comparing the respective weights as outlined below.

$$\text{Contrast 1} \quad a_1 \quad a_2 \quad a_3$$
$$\text{and Contrast 2} \quad b_1 \quad b_2 \quad b_3$$

will be independent if

$$a_1b_1 + a_2b_2 + a_3b_3 = 0$$

(more generally, if

$$\Sigma a_ib_i = 0)$$

In this situation the questions posed are totally unrelated to one another. Contrasts of this type are said to be orthogonal.

If, for instance, in the previous example we had decided to deliberately ask the question, "Does dosage level make a difference?" (that is, ignore the dummy treatment and contrast the two active treatments), we could have used the following weights:

CONTROL	ACTIVE (low)	ACTIVE (high)
0	-1	$+1$

When these are compared with the dummy/active weights, that is,

CONTROL	ACTIVE (low)	ACTIVE (high)
2	-1	-1

it is immediately clear that the two questions are independent and that the between-treatments variation can be split up to provide a separate, legitimate answer to each question.

The idea of defining linear contrasts can be used to answer a wide variety of specific questions, including testing for trends in a set of results. Suppose, for instance, that in an experiment comparing responses to different dosage levels, the doses administered were of sizes spaced at equal intervals (for example, 5 mg, 10 mg, 15 mg). A logical question would be, "Does the patient's response increase (or decrease) uniformly with the dose?" This could be defined by the following sets of weights and tested in the usual way.

THREE EQUALLY SPACED TREATMENTS

$$-1 \qquad 0 \qquad +1$$

FOUR EQUALLY SPACED TREATMENTS

$$-3 \qquad -1 \qquad +1 \qquad +3$$

and so forth.

Note that these weights total to zero. Also note that they reflect the consistent trend from low to high and the equal spacing between the treatments.

Such a contrast will again have 1 degree of freedom. If the additional between-treatments variation not accounted for by this trend is not significant, then the linear response to dosage levels (assuming it is significant) is the only pattern evident in the data. If, however, it is significant, then other trends are also present.

Key Terms

Treatment effect

Within-treatment variation

Between-treatment variation

Analysis of variance

F tables

Least significant difference

Tukey's multiple comparison test

Linear contrast

PROBLEMS

1. In an experiment to investigate the effects of differing doses of insulin on blood sugar levels, 18 rabbits were randomly allocated to three treatment groups (A, B and C) so that each treatment was administered to six rabbits. Initial, pre-treatment blood sugar levels were measured for each rabbit, and the appropriate insulin treatment administered. After five hours the blood sugar levels were remeasured and the percentage decrease in blood sugar was calculated.

The results obtained for each treatment are outlined below. Test for evidence of differences between the treatments.

TREATMENT		
A	B	C
13.6	19.2	23.8
17.4	23.9	28.0
21.8	26.0	31.2

TREATMENT

A	B	C
14.7	16.4	16.2
16.3	19.5	20.4
21.2	22.1	24.2

2. To evaluate the effect of aerobic training on aerobic performance, athletes from four disciplines were tested on a treadmill sloped at 5 degrees and were subjected to a graded workload. The athletes were tested until the point of subjective exhaustion was reached, and their maximum oxygen output (in l/min) during the test was recorded. The results are given below. Analyze and interpret.

PENTATHLON	MARATHON	FOOTBALL	CANOEING
5.3	4.3	4.5	4.6
5.9	4.5	4.4	5.0
4.8	5.0	3.9	5.3
4.5	3.8	5.2	5.8
6.1	5.4	4.1	5.5
5.5	4.7	5.0	6.0
5.0	4.3	5.4	5.1
5.6	5.2	4.7	4.8
4.6	4.8	4.2	5.6
5.3	4.1	4.9	5.3

3. As a part of a study of the functioning of the hypothalamic/pituitary/gonadal system in psychosis, plasma concentrations of luteinizing hormone (LH) were measured in 12 young males demonstrating symptoms of schizophrenia, 12 young males demonstrating symptoms of mania, and a comparable group of 12 controls. Test the data for any evidence of difference among these three groups.

LH (IU/l)

Mania	Schizophrenia	Control
2.9	2.7	2.6
3.8	3.9	2.2
3.3	1.3	3.9
2.1	2.0	3.2
5.0	4.1	1.4
4.2	2.9	0.7
2.3	0.9	2.4

LH (IU/l)

Mania	Schizophrenia	Control
5.4	2.3	3.4
3.7	1.8	1.9
3.2	3.3	4.2
4.4	2.4	2.9
3.5	3.0	2.2

4. The creatinine levels of 30 patients (10 per group) with low renin, normal renin and high renin hypertension (as established from measurement of concurrent plasma renin activity in untreated patients with essential hypertension) were measured. Use these results to test the hypothesis that creatinine levels are influenced by the extent of plasma renin activity in such patients.

CREATININE (mg/dl)

Low renin	Normal renin	High renin
1.45	1.53	1.60
1.20	1.25	1.34
1.63	1.36	1.17
1.85	1.18	1.30
1.41	1.73	1.12
1.08	1.41	1.43
1.56	0.96	1.21
1.49	1.34	0.83
1.35	1.04	1.23
1.73	1.61	0.91

5. The emergence of choloroquinine-resistant malaria has caused a need for new anti-malarial drugs. In an investigation of potential treatments, 40 patients with significant malarial parasite *(P. falciparum)* counts were randomly assigned to four treatment groups, and the time from initiation of treatment until clearance of parasites (as estimated by examining blood films every four hours) was recorded for each patient.

PARASITE CLEARANCE TIME (hours)

Oral quinine	Intramuscular quinine	Intramuscular quinghaosu	Oral quinghaosu
100	120	104	80
120	104	80	68
92	84	72	60
80	104	92	92

PARASITE CLEARANCE TIME (hours)

Oral quinine	Intramuscular quinine	Intramuscular quinghaosu	Oral quinghaosu
128	116	88	84
104	80	112	48
76	92	68	72
96	108	80	96
132	100	52	40
104	136	60	56

Use this data to test for any evidence of differences between drugs (Quinine vs. Quinghaosu) and any evidence of differences between routes of administration (intramuscular vs. oral). Are these independent questions?

6. In a study of the impact of smoking on general cardiovascular fitness, six non-smokers, six light smokers, six moderate smokers, and six heavy smokers were subjected to a period of sustained physical exercise. After three minutes' rest their heart rates were measured. The results are presented below.

NONSMOKERS	LIGHT SMOKERS	MODERATE SMOKERS	HEAVY SMOKERS
69	55	66	91
52	60	81	72
71	78	70	81
58	58	77	67
59	62	57	95
65	66	79	84

Test the hypothesis that a heavier smoking habit should be reflected in a higher heart rate.

7. As part of an investigation of the long-term impact of diet on cardiovascular disease, total cholesterol (in nmols/l) was measured for 48 volunteers who adhered to 4 different dietary lifestyles (12 subjects/lifestyle); the lifestyles were vegan (no meat or milk products), vegetarian (milk products but no meat), fish (fish and milk products but no beef, pork, chicken, and so forth), and meat (unrestricted animal products). You hypothesize that total cholesterol levels should reflect the extent to which individuals use animal products in their diets. Use the experimental evidence to investigate this hypothesis.

VEGAN	VEGETARIAN	FISH	MEAT
5.1	4.8	4.9	4.3
4.2	3.8	6.4	4.8
3.9	4.2	5.1	5.7
3.0	5.5	3.7	5.9
5.5	4.7	3.9	4.6
4.5	5.3	4.9	5.8
3.4	5.1	5.3	4.9
4.6	4.3	5.4	6.7
3.7	3.5	6.0	5.4
4.1	5.9	4.8	6.3
5.9	6.4	5.7	4.0
4.7	4.9	4.3	5.2

Analyzing Qualitative Data

QUANTITATIVE AND QUALITATIVE DATA

In the very first section of this book we considered the different types of measurements that can be encountered in the life sciences. The most dramatic distinction that we encountered was between quantitative (or continuous) measurements and qualitative (or categorical) measurements. To recap, quantitative data involves locating individuals on some sort of continuous scale (such as blood pressure measurements or weight measurements). Our whole strategy for handling data of this type is based on the premise that the data tends to occur in a basic pattern (that is, the Normal distribution discussed in Chapter 2) that can be described by such characteristics as the mean, standard deviation, and standard error. In Chapters 2 through 5 we concentrated on learning how to explore quantitative data, principally by expressing differences in terms of standard deviations or standard errors (depending on whether we were evaluating individuals or means), that is, standardizing them.

Qualitative data, however, represents a completely different type of measurement. In this, results are simply placed in different categories (such as male or female, or blood type O, A, B, or AB) rather than located on a scale. Here the idea of a Normal pattern described by a mean and standard deviation is, of course, usually quite inappropriate (although there may be certain situations in which it is a useful approximation; see Chapter 2). It is as patently silly to try to average blood types as it is sensible to average blood pressures. To summarize qualitative information, we often quote the percentage of individuals falling into each category (percentage blood type O, and so forth). In many ways this summarizes our information much more completely than the equivalent act of calculating a mean and standard deviation for a quantitative variable, since it effectively defines all the results we have observed.

Since qualitative measurements are intrinsically quite different from quantitative measurements, it should come as no surprise that the statistical methods we employ to study them are also quite different. In general we can say that these methods concentrate on comparing the percentages, or more precisely, the actual numbers of individuals, in the various categories, whereas the quantitative methods concentrate on comparing means. (Qualitative variables are usually very easy to observe and record. Their statistical analysis is correspondingly simple and usually involves much less calculation than the equivalent analysis for a set of quantitative data.)

━━ OBSERVED AND EXPECTED RESULTS ━━━━━━━━━━━

Medical genetics is one field in which qualitative data and categorical observations are of fundamental importance. Many inherited characteristics are, of course, controlled by the actions of many genes acting together and are essentially quantitative in nature. However, many of the most important inherited conditions are controlled by the influence of single genes. The resulting phenotypes (or genotypes, if they can be distinguished) are clearly qualitative in nature. Considering the results of a simple medical genetics study provides an ideal framework for introducing and developing a general method for analyzing qualitative information.

Cystic fibrosis is an inherited condition proven to be controlled by the action of a single recessive gene, c. Sufferers of the disease, therefore, possess the homozygous genotype, cc. For an offspring to be at risk of the disease, both parents must be carriers of the disease and possess the heterozygous genotype Cc.

If both partners are, in fact, carriers of the disease, the basic Mendelian theory of random gene assortment suggests that on average we would expect to see 25% of their children "normal," 50% unaffected but carrying the disease, and 25% explicitly affected by cystic fibrosis. In other words, we would expect to see, on average, 75% of the offspring unaffected by cystic fibrosis and 25% suffering from the disease (Figure 6-1).

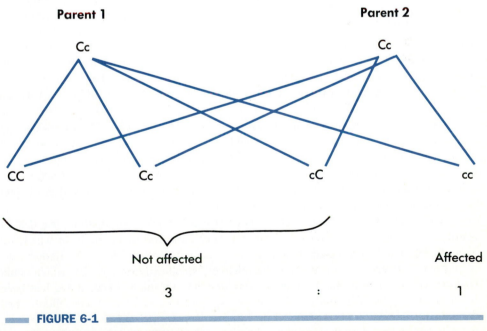

FIGURE 6-1

Inheritance of cystic fibrosis.

Suppose that we have been following the pregnancies of 86 couples known to be carriers of cystic fibrosis. Of the 86 children (we will assume, for simplicity, that none of the pregnancies result in multiple births), 16 were afflicted with cystic fibrosis; the other 70 were not. One natural question to ask is whether the results we observed in practice are consistent with what our knowledge of genetic theory suggested.

If the basic Mendelian theory was correct, and the genes were randomly mixed at mating, we would have expected to find 25% of the children affected and 75% unaffected. For a study based on 86 children we would, therefore, have expected to find 21.5 (25% of 86) children affected and 64.5 children unaffected. It would be unreasonable to demand that the results we actually observe in practice be identical to those we expected to see on the basis of our theory. They should, however, be very close to what we expect, if that theory is correct. If, on the other hand, they differ very markedly from what we expected, then surely this has to be a strong indication that the basic theory may well be wrong. (If theory clashes with reality, then our faith in the theory has to be severely shaken.)

Checking the agreement between the results we observe in practice and the results we expected to see on the basis of some theory or hypothesis thus gives us a very practical method of investigating or testing the probable correctness (or otherwise) of our basic hypothesis and forms the cornerstone of the way we approach and test questions involving qualitative or categorical measurements.

THE CHI SQUARED MEASURE

Having developed the idea that the level of agreement between the number of individuals observed in practice and the number of individuals predicted by some theory or hypothesis is a very sensible indication of how the hypothesis stands up to scrutiny (with a high level of disagreement suggesting that the hypothesis looks pretty unimpressive), the challenge is now to come up with a reasonable measure of the extent of the agreement or disagreement. To help us develop such a measure, O will denote the **observed values** of a qualitative variable and E will denote the **expected values** (the values predicted by some theory or hypothesis). An obvious measure of the agreement between our observed and expected values is, therefore, $(O - E)$, the difference between them.

Clearly, however, we would be just as interested to find observed values that are much smaller than we expected as we would be to find observed values that are much bigger than expected. It is the magnitude of the difference rather than its sign that tells us how good the agreement between the observed and expected values is. Luckily, we can remove the complication of sign by the very simple device we employed in Chapter 1, namely squaring the differences, that is, $(O - E)^2$.

We should, however, interpret this new measure of agreement with a great deal of caution. Suppose we had been involved in a very large follow-up study in which we expected to find 320 affected offspring. If we actually found 330 such children, we would not feel very concerned about the relatively small difference of 10, which could easily have arisen from natural random variation. On the other hand, if we had been involved in a very small study in which we expected to find 15 affected children, and we actually found 25, we might well be extremely concerned. After all, the difference of 10 is now relatively large and represents more than a 66% increase in what our Mendelian theory tells us to expect. It is fairly clear, therefore, that only a measure of agreement that reflects the relative size of the differences can be expected to lead to a sensible conclusion. Such a measure is obtained by calculating our differences relative to the size of the expected value (we use the expected value since, as always, we start off believing the null or initial hypothesis to be true, and the expected value is the value that is defined by this hypothesis):

$$(O - E)^2 / E$$

This represents the agreement between each individual expected value and the corresponding observed value. That is, it provides us with a measure of the level of agreement for each category of the categorical measurement. To obtain an overall measure of agreement, we simply sum all the individual agreement values over all the categories to give our final measure:

$$\Sigma\, (O - E)^2 / E$$

This value is known as the χ^2 or **Chi squared** (pronounced ki squared) statistic, and it measures the agreement between experimentally obtained (observed) results and the (expected) results suggested by a theory or hypothesis. Small values of χ^2 show good agreement and suggest that the hypothesis is plausible and should not be rejected. Large values of χ^2 show poor agreement and suggest that the basic hypothesis is probably wrong and should be rejected. What we are doing, therefore, is exactly the same in principle as what we did in Chapters 4 and 5 when we calculated a value of t or F and used it to test a hypothesis about the means of two or more groups.

We now face a problem similar to the problem we encountered with our t and F values. What value of χ^2 is so big that we would consider it to be very strong evidence that our original hypothesis must be wrong? As with most statistical tests, tables of χ^2 values that would be exceeded very rarely when the null hypothesis is correct have been calculated, and a set is presented in table T/5 of the appendix.

Before we decide on the exact critical value of χ^2 we will use to test our hypothesis, we should consider one simple but important fact. Our χ^2 value has been obtained by adding together values obtained from a number of differences. The more differences involved in our final total, the larger we must expect our final χ^2 value to be, all other things being equal. Therefore, any final decision we make on our overall χ^2 measure of agreement must take some account of the number of differences or categories involved in its calculation.

As you may anticipate, the appropriate value of χ^2 is determined, not by the number of differences (or expected values) involved, but by the number of truly independent expected values involved (the degrees of freedom). In our genetics example, we have 2 expected values and 2 differences. However, we have only 1 degree of freedom. Once we know that we expect 75% of children to be unaffected, it follows automatically that we have to expect 25% to be affected. (If we have 86 children in total and we expect to see 64.5 unaffected, we have no choice but to expect to see 21.5 affected). We, therefore, have only one truly independent expected value and consequently, only one degree of freedom. In general (with some reservations we will discuss in the section on Goodness of Fit Tests), the degrees of freedom are given by the number of categories or expected values -1.

If we consult a table of χ^2 values we will see that the standard 5% (p = .05) value of χ^2 for 1 degree of freedom is 3.84. (Very observant readers may note that 3.84 =

1.96^2). A χ^2 value as large as or larger than this will only happen five times in every 100 if our basic hypothesis is correct. If our calculations produce a result larger than this, it would seem more sensible to conclude that our basic null hypothesis did not match reality, was therefore in error and must be rejected.

We are, of course, still faced with the problem of mistaken conclusions that we discussed in Chapter 4. Quite large differences between observed and expected values can sometimes arise purely from chance. In fact, differences that give rise to χ^2 values of 3.84 or larger will probably arise purely from natural variation about five times in every 100 investigations. In these situations, we would mistakenly (but understandably) conclude that our null hypothesis was incorrect. If, however, we seek to guard against these Type 1 errors by setting more severe standards and using larger values of χ^2 (say, the .01 value of 6.63, for 1 degree of freedom) as our decision value, we will markedly increase the number of times we fail to pick up important discrepancies between theory and reality (Type II errors). As before, the use of the .05 value as the basic decision or critical value is generally accepted as achieving a reasonable balance between the two types of errors.

Let us return to our cystic fibrosis example. Remember that the 86 children turned out to consist of 16 affected children and 70 unaffected children. Are these results consistent with the basic theory of random gene assortment, or are there serious discrepancies that require investigation?

	CATEGORIES	
	Affected	**Unaffected**
Observed results	16	70
Expected results	21.5	64.5
$O - E$	-5.5	$+5.5$
$(O - E)^2$	30.25	30.25
$\dfrac{(O - E)^2}{E}$	$\dfrac{30.25}{21.5}$	$\dfrac{30.25}{64.5}$
	$= 1.407$	$= 0.469$
$\chi^2 = \Sigma\,(O - E)^2 / E$	$= 1.407 \qquad +$	0.469
	$= 1.876$	
Degrees of freedom	$=$ Number of categories $- 1$	
	$= 2 - 1 = 1$	

Consulting a set of χ^2 tables for one degree of freedom shows that the .05 critical value for χ^2 is 3.84. The calculated value of 1.876 does not exceed this. We, therefore, conclude that the differences between the observed and expected values are small and could quite easily have arisen from simple random variation. There is, therefore, no evidence to justify doubting the concept of random gene assortment. (More formally, we cannot reject the null hypothesis of random gene assortment. Remember that we can never actually prove the null hypothesis, any more than we can or should have to prove a person innocent. We can simply note that there is no evidence to disprove it.)

— YATES CORRECTION

There is one feature of the cystic fibrosis data that may have concerned you. For a study size of 86 children, we expect to find, on the basis of Mendelian theory, 21.5 children affected and 64.5 not affected. Common sense tells us that, even if we had no random variation at all, these expected results could never be reproduced exactly by the observed results. The closest possible results are 21 affected and 65 unaffected (or 22 and 64). Results like these would give a χ^2 value of

$$\frac{(21 - 21.5)^2}{21.5} + \frac{(65 - 64.5)^2}{64.5}$$

$$= \frac{(.5)^2}{21.5} + \frac{(.5)^2}{64.5}$$

$$= .0155$$

Even though our results follow the null hypothesis as closely as is possible in practice, we still obtain a definite (albeit small) value. Since the χ^2 statistic is designed to measure variations from the null hypothesis, this seems rather unreasonable. In effect the χ^2 value always slightly overestimates the differences between our expected and observed results simply because the observed results must, by their nature, be discrete whole numbers and cannot, even in the ideal case, perfectly match the expected (continuous) values.

We can correct for this overestimation very easily, by decreasing the absolute values of our $(O - E)$ differences by .5. (We use .5 because it is the maximum possible discrepancy between an expected value and the corresponding "ideal" observed value. By correcting for the worst possible case, we ensure that all other situations are also corrected.) This means that a value of -5.5 becomes -5.0, and a value of $+5.5$ becomes $+5.0$.

Recalculating the cystic fibrosis example with this correction incorporated gives us

$$= \frac{(-5.0)^2}{21.5} + \frac{(+ 5.0)^2}{64.5}$$

$$= \frac{25}{21.5} + \frac{25}{64.5}$$

$$= 1.550$$

The Chi squared value is slightly reduced but our conclusions are, of course, unaffected.

This correction is known as **Yates' correction,** or sometimes as the continuity correction. It usually has only a fairly minor influence on the final χ^2 value and is normally only used in a χ^2 calculation when the degrees of freedom equal 1. Chi squared calculations involving two or more degrees of freedom are always calculated using the original, uncorrected differences.

It must be pointed out that the use of Yates' correction is somewhat controversial. Since it corrects for the maximum possible discrepancy (.5) between a continuous expected value and the closest possible discrete observed value, it also inevitably overcorrects in other, less dramatic situations. As a result the Chi squared test with Yates' correction tends to be somewhat conservative and loses a little power. There is a body of statistical opinion that argues that, for this reason, it should never be used. However, the majority opinion is that it is preferable to err on the side of caution and that therefore Yates' correction should always be used when one degree of freedom is involved. (There is universal agreement that its use is quite unnecessary with two or more degrees of freedom).

GOODNESS OF FIT TESTS

The test we have just carried out is known as a **Goodness of Fit Test.** In it we have used a particular theory, belief, or hypothesis to arrive at the pattern of results we would have expected to find if the theory or hypothesis were true. The validity of this hypothesis is tested by checking how closely the results we observe in practice actually fit the pattern we expect in theory. If the fit is good (that is, the agreement is close and the χ^2 value is small), then we accept the theory (null hypothesis) as a reasonable explanation of how reality behaves. If the fit is poor (there is marked disagreement and the χ^2 value is large), then common sense tells us that the theory we have employed is an unacceptable description of reality. We would therefore reject the null hypothesis (the basic theory) and seek a more reasonable explanation of our observed results.

Goodness of fit tests are, of course, not confined to two-category situations. Exactly the same basic principles are employed if we wish to test a hypothesis that seeks to explain three or more categories. Suppose, in our cystic fibrosis example, we were able to determine the genotypes of the children involved, that is, we could categorize them as "normal," "carriers," or "affected." On the basis of Mendelian theory, we know that we would expect to see, on average, 25%, 50%, and 25% of children in these respective categories.

	CATEGORY		
	NORMAL	**CARRIER**	**AFFECTED**
GENOTYPE	CC	Cc	cc
EXPECTED RESULTS	21.5	43.0	21.5
OBSERVED RESULTS	25	45	16
$(O - E)$	+ 3.5	+ 2.0	− 5.5
$(O - E)^2$	12.25	4.00	30.25
$\dfrac{(O - E)^2}{E}$	$\dfrac{12.25}{21.5}$	$\dfrac{4.00}{43.0}$	$\dfrac{30.25}{21.5}$
	= 0.570	.093	1.407
$\Sigma(O - E)^2/E$	= 0.570 +	0.093 +	1.407
	= 2.070		

Degrees of freedom   = Number of categories − 1

= 3 − 1 = 2

.05 critical value of χ^2 with 2 df = 5.99

It is clear that the calculated χ^2 value does not exceed the tabulated .05 critical value. The expected results, based on Mendelian theory, give a good fit to the observed results actually obtained in practice. We are therefore happy to accept the concept of random gene assortment as the mechanism controlling the inheritance of cystic fibrosis.

Note that we did not employ Yates' Correction (since the degrees of freedom involved were greater than 1) and that the degrees of freedom were 2, not 3. Just as before, if we expect 25% (21.5 out of 86) of the children to be normal and 50% (43.0 out of 86) to be carriers, then we must expect 25% (21.5 out of 86) to be affected by the disease. Only two, not all three, of the categories are truly independent.

If the theory being tested predicts the expected values independently of the observed results, then the degrees of freedom are always given by the number of categories less one. (The 25% normal, 50% carrier, 25% affected prediction is based entirely on the logic of random gene assortment and could have been stated months before any of the babies were born.) However, there are many situations in which we have to make some use of the observed results to enable us to predict expected values.

In Chapter 2 we discussed the idea that data tends to follow basic, well-known patterns. Continuous measurements, for instance, frequently follow the Normal distribution, and we saw in Chapter 2 that counts frequently follow a Poisson distribution. The properties of these distributions are well-known and easily described. Hence, it is quite straightforward to work out the pattern of results we would expect to find if the data really did follow a particular distribution. When a set of results is collected, therefore, it is often useful to see if the pattern of results we observed in practice really do match the pattern or distribution we expected or predicted, using a goodness of fit test of the type we just developed.

The number of bacterial colonies observed on a set of microscope slides might, for instance, reasonably be expected to follow a Poisson distribution. The Poisson distribution does, however, assume that the events involved are independent. In the microscope slide situation this means that the colonies are scattered randomly about the slides without any tendency to clump together (or, conversely, to repel one another). When a set of counting data is collected, it is a good idea to check the goodness of fit to a Poisson distribution. This will confirm whether this basic assumption of independence holds in practice.

In Chapter 2 we saw that the probability of a zero count (that is, the chance of encountering a microscope slide with no colonies on it) was given by

$$p(0) = e^{-\lambda}$$

where λ is the mean of the Poisson distribution; the probability of a count of 1 was given by

$$p(1) = \lambda \, e^{-\lambda} \text{ or } \lambda \, p(0), \text{ and so forth.}$$

These probabilities define the pattern of results we would expect to see if a Poisson distribution were followed exactly.

To calculate them, however, we need to know λ, the mean parameter. We have to estimate this by calculating the sample mean (there is, after all, no abstract or theoretical basis for deciding on the average number of bacterial colonies we would expect to see on a microscope slide).

If 40 slides yield a total of 50 colonies, then the mean number of colonies/slide is $50/40 = 1.25$.

The probability of observing no colonies on a slide is therefore

$$p(0) = e^{-1.25}$$

$$= .286$$

(that is, we would expect just over 1 slide in every 4 to have no colonies at all, if the results really do exhibit the Poisson pattern).

Out of our total of 40 slides we would, therefore, expect to see

$$40 \times .286 = 11.44$$

slides with no colonies present. If we actually observed 14 such slides, we calculate the agreement (goodness of fit) between the two in the usual way.

OBSERVED	EXPECTED	(O − E)	(O − E)²	(O − E)²/E
14	11.44	2.56	6.5536	.573

This agreement is calculated for each category (that is, each count situation, 0, 1, 2, and so forth) and then totalled to give the overall χ^2 statistic, which is tested as before.

The potential number of categories in a Poisson distribution is, in theory, unlimited, but most of these are very unlikely to occur in practice. In Chapter 2 we saw that we normally curtail the number of Poisson categories at some point by calculating, say, $p(4+)$, the probability of observing a count of four or more. Just where you place this cutoff point is important when doing a goodness of fit test.

The formula we have used, $\Sigma(O - E)^2/E$, is actually an approximation of a slightly more accurate but substantially more complicated formula. This approximation starts to break down if the expected value falls below 5.0. The categories should, therefore, be curtailed so that the smallest expected value is never less than 5.0.

To see this idea in action, let's consider the pattern we would expect to see for 40 slides with an average of 1.25 colonies/slide.

$$p(0) \quad = e^{-1.25} \quad = .286, \text{ that is, expect 11.44 out of 40}$$

$$p(1) \quad = 1.25\ p(0) = .358, \quad '' \quad 14.32 \quad ''$$

$$p(2) \quad = \frac{1.25}{2}\ p(1) = .224, \quad '' \quad 8.96 \quad ''$$

$$p(3+) = 1 - [p(0) + p(1) + p(2)]$$

$$= .132 \quad '' \quad 5.28 \quad ''$$

By curtailing the categorization at 3 or more colonies per slide (a total of 4 categories), we have ensured that none of the expected values will fall below 5 and that the Chi squared test can be legitimately used to measure the goodness of fit between expectation and reality.

If, by contrast, the categorization had been restricted at 4 or more colonies per slide, that is, $p(0)$, $p(1)$, $p(2)$ as above, and

$$p(3) = \frac{1.25}{3}\ p(2) = .093, \text{ that is, expect 3.72 out of 40}$$

$$p(4+) = 1 - [p(0) +. \ . \ . \ .+ p(3)]$$

$$= .039 \qquad '' \qquad 1.56 \qquad ''$$

(a total of 5 categories), we would have ended up with two categories with expected values less than 5.0. (There may be occasions when it will prove necessary to combine categories at both ends of the distribution. If, for instance, the expected number of zero count situations were less than 5, it would be wise to combine the 0 and 1 categories (that is, form a "1 or less" category) to ensure the absence of an expected value less than 5.0.)

One very important point to note is that we have been forced to use our data to estimate a parameter of the distribution, that is, we have, by sheer necessity, had to use the sample mean as our measure of λ. As always, using the data once to estimate an unknown population parameter reduces the degrees of freedom available to us by 1. (More generally, the degrees of freedom are reduced by the number of parameters estimated from the data.)

The general rule for goodness of fit degrees of freedom is, therefore:

$\qquad$ df = Number of categories

$\qquad\qquad$ − Number of parameters estimated from sample data

$\qquad$ −1.

The degrees of freedom involved in the four category situation is therefore

$$4 - 1 - 1 = 2$$

Suppose the numbers of slides we actually observed in practice were as follows.

	OBSERVED NUMBER OF SLIDES	EXPECTED NUMBER OF SLIDES	$(O - E)$	$(O - E)^2$	$(O - E)^2/E$
0 colonies	14	11.44	2.56	6.5536	0.573
1 colony	12	14.32	−2.32	5.3824	0.376
2 colonies	6	8.96	−2.96	8.7616	0.978
3+ colonies	8	5.28	2.72	7.3984	1.401
TOTAL	40	40.00			3.328

$$\chi^2 = 3.328 \text{ with 2 df, N.S.}$$

The .05 critical value of χ^2 with 2 df is 5.99. The calculated value of 3.328 does not exceed this. There is therefore good agreement (or fit) between the observed and expected values. We have no reason to doubt that the data is behaving as a Poisson distribution.

The Normal distribution plays a particularly important role in the analysis of continuous data, since the ideas and tests we have developed for analyzing data of this type (confidence limits, t testing, and so forth) all assume that this pattern is a reasonable way to describe a set of continuous measurements. Having some way to check the reasonableness of this assumption could, therefore, be very useful, and the goodness of fit test seems to address this general need.

An obvious problem is that the Normal distribution is continuous, not categorical. An ad hoc solution is to arbitrarily split the distribution into a number of categories, such as observations falling more than 1 standard deviation above the mean, observations falling between 1 standard deviation above the mean and the mean itself, and so forth.

From our knowledge of the Normal distribution, as discussed in Chapter 2, we know, for instance, that 34.13% of results would be expected to lie between the mean and +1 standard deviation. It follows that 15.87% must lie above +1 standard deviation and that these percentages would also be expected in the corresponding categories below the mean (Figure 6-2). From the total sample size we can, therefore, work out the expected numbers in each of these four categories, if the data follows a Normal distribution.

Another obvious problem is that we have no theoretical basis for predicting the mean and standard deviation of our data. We will just have to look at our data and use the sample mean and standard deviation as the most reasonable estimate of what these values should be. (This unavoidable dependence on values that are "educated guesses" means that we should really use *t* tables, rather than Normal tables, to obtain our expected values. However, when the data set is reasonably large, with more than, say, 40 individuals, there will be very little difference. For data sets smaller than this, goodness of fit testing in this way is probably a waste of time, since the limited evidence available will result in a test that will be unable to detect anything but very extreme departures from normality.)

Once the mean and standard deviation have been calculated, the number of results actually falling into these four categories can be determined. The level of agreement between these observed numbers and the numbers expected under the null hypothesis of the Normal distribution can then be tested using the usual Chi squared statistic. A significant value is clear evidence that the set of results under examination does not conform to a Normal distribution.

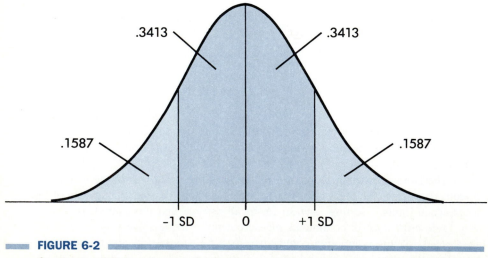

FIGURE 6-2

Goodness of fit to a normal distribution.

Remember that, in a goodness of fit test, the Chi squared value has

Number of categories − (number of parameters estimated from data) − 1

degrees of freedom. When testing for normality the mean and standard deviation (two parameters) will inevitably have to be estimated from the data. In the four-category situation we have discussed here, the calculated value of Chi squared should therefore be tested with 1 degree of freedom.

Any number of categories can, of course, be defined. The more categories used, the more sensitive the test will be to potential deviations from normality. Do remember, though, that the expected values should all be greater than five for the Chi squared statistic to be accurate. This restriction will effectively determine the number of categories that can be used.

─── ASSOCIATION BETWEEN QUALITATIVE VARIABLES ───

Testing the fit of a set of data to some hypothesized distribution or pattern provides a very useful introduction to the concept of observed and expected results, ideas that form the cornerstone of our methods of handling qualitative information. Much more frequently, however, when we collect qualitative information in a health study, we are interested in exploring the question, "Is there any connection or association between two particular qualitative measurements?" We might pose the question, for instance, "Is there any connection or association between an individual's gender (male or female and, hence, a categorical or qualitative variable), and their smoking habits (defined as smoker or nonsmoker and, again, a qualitative variable)?" Another way to phrase exactly the same question would be, "Is there any evidence to suggest that a higher proportion of males than females smoke or vice versa?" or, "Does the pattern of smoking we see in males differ markedly from the pattern we see in females?"

How could we attempt to answer questions of this type? A simple and logical way to start is to display the results you have obtained. With qualitative data this can be done very easily by presenting the results in the form of a table, known as a **contingency table**. Consider for a moment the results of a hypothetical survey of smoking habits in which 100 men and 100 women were asked to classify themselves as smokers or nonsmokers. The following contingency table summarizes the survey results in a very comprehensive yet accessible way.

		GENDER		
		Male	**Female**	**Total**
Smoker?	**Yes**	54	32	86
	No	46	68	114
		100	100	200

A table of this type, consisting of two rows (the smoking categories) and two columns (the gender categories), is referred to as a 2 × 2 contingency table or simply a 2 × 2 (read "2 by 2") table.

Clearly the samples of men and women have different observed or estimated smoking percentages (54% and 32% respectively), but what does this really tell us? At the heart of a test of association is the question, "Do these figures provide real evidence that smoking habits differ between the sexes, that is, there really is an association, or could there simply be random variation about a common smoking pattern, that is, there is no association, and smoking is equally prevalent in the two genders?"

Testing for Association

In the section on Observed and Expected Results we learned that tests of qualitative variables are based on measuring the agreement between expected results based on some hypothesis or theory, and observed results actually obtained in practice. If the agreement is good (that is, the Chi squared value is small), then the hypothesis is regarded as believable. If the agreement is poor (that is, the Chi squared value is large), then the hypothesis is regarded as unbelievable and rejected as an acceptable explanation of reality. Exactly the same principles are employed to test for association between qualitative variables.

The experimental or investigative results obtained in practice, such as those quoted in the smoking/gender contingency table, are clearly the observed values. What is a lot less clear, however, is exactly what the expected values are. Remember that the expected values are defined by the appropriate null hypothesis, that is, that no association exists between the two qualitative variables. In the above example, the null hypothesis is that smoking habits are not influenced by an individual's gender, that is, that in reality the percentage of smokers is the same in both genders.

The question is, "What is this percentage?" If males and females really have the same smoking habits, then there is no need to quote their results separately. Since gender is quite irrelevant if the null hypothesis is true, it would be far more sensible to simply quote the combined, overall result, that is, 86 smokers out of a total of 200, a percentage of 43. This is the best (indeed, the only) indication we have of the smoking rate, if the null hypothesis is true. Since this estimated percentage should hold for both genders (remember, the null hypothesis asserts that the two genders behave identically) we would, therefore, expect to find:

$$100 \times \frac{86}{200} = 43$$

smokers among the 100 males, and

$$100 \times \frac{86}{200} = 43$$

smokers among the 100 females, if no gender association exists.

This approach to calculating expected values is often presented as the following general rule:

The expected value of any cell in a contingency table is given by:

$$\frac{\text{Appropriate row total} \times \text{appropriate column total}}{\text{Overall total}}$$

This is simply a more general restatement of the argument presented previously, in which we used the smokers' row total (86) and overall total (200) to find the overall smoking rate (43%) and applied this to the male column total (100) to find the expected number of male smokers.

A moment's thought will confirm that, in this situation, we only need to calculate explicitly one expected value to be able to determine all four expected values. If we expect to see 43 male smokers, we must expect to see 57 male nonsmokers, since there are 100 males in the study. If we expect to see 43 male smokers, we must expect to see 43 female smokers, since there are 86 smokers in the study. Similarly, expecting 43 female smokers means we have to expect 57 female nonsmokers. Any one expected value explicitly and totally defines the other three expected values in the table. This is a clear indication that only one degree of freedom is involved in the analysis of a 2 × 2 table.

As a general rule the degrees of freedom involved in the analysis of any contingency table are given by:

$$df = (\text{Number of rows} - 1) \times (\text{number of columns} - 1)$$

Now that we possess both observed and expected results, we are in a position to carry out a Chi squared test of association.

		Male	**Female**	**Total**
		GENDER		
Smoker?	**Yes**	54 (43.0)	32 (43.0)	86
	No	46 (57.0)	68 (57.0)	114
		100	100	200

where ( ) denotes the expected values.

$$\chi^2 = \Sigma(O - E)^2/E$$

$$= \frac{(54 - 43)^2}{43} + \frac{(32 - 43)^2}{43} + \frac{(46 - 57)^2}{57} + \frac{(68 - 57)^2}{57}$$

$$= \frac{(11)^2}{43} + \frac{(-11)^2}{43} + \frac{(-11)^2}{57} + \frac{(11)^2}{57}$$

Because only 1 degree of freedom is involved, that is,

$$df = (2 - 1) \times (2 - 1) = 1$$

we should employ Yates' correction and reduce the absolute value of all differences by .5.

$$= \frac{(10.5)^2}{43.0} + \frac{(-10.5)^2}{43.0} + \frac{(-10.5)^2}{57.0} + \frac{(10.5)^2}{57.0}$$

$$= 2.564 + 2.564 + 1.934 + 1.934$$

$$= 8.996$$

The calculated value of χ^2 exceeds the .05 χ^2 critical value of 3.84 and also the .01 χ^2 critical value of 6.63. There is therefore very strong evidence that the expected values (based on a null hypothesis of no difference in smoking habits) explain or fit the observed results very poorly. Therefore we reject this null hypothesis and conclude that a real association exists between gender and smoking habits. Since the observed number of male smokers exceeds the number we would have expected to see if males and females behaved identically we can conclude, to be somewhat more precise, that males are significantly more likely to smoke than are females.

$$\chi^2 = 8.996 \text{ with 1 df, p} < .01$$

It is worth comparing this use of the Chi squared test with our use of the *t* test to compare the means of two groups in Chapter 4. In that case we were, in one sense, testing for any evidence of an association between a continuous variable (as characterized by its mean values) and a categorical variable (group 1 or group 2, Treatment A or Treatment B, and so forth). The Chi squared test is, therefore, the equivalent of the *t* test when the outcome we are interested in is qualitative rather than quantitative.

COMPARISON	OUTCOME	TEST
Treatment A vs. Treatment B	Quantitative (i.e. blood pressure levels)	*t* test
Treatment A vs. Treatment B	Qualitative (i.e. survival/ death)	χ^2 test

As we noted in the section on Goodness of Fit Tests, the χ^2 statistic is based on an approximation that starts to break down if the expected value falls below 5.0. If this happens, the Chi squared test cannot be safely used. An alternative test, known as **Fisher's Exact Test,** may be used in situations like this. This test directly calculates the probability of the observed results, or even more unusual results, arising if the null hypothesis is true. If this probability is less than .05 we would conclude that the observed results do indeed differ dramatically from the expected results predicted by the null hypothesis of no association and that this hypothesis has to be rejected. Fisher's Exact Test is, however, rather tedious to calculate, and we will not consider it in any detail.

General Contingency Tables

Qualitative variables can, in general, have more than two categories (such as blood type or ethnic group). The method we developed in the previous section to test for association is quite general and need not be restricted to situations that can be described by 2 × 2 tables. It can be applied just as easily to 2 × 3, 2 × 4, 3 × 3, 3 × 4, or other contingency tables and, hence, used to test for association between qualitative variables with any number of categories.

For instance, smoking habits might have been recorded in more detail by using three categories: nonsmoker, light smoker, and heavy smoker, using some accepted definition. The observed results are presented in the following 2 × 3 contingency table.

		SMOKING HABIT			
		None	Light	Heavy	
	Male	46	30	24	100
Gender					
	Female	68	20	12	100
		114	50	36	200

The expected values for the null hypothesis of no association are calculated exactly as we outlined in the section on Testing for Association, by using the appropriate row and column totals. (Again, if gender bears no relationship to smoking habits, the overall proportion of heavy smokers in the study offers the most reasonable basis for predicting the number of males who are likely to be heavy smokers. If 18% of the people in the study are heavy smokers, and men and women behave identically, then we expect to see 18 out of 100 males in this category.)

That is,

$$\text{Expected number of male heavy smokers} = \frac{\text{Number of males} \times \text{number of heavy smokers}}{\text{overall total}}$$

$$= \frac{100 \times 36}{200}$$

$$= 18.0$$

This yields the following contingency table of observed and corresponding expected values:

		SMOKING HABIT			
		None	**Light**	**Heavy**	**Total**
	Male	46	30	24	100
		(57.0)	(25.0)	(18.0)	
Gender					
	Female	68	20	12	100
		(57.0)	(25.0)	(18.0)	
		114	50	36	200

The degrees of freedom involved are given, as before, by:

$$df = (\text{Number of rows} - 1) \times (\text{number of columns} - 1)$$

$$= (2 - 1) \times (3 - 1)$$

$$= 2$$

That is, we need to be able to calculate only any two expected values from different columns to be able to specify all others exactly. (Defining the expected number of male heavy smokers and male light smokers automatically defines all the other expected values in the table. Try it and see.)

The χ^2 statistic is calculated exactly as before.

That is,

$$\chi^2 = \Sigma(O - E)^2/E$$

$$= \frac{(46 - 57)^2}{57.0} + \frac{(30 - 25)^2}{25.0} + \frac{(24 - 18)^2}{18.0}$$

$$+ \frac{(68 - 57)^2}{57.0} + \frac{(20 - 25)^2}{25.0} + \frac{(12 - 18)^2}{18.0}$$

Since the degrees of freedom involved are greater than one, there is no need to employ Yates' correction.

$$\chi^2 = 2.123 + 1.000 + 2.000 + 2.123 + 1.000 + 2.000$$

$$= 10.246$$

This exceeds the .05 and .01 χ^2 critical values for two df (5.99 and 9.21 respectively). Therefore we conclude that there is a significant association between gender and smoking habits with more males (and fewer females) smoking than we expected.

The only problem with more complex contingency tables is that it becomes more difficult to decide exactly what your results mean. For instance, does the association mean that the genders differ in their tendency to smoke or not smoke, but that male smokers and female smokers do not differ in their tendency to be light or heavy smokers? Does it, alternatively, mean that they differ on all aspects of smoking, that is, that males are more likely to smoke and also, if they smoke, are more likely to be heavy smokers?

This lack of clarity is reminiscent of the problems of interpretation posed by analysis of variance (see Chapter 5). Indeed, just as we can view a 2 × 2 Chi squared test as the parallel of a *t* test when the outcome is qualitative, so we can view a 2 × *k* Chi squared test as the parallel of an analysis of variance. If the analysis of variance confirms the existence of differences between several group means, then exploring the nature of these differences involves basically carrying out a series of *t* tests (with the decision values suitably stiffened to control the type I error). In the analysis of variance situation this followup or supplementary testing can be carried out in a very elegant way (Tukey's multiple comparison test discussed in Chapter 5). In the simpler world of qualitative measurements this level of sophistication is unfortunately not available, and a more direct approach will have to suffice.

If the Chi squared test of the full contingency table confirms that some association does exist between the qualitative variables, one possible solution would be to carry out a series of 2 × 2 tests to explore specific hypotheses of particular interest. These hypotheses should be stated before the overall analysis is carried out, and should be a logical reflection of what the contingency table is actually describing. For this data set the questions, "Does the tendency to smoke at all differ between the sexes?" and "For smokers, does the tendency to be a heavy smoker differ between the sexes?" would be meaningful and logical, and could be used as the basis for two supplementary 2 × 2 tests.

Once again, however, the more tests that are carried out, the greater the risk we run of rejecting the null hypothesis when it is, in fact, correct, that is, no real association exists (a Type I error). To maintain the overall error rate of .05, we can use exactly the same procedure we employed in Chapter 5, namely dividing the overall error rate (.05) by the number of supplementary tests we wish to carry out (2) and using this to define our basic critical value for testing these supplementary questions, that is,

$$.05/2 = .025$$

.025 critical value of χ^2 with 1 df = 5.02

We have, in fact, already tested the first supplementary question in the previous section. The Chi squared value obtained there exceeded a 5.02 critical value, and confirms that the genders differ in their tendency to smoke or not smoke.

The second supplementary question is described in the following table.

		SMOKING HABIT		
		Light	**Heavy**	**Total**
	Male	30	24	54
		(31.40)	(22.60)	
Gender				
	Female	20	12	32
		(18.60)	(13.40)	
		50	36	86

This leads to an χ^2 value (using Yates' correction) of

$$\frac{(-.9)^2}{31.4} + \frac{(.9)^2}{22.6} + \frac{(.9)^2}{18.6} + \frac{(-.9)^2}{13.4}$$

$$= 0.166 \text{ with 1 df}$$

This clearly does not exceed a 5.02 critical value. Therefore, we conclude that, for individuals who smoke, their gender does not influence their tendency to be light or heavy smokers.

The overall conclusion, that gender is associated with smoking habit, can, therefore, be interpreted more precisely as "males are more likely to smoke than females, but once a person starts smoking, their gender does not affect their tendency to become a heavy smoker."

The Analysis of Paired Qualitative Data

In the two previous sections we pointed out that the Chi squared analysis of 2×2 tables and more general contingency tables can be considered the qualitative equivalent of the *t* test and analysis of variance, respectively. Back in Chapter 4 we commented that, when we compare two treatments, the differences (if they exist) will show up more clearly if we constructed our experiments around pairs that were inherently similar to one another. The consequence of this is that the method of analysis has to concentrate quite specifically on the differences between the pairs, that is, the paired *t* test, rather than the more general analysis provided by the independent *t* test.

Exactly the same principles apply to qualitative data. It is often more sensitive to investigate effects (such as the influence of gender on smoking habits) by grouping the subjects into matched pairs (such as males and females who are inherently very alike) rather than simply studying, say, smoking habits in two unrelated groups of individuals (such as separate unpaired groups of males and females).

Age and socioeconomic status both could influence smoking habits, and the wide variation in these factors that might be present in a typical study group could well obscure any potential gender effect. In order to control for this we might form pairs of males and females, with the partners in each pair being of the same age and socioeconomic status. Any differences in smoking behavior within such a pair, therefore would likely be due solely to the difference in gender and certainly could not reflect the influence of age or socioeconomic factors.

In such a paired study there are four possible outcomes that might be observed for any male/female pairing:

Male smokes/female smokes	$++$
Male smokes/female does not smoke	$+-$
Male does not smoke/female smokes	$-+$
Male does not smoke/female does not smoke	$--$

Note that only the $+-$ and $-+$ results tell us anything about potential differences between the two "treatments" (male and female) being compared. The analysis of paired qualitative data, therefore, concentrates solely on the numbers of $+-$ and $-+$ results observed.

If there is no inherent tendency for one gender to be more prone to smoke than the other, then we would expect the $+-$ and $-+$ results to be equal or certainly very similar to one another,
that is,

$$\text{no true gender effect} \qquad (+-) - (-+) = 0$$

If there is a real tendency for one gender to be more likely to smoke than the other, then we would expect one of these to be large and the other to be small, that is, we would expect the difference between the two to be large. Since we are interested in a difference in either direction, it is more convenient to work with the squared differences, rather than worry about which way we should calculate the difference.
That is,

$$\text{true sex effect} \qquad [(+-) - (-+)]^2 \text{ is large}$$

As with the observed and expected values of the section on the Chi squared Measure, we must be careful when assessing the size of the difference. A difference of 10 could appear to be dramatically important if $(+-) = 11$ and $(-+) = 1$, or totally trivial if $(+-) = 1,011$ and $(-+) = 1,001$. To reach a meaningful conclusion, we must judge the size of the difference relative to the total number of such observations recorded.

Differences between paired qualitative results are, therefore, tested by calculating the test statistic:

$$\frac{[(+-) - (-+)]^2}{(+-) + (-+)}$$

A small value suggests that the two categories being compared could well behave in a similar way and that the null hypothesis of no effect or association should not be rejected. A large value suggests that a real association is present and that the null hypothesis should be rejected.

In view of the general form of the paired formula, and the fact that there are two categories ($+-$ and $-+$) involved, it should come as no surprise to learn that it is tested for significance using the appropriate Chi squared critical value with 1 degree of freedom.

Since the paired Chi squared test, also called **McNemar's test,** involves 1 degree of freedom it should really incorporate Yates' correction. This can be done very simply by reducing the absolute value of the difference by 1. (Instead of reducing each of the two categories by .5, it is easier and quicker to do this all at once).

$$\frac{[|(+-) - (-+)| - 1]^2}{(+-) + (-+)}$$

Suppose that the gender/smoking study was based on 100 pairs of matched male and female subjects, and the following results were obtained:

Male smokes/female smokes	(++) 24
Male smokes/female does not smoke	(+−) 26
Male does not smoke/female smokes	(−+) 10
Male does not smoke/female does not smoke	(−−) 40

In other words, 36 couples show a gender-related disparity in their smoking habits, with the male partner being more likely to be a smoker than the female partner. (In 26 couples the male was the smoker compared with only 10 couples in which the female was the smoker.)

The appropriate test statistic for male/female differences (that is an association between smoking and sex) is

$$\frac{(|26 - 10| - 1)^2}{26 + 10} = \frac{15^2}{36} = \frac{225}{36}$$

$$= 6.250$$

This exceeds the .05 critical value for Chi squared with 1 DF of 3.84 and also the .025 critical value of 5.02. We therefore reject the null hypothesis and conclude that a real difference in smoking habits exists between males and females. More specifically, we conclude that males are more likely to smoke than females.

$$\chi^2 = 6.250 \text{ with 1 df, } p < .025$$

—— **Key Terms** ——————————————————————

Observed values	**Goodness of fit test**
Expected values	**Contingency table**
Chi squared	**Fisher's exact test**
Yates' correction	**McNemar's test**

── **PROBLEMS** ──────────────────────────

1. A study of stress in the industrial workplace was carried out by interviewing 1342 employees of a major New Zealand manufacturing firm. The authors argued that this company was representative of New Zealand industry as a whole since the socioeconomic composition of its work force was similar to that of the New Zealand work force as a whole. The breakdown of workers by socioeconomic group as actually encountered in the study is given below, as are the known percentages for the New Zealand work force as a whole.

SOCIOECONOMIC GROUP	STUDY WORK FORCE (n)	NEW ZEALAND WORK FORCE (%)
1	59	5.8
2	154	10.3
3	346	24.3
4	254	29.3
5	358	17.2
6	171	13.1

Is it reasonable to claim that the socioeconomic distribution within this company is comparable to that of the country's work force as a whole?

2. As part of a national survey of the natural history of bacteriuria in elderly males, 60 nursing homes across Canada each were asked to randomly select a sample of 10 male residents, collect urine samples, and have these tested for bacteriuria. The results reported by the 60 nursing homes are presented below.

Number of nursing homes reporting 0 infected individuals	1
1	7
2	6
3	12
4	18
5	5
6	8
7	3
8 or more	0

What distribution would you expect these results to follow? Test to see if the observed results do in fact follow this distribution.

3. To determine the density of bacteria in a particular solution, diluted samples of the solution were placed on 60 culture plates. After a period the bacterial colonies on each plate were counted. The results are presented below. Is it reasonable to assume that these counts follow a Poisson distribution?

NUMBER OF COLONIES OBSERVED/PLATE	NUMBER OF SUCH PLATES OBSERVED
0	5
1	19
2	18
3	9
4	7
5	1
6	0
7	0
8	1
	60

4. Population statistics indicate that the chances of a newborn being male are .52. A survey of 50 quadruplet births yielded the following pattern of gender outcomes.

0 males	5
1 male	14
2 males	14
3 males	10
4 males	7

Test if these results follow a Binomial distribution. What does this imply about the genders of the children in quadruplet births?

5. A total of 268 patients visiting an emergency room complaining of an apparently sprained ankle were examined and any evidence of bruising noted. Radiographs were then taken to determine if the ankles were fractured. The results of the study are presented below. Is there any evidence that a fracture is associated with bruising?

		BRUISING PRESENT?	
		Yes	No
FRACTURE?	Yes	38	7
	No	151	72

6. In a study of the potential role of drug therapy in the treatment of bladder instability in the elderly, 19 incontinent elderly patients received Imipramine, and 14 received a placebo treatment. Of the Imipramine patients, 14 became dry after treatment, compared with only 6 of the placebo patients. Is there any evidence of genuine treatment differences?

7. In a study to determine whether the type of bed surface employed influences the incidence of decubitus ulcers in elderly orthopaedic patients, the following results were obtained.

	LOW AIR-LOSS BED	WATERBED	RIPPLE MATTRESS
Decubitus ulcers observed	7	13	25
No decubitus ulcers observed	49	58	47

Does the evidence suggest that bed surface does have an influence on the incidence of decubitus ulcers?

8. Asthma presents a significant pediatric health problem in New Zealand. The extent and nature of the problem, however, differ between New Zealand's two major ethnic groups, Europeans and Polynesians. A total of 355 children (199 European and 156 Polynesian) discharged from the pediatric ward of a New Zealand hospital, following admission for asthma, were enrolled in a study to explore the influence of social and other factors on ethnic differences.

   a. The existence of maternal asthma was reported for 73 of the Polynesian children and 66 of the European children. Do these figures suggest any difference in the prevalence of maternal asthma in the two groups of asthmatics?

   b. The children's families were classified by socioeconomic level. (I = highest income level, and so forth) Test whether these results indicate an association between socioeconomic status and ethnic group, and comment on the nature of the association, if any.

	SOCIOECONOMIC GROUP				
	I	II	III	IV	V
European	70	55	36	17	21
Polynesian	18	20	40	53	25

9. A total of 183 men enrolled in a health education course aimed at educating them to the risk factors associated with coronary thrombosis. At the start of the course 26 individuals were found to be hypertensive (defined as a diastolic BP over 100 mm Hg). After a 2-year followup period, 12 of these individuals had become non-hypertensive. However, this followup screening also revealed that 3 individuals had developed hypertension for the first time.

   Does this evidence suggest that, on balance, the course has been successful in promoting a lifestyle that has a real influence on hypertension?

CHAPTER 7

Relationships Between Continuous Measurements

The purpose of research is to expand our understanding of why the world (or a specific portion of it) is the way it is and how it might be changed. This understanding depends on our ability to explore and test relationships, and much of this textbook is designed to give us the skills needed to do exactly that. In Chapters 4 and 5 (*t* testing and analysis of variance) we developed strategies for testing possible relationships between qualitative and quantitative variables. (For example, does the type of hypertension therapy that a patient receives actually have any effect on the patient's blood pressure?) In Chapter 6 (Chi squared testing) we developed a strategy for testing possible relationships between two qualitative variables. (For example, does a person's gender have any connection to that person's risk of becoming a smoker?) In this chapter we will explore the issues involved in describing and testing the third possible type of relationship, a relationship between two quantitative variables.

In many activities, including biostatistics, the simplest approach often can be the most informative. In our study of patterns and distributions in Chapter 2 we saw that a simple graphic plot (a histogram) of a single set of continuous measurements illustrated very clearly all the major features of the pattern followed by that data and led us toward an understanding of the Normal distribution. Similarly, a simple graphic plot of two continuous measurements will give us much useful information about the relationship between them. Such a plot is known as a **scatter diagram.**

A scatter diagram consists of a graph with a horizontal axis and a vertical axis representing the two variables under study. The horizontal axis is usually called the *x* axis, and the vertical axis the *y* axis, although which variable we call *x* and which we call *y* is usually immaterial. (When a dependence relationship exists between the two variables, then there is a well-established convention for the way the results should be plotted. This situation will be discussed in the section on Regression Analysis). The final diagram is obtained by plotting the individuals for which information is available as points at the appropriate locations on the diagram.

Suppose, for instance, we had data available on the heights and weights of a number of patients.

PATIENT	HEIGHT (in)	WEIGHT (lb)
1	68	155
2	73	178
3	70	175

and so forth

A scatter diagram of such a set of results might look something like Figure 7-1.

Even a glance at this scatter diagram tells us there is a strong, clear-cut (although certainly not perfect) relationship between these two variables. More precisely, it strongly suggests to us that, as height increases, weight tends to increase also, reflecting a tendency for the body to remain in proportion. A relationship in which both measurements tend to increase (or decrease) together is called a positive relationship.

Let's pause just for a moment and ask ourselves the seemingly obvious question, "Why does a glance at Figure 7-1 make me believe that height and weight are indeed related to one another?" What we are subconsciously doing when we look at a scatter diagram and decide whether there is "clear-cut" evidence of a relationship, is asking ourselves how closely the points resemble a straight line pattern (a straight line is, after all, the simplest and most obvious relationship there is). The more marked such a

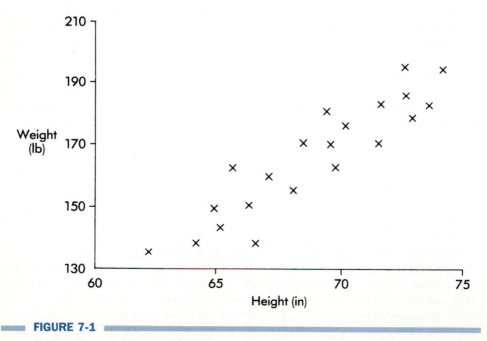

FIGURE 7-1

Scattergram—weight and height.

straight line tendency is, the more striking we feel the evidence must be in favor of a relationship or connection between the two variables. Glancing at Figure 7-1 we can easily envisage a straight line running through the swarm of data points and summarizing the theme of the relationship, "tall people tend to weigh more than short people!" In reality the relationship is not as clear cut as this. There are deviations from this overall summary (the points do not all lie on this imaginary line but scatter around it). However, perfection is too much to expect in the real world, and we would view the fact that the scatter around the imaginary line seems fairly small as convincing evidence of the relationship.

Relationships can of course be negative, as well as positive. A scatter diagram of body weight plotted against a measure of body flexibility might look like Figure 7-2.

Here we would think that there is fairly clear visual evidence of a negative relationship, in which an increase in body weight tends to be matched by a decrease in body flexibility. We would also subjectively think that the pattern is substantially less clear cut than was the weight/height relationship. While the general straight line tendency can still be seen, there is obviously much more fluctuation around this straight line than was the case in the previous example. This, at the very least, would make us think that the weight/flexibility relationship is not as strong as the weight/height relationship and should make us somewhat more cautious about asserting that there really is a weight/flexibility relationship.

There are also situations in which no pattern of any sort may be evident. Suppose we plot a scatter diagram of body weight against reaction time to a visual stimulus (Figure 7-3).

Here there is no visual evidence of any pattern in the data or, hence, of any relationship between weight and reaction time. In terms of our mental picture, there appears to be no logical way to insert a straight line through this bundle of points. At best we

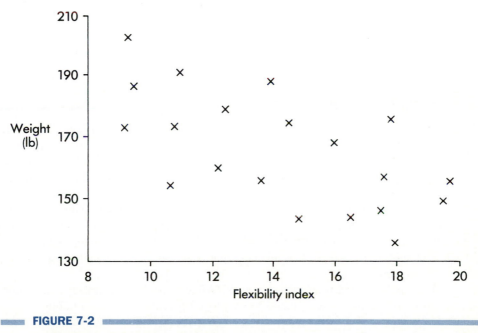

FIGURE 7-2

Scattergram—weight and flexibility.

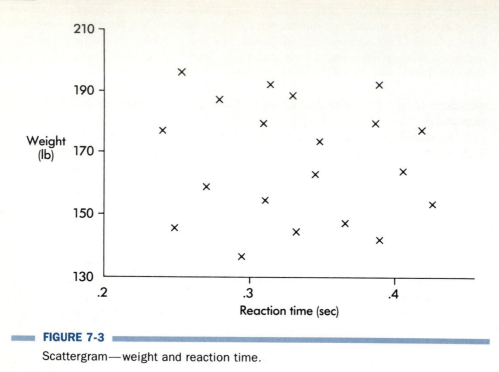

FIGURE 7-3

Scattergram—weight and reaction time.

might envisage a straight line running horizontally through the center of the points, sloping neither up nor down, and still, of course, showing an absence of any pattern or relationship.

The simple device of the scatter diagram can, therefore, provide a lot of useful information very painlessly and should always be plotted when one is investigating potential relationships between two quantitative variables.

MEASURING CO-VARIATION

Although a scatter diagram is an extremely useful, indeed invaluable, device for obtaining an insight into the relationship between two continuous variables, it is clearly insufficient on its own to answer all the questions we might wish to ask. (As we saw, interpreting a scatter diagram does involve subjective decisions about the nature, strength, and existence of a relationship).

To obtain a more objective description of a relationship under study, we need to be able to calculate a numerical value that would in some way measure the strength of the relationship. Such a measure would also be of great help in deciding whether any genuine relationship exists at all, something that is not always obvious from a glance at a scatter diagram.

How might we construct such a measure? What basically concerns us is measuring the tendency of our two measurements to vary together (either rising and falling together or one tending to rise as the other falls). This pattern of joint variation is formally known as **co-variation.**

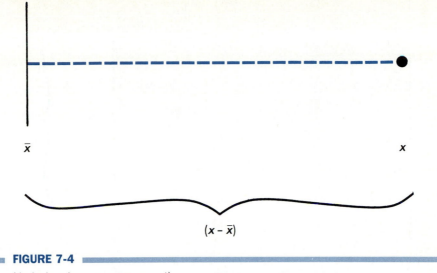

$(x - \bar{x})$

FIGURE 7-4

Variation (one measurement).

Co-variation is clearly closely related to the individual variation of the two variables involved. This gives us an immediate suggestion of a way we might measure co-variation. When measuring the variation of a single variable, we assess the overall degree of variation among our results by measuring how much each individual value varies from the reference value of the overall mean (as discussed in Chapter 1 and illustrated in Figure 7-4).

We can measure the co-variation of two variables in the same basic way, by measuring individual values relative to a reference point defined by the mean values of the two variables involved (that is, relative to a mythical "typical individual" located in the center of the scatter diagram), as shown in Figure 7-5.

Since co-variation simply measures the joint variation of two variables, we can obtain a single measure of how much co-variation any individual exhibits by simply multiplying these two individual deviations.

$$(x - \bar{x})(y - \bar{y})$$

For instance, suppose that the "average" or reference individual in the height/weight study has a height of 5 ft 10 in and a weight of 172 lbs. An individual with a height of 6 ft 0 in and a weight of 178 lbs is 2 inches above the average height and 6 lbs above the average weight, and contributes a co-variation of 12 inch lbs. To obtain an overall measure of co-variation, we add the results for all individuals together and divide by $n - 1$ to give our final "average" measure (in a manner exactly analogous to our calculation of the variance in Chapter 1).

$$\frac{\Sigma(x - \bar{x})(y - \bar{y})}{n - 1}$$

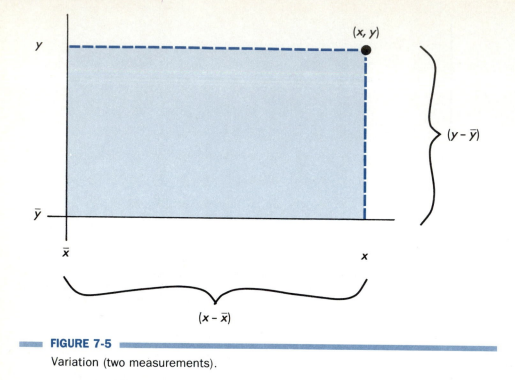

FIGURE 7-5

Variation (two measurements).

(You may wonder why we divide by $n - 1$ and not $n - 2$, since two means are involved. Remember that co-variation is measured around a single reference point or individual. The only difference from our discussion of the variance is that variation can now take place in two directions simultaneously, rather than just one. Don't worry, we will indeed have to pay the appropriate price for estimating two means, as we shall see in the section on Testing the Correlation Coefficient.)

Our "average" measure of variation was known as the variance. Similarly, the "average" measure of co-variation is known as the **covariance.** Unlike the variance, however, the covariance can take a positive or negative value. The reason for this should be obvious if you consider the way it has been calculated.

Consider a scatter diagram split into four quadrants around the mean individual or reference point (Figure 7-6). When the individual values scatter around a line running from the lower left quadrant to the upper right quadrant (line AB) the co-variation contributions of these individuals will almost all be positive (either both values above average in quadrant A or both values below average in quadrant B, resulting in a positive product in both cases). The consequence will be a large positive covariance (corresponding to a positive relationship) when the individual contributions are summed and averaged. When the individuals are scattered around a line (such as CD) that typifies a negative relationship, the covariation contributions will almost always be negative (with one value above average and one below average, resulting in a negative product) and will produce a large negative covariance. When no clear pattern is evident, individuals will be scattered in all four quadrants, with a mixture of positive and negative covariation

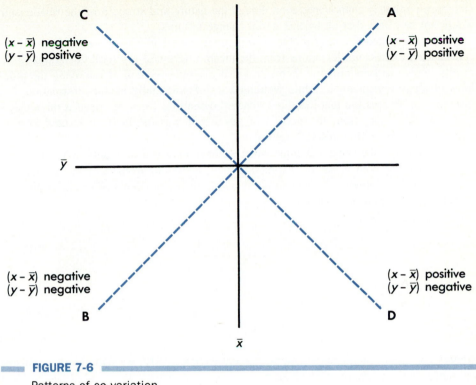

C   A

$(x - \bar{x})$ negative
$(y - \bar{y})$ positive

$(x - \bar{x})$ positive
$(y - \bar{y})$ positive

$\bar{y}$

$(x - \bar{x})$ negative
$(y - \bar{y})$ negative

$(x - \bar{x})$ positive
$(y - \bar{y})$ negative

B   D

$\bar{x}$

FIGURE 7-6

Patterns of co-variation.

contributions and a covariance that will be very close to zero.

(Thoughtful readers will note that if you, for some inexplicable reason, tried to calculate the covariance of a variable with itself, you would simply end up with the variance of that variable.)

$$\frac{\Sigma(x - \bar{x})(x - \bar{x})}{n - 1} = \frac{\Sigma(x - \bar{x})^2}{n - 1}$$

THE CORRELATION COEFFICIENT

Unfortunately, the covariance has severe limitations as a practical tool for measuring and describing relationships between variables. Because it combines information from two variables, it measures co-variation in rather strange composite units. In the case of our previous height/weight example, the covariance would be measured in inches × pounds. Clearly, composite units of this type make little sense. Furthermore, since the units employed and the magnitude of the values involved will differ from one study to another, it is impossible to form any idea of what constitutes a large or small covariance. Even within the same study a change in measurement scale from inches and pounds to meters and kilograms (equally appropriate units of measurement and ones that would be the units of choice in many countries) would produce a vastly different (in fact, much smaller) covariance, even though the nature and strength of the relationship would not be affected at all by this trivial change in units.

We encountered similar problems with single variables in Chapter 2. In those circumstances, expressing deviations from the mean in terms of standard deviations (technically known as standardizing them) provided a simple method of removing the problems of measurement units, while yielding a lot of potentially useful information. For instance, an observation that lies three standard deviations from the mean is obviously a very unusual observation, irrespective of whether it was originally measured in kilograms, cubits, or milligrams/100 milliliters.

We can use the same technique when we have two measurements rather than one. By standardizing both variables (expressing their deviations from the mean in terms of their standard deviations), we can similarly avoid all the problems of trying to interpret unit-dependent covariance values.

That is, if we replace

$$(x - \bar{x}) \text{ by } \frac{(x - \bar{x})}{s_x}$$

and

$$(y - \bar{y}) \text{ by } \frac{(y - \bar{y})}{s_y}$$

when s_x, s_y are the standard deviations of x and y respectively, we will obtain a unit-free or standardized, measure of covariation.

$$\text{Standardized covariance} = \frac{1}{n-1} \frac{\Sigma (x - \bar{x})(y - \bar{y})}{s_x \quad s_y}$$

The standardized covariance is more commonly known as the **correlation coefficient** and is often referred to in statistical shorthand notation as r.

Calculating the Correlation Coefficient

By examining the formula we have just derived

$$r = \frac{1}{n-1} \frac{\Sigma (x - \bar{x})(y - \bar{y})}{s_x \quad s_y}$$

it becomes easy to appreciate the standardization rational that underlies the correlation coefficient (that is, that it is simply a standardized covariance). However, just as the basic obvious formula for the variance has an alternate, intuitively less obvious version that is, in practice, much easier to calculate, so too has the covariance, and consequently, the correlation coefficient.

Since $s = \sqrt{\text{Variance}}$

$$r = \frac{1}{n-1} \frac{\Sigma (x - \bar{x})(y - \bar{y})}{\sqrt{\text{variance}_x \times \text{variance}_y}}$$

$$= \frac{1}{n-1} \frac{\Sigma\,(x-\bar{x})(y-\bar{y})}{\sqrt{\dfrac{\Sigma(x-\bar{x})^2}{n-1}\dfrac{\Sigma(y-\bar{y})^2}{n-1}}}$$

$$= \frac{\Sigma\,(x-\bar{x})(y-\bar{y})}{\sqrt{[\Sigma(x-\bar{x})^2\,\Sigma(y-\bar{y})^2]}}$$

Just as

$$\Sigma(x-\bar{x})^2 = \Sigma x^2 - (\Sigma x)^2/n$$

similarly

$$\Sigma(x-\bar{x})(y-\bar{y}) = \Sigma xy - (\Sigma x)(\Sigma y)/n$$

The version of the correlation coefficient formula most easy to compute is, therefore:

$$r = \frac{\Sigma xy - (\Sigma x)(\Sigma y)/n}{\sqrt{(\Sigma x^2 - (\Sigma x)^2/n)(\Sigma y^2 - (\Sigma y)^2/n)}}$$

To see the correlation coefficient in action, let's attempt to measure the relationship between the following two sets of measurements, the heights (in inches) and weights (in pounds) of 10 women being monitored as part of an investigation of contraceptive pill usage.

HEIGHT (in) x	WEIGHT (lb) y	x^2	y^2	xy
64	130	4096	16900	8320
65	148	4225	21904	9620
71	180	5041	32400	12780
67	175	4489	30625	11725
63	120	3969	14400	7560
62	127	3844	16129	7874
67	141	4489	19881	9447
64	118	4096	13924	7552
65	120	4225	14400	7800
64	119	4096	14161	7616

$\Sigma x = 652$ $\Sigma y = 1378$ $\Sigma x^2 = 42570$ $\Sigma y^2 = 194724$ $\Sigma xy = 90294$

$$r = \frac{90294 - 652 \times 1378/10}{\sqrt{[(42{,}570 - 652^2/10)(194{,}724 - 1378^2/10)]}}$$

$$= \frac{448.4}{\sqrt{59.6 \times 4835.6}}$$

$$= +\ .8353$$

Understanding the Correlation Coefficient

The correlation coefficient has one important practical advantage over the other statistics that we have encountered so far. It can take only values between the limits of $+1$ and -1 (with the consequent assurance that a value outside these limits must be the result of calculation error).

A value of $+1$ (Figure 7-7, *A*) indicates a perfect relationship in which the two variables rise and fall together in complete unison. We have intuitively accepted the straight line as a reference pattern, and the calculation of the covariance and correlation are similarly based on the straight line as the ideal perfect relationship.

Similarly a value of -1 (Figure 7-7, *B*) indicates a perfect relationship in which the value of one variable rises as the other falls.

By contrast, a value of 0 (Figure 7-7, *C*) indicates the complete absence of any relationship between the two variables, that is, the value of one variable has no influence on the value taken by the other variable.

Naturally, in real life we would never expect to observe these ideal clear-cut relationships. We are much more likely, for example, to observe relationships that are strong, even if not perfect (Figure 7-8, *A*), nonexistent relationships that have correlations close to, but not exactly, zero (Figure 7-8, *B*) and perhaps the most intriguing and perplexing of all, weak correlations (Figure 7-8, *C*) that may indicate real, but far from perfect, relationships or may be simply quirks of numerical fate.

The sign of a correlation coefficient therefore defines the nature or direction of the relationship, while the absolute value of the correlation coefficient measures the strength or clarity of the relationship on a scale of 0 (no relationship whatsoever) to 1 (perfect correspondence between the measurements). One very useful way to interpret the strength of a relationship is to appreciate that the square of the correlation (this, of course, eliminates the sign and comments solely on its size) measures the proportion of the variation in the two variables that is common to, or shared by, both variables. The

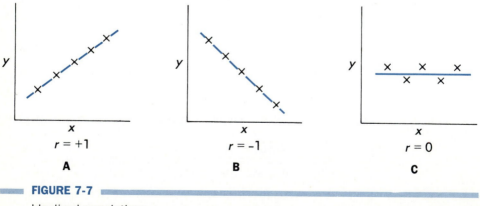

$r = +1$

A

$r = -1$

B

$r = 0$

C

FIGURE 7-7

Idealized correlations.

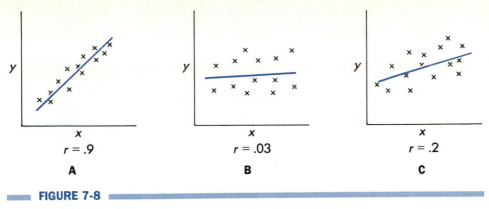

FIGURE 7-8

Realistic correlations.

higher this is, by definition, the stronger the relationship between the two variables. For instance the height/weight correlation of .8353 tells us that the relationship is a positive one (with height and weight tending to increase or decrease together) and that the two variables have 69.77% (.6977 or .8353^2) of their variation in common (reflecting, no doubt, the general tendency for the body to remain in proportion).

Testing the Correlation Coefficient

Saying that small correlation coefficients are dubious evidence of a relationship (in other words, quite possibly consistent with a null hypothesis of no real relationship) while large correlation coefficients are convincing evidence of the existence of a relationship is, in many ways, stating the obvious. It simply begs the question "Just how can we decide when a correlation coefficient is large enough to impress us?" Take the height/weight correlation of .8353 that we calculated earlier in this chapter. Is this unequivocal evidence of a height/weight connection, or might it be nothing more than a coincidence, always a possibility in studies based on small sample sizes?

The crunch question we must answer is, "Is the value of r we have calculated so large that it is impossible (or, more realistically, very unlikely) that it could have arisen from purely accidental causes?" If this is so, then we must conclude that some real relationship exists between the two variables.

Using the formal jargon we developed in Chapter 4, we wish to test the null hypothesis that no relationship exists between the two variables. If the calculated value of our test statistic (r) is so large that it is extremely unlikely (that is, the chance is less than 1 in 20 or 5%) that the null hypothesis is true, then we must reject it in favor of the alternate hypothesis that a real relationship does exist between our variables.

Critical values of r are available to aid in this decision and are given in table T/6 of the Appendix. As usual this table lists the basic .05 critical value and also some more severe (usually .01 and .001) critical values. If the calculated sample value of r exceeds the .05 critical value, there is less than a .05 chance that a value of that size arose accidentally, and we would conclude that it is more likely that a relationship does exist between the two variables. (In formal terms, we are prepared to run a .05 risk of making a Type I error in our conclusions.) Naturally, if the calculated value exceeds some of the more severe critical values, this should be mentioned in any report. (Please appreciate that, when checking a calculated value of r against a set of tables, the sign is ignored since it is the absolute value of r that comments on the strength and possible existence of the relationship.)

The exact size of the critical value used to check the calculated sample value depends, as in most statistical tests, on the sample size or, more precisely, the degrees of freedom. For correlation coefficients, the degrees of freedom are given by:

$$df = \text{Sample size} - 2$$

This should not surprise us at all. Calculation of the correlation coefficient requires the estimation of two reference means from the sample data and, as we mentioned earlier, the appropriate price has to be paid.

Should tables of the correlation coefficient not be available, it is possible to test for significance using a t test. This involves simply calculating how many standard errors away from zero (the null hypothesis position of no relationship) our calculated correlation coefficient lies. To do this we need to know the standard error of a correlation coefficient. (Remember that if we could collect fresh samples and calculate their correlation coefficients, we would certainly observe a range of r values. The standard error of a correlation coefficient measures the likely extent of this variation, just as the standard error of a mean enables us to determine the range of means we might observe in repeated sampling.)

$$SE\ (r) = \sqrt{\frac{1 - r^2}{n - 2}}$$

The formula for the standard error of r, although it seems unusual at first glance, is actually very reasonable. The top line measures the proportion of variation that is not shared by the two variables and, hence, is not involved in the correlation. Therefore, it reflects how much the correlation coefficient can fluctuate because of random variation about the true correlation pattern. Note that this is largest when $r = 0$ and the amount of "non-correlation" variation is maximum. It is smallest (in fact, nonexistent) when $r = \pm 1$. In a perfect relationship, random variation is a dirty (or, more precisely, irrelevant) word. The bottom line contains the degrees of freedom involved in the correlation calculation. It is, of course, directly related to the sample size and reflects the fact that the larger the sample available to us, the more reliable our correlation coefficient will be.

The r value, expressed in standard errors, is, therefore

$$t = \frac{r - 0}{\sqrt{\dfrac{1 - r^2}{n - 2}}}$$

$$= r \sqrt{\frac{n - 2}{1 - r^2}}$$

This is tested by using t tables with $n - 2$ df

The height/weight correlation of .8353 results in a value of

$$t = .8353 \sqrt{\frac{8}{.3023}}$$

$$= 4.297 \text{ with 8 df, } p < .01$$

Unlike the F and χ^2 tests (which are both based on squared values), the t test and the correlation coefficient both retain a clear indication of the direction of the relationship under investigation, and hence both lend themselves to one-tailed testing, if appropriate. We have assumed so far that the general alternative hypothesis (there is a relationship) is the one we are interested in exploring, and this is indeed usually the case. However, there may be situations in which exploring a specific alternative hypothesis makes more sense. (For instance, exploring the specific alternative hypothesis that there is a positive relationship between height and weight has a great deal of logical merit.) If a one-tailed test of the correlation coefficient is appropriate, then the procedure introduced in Chapter 4, that is, using the appropriate .10 two-tailed critical value as a .05 one-tailed critical value, is the approach to use.

Some Notes of Caution

We commented earlier that we recognize the presence of a relationship because there appears to be an underlying straight line in a scatter diagram. The human eye can, of course, recognize patterns more subtle than this. Unfortunately the correlation coefficient cannot. The correlation coefficient is designed to measure linear relationships and linear relationships only. When the relationships involved are curved (as quite frequently happens), then the correlation coefficient can be highly misleading.

The calculation of a correlation coefficient for a set of results such as those depicted in Figure 7-9 might well apparently indicate that no relationship exists when it is clear from even a glance at the scatter diagram that a real, if somewhat complex, relationship does exist. Techniques are available for the study of curvilinear relationships

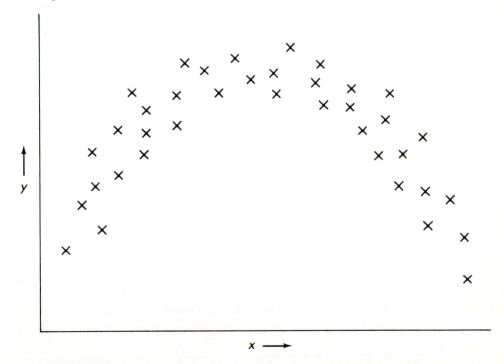

FIGURE 7-9

Curvilinear relationship.

and are based on the ideas we will discuss in Chapter 13. (A glance at a scatter diagram will usually tell you if the relationship is approximately linear, an important reason why drawing a scatter diagram is well worth the minimal time and effort involved.)

Finally, both variables should be approximately normal for the statistical test of *r* to be valid. (They should, ideally, show a pattern of joint normality in which the swarm of points on the scatter diagram exhibits a symmetrical egg shape). This requirement of normality holds, of course, for all the statistical tests we have discussed (with the exception of the Chi squared test, since qualitative variables cannot, of course, have a Normal distribution). There are quite simple methods (known as nonparametric methods) for the analysis of quantitative data that does not follow a Normal distribution, and a nonparametric equivalent of the correlation coefficient will be discussed in Chapter 14. These techniques cannot, however, solve the problems posed by nonlinear relationships.

REGRESSION ANALYSIS
Dependence Relationships

In the previous section we discussed how we might tackle the problem of determining if there was any evidence of a relationship between two continuous variables and how we might measure just how strong that relationship was. At no stage did we make any distinction between the two variables involved. Height and weight are, for example, both simply continuous measurements that vary from individual to individual and that, we believe, may well behave in the same general way.

There are many situations, however, in which we want to study the relationship between two continuous variables for which there is a clear, logical distinction in the way they can relate to one another. Consider the exploration of a relationship between age and blood pressure, for instance. Here the relationship is not a symmetrical one in the logical sense. A change in an individual's age could well produce a change in the individual's blood pressure. However, a change in blood pressure would not possibly produce a change in the patient's age. One of the variables (blood pressure) depends on, or is influenced by (directly or indirectly), the other one (age). The variable that depends on the other one is called, quite logically, the **dependent variable.** The variable that influences the other one is called, not quite so logically, the **independent variable.** (The independent variable is sometimes called, much more meaningfully, the explanatory variable.)

There are also situations in which there is no logical distinction between the two variables, but circumstances or research objectives create the dependent/independent relationship. As we commented earlier, no logical distinction exists between height and weight. Both are simply numerical comments on certain aspects of body stature. Suppose, however, that our research objective was to develop a way of predicting an individual's weight from his height. Our ability to predict an individual's weight depends on our having a knowledge of his height, and also, of course, on being able to quantify and describe the nature of the weight/height relationship. In other words, in this situation, weight is a dependent variable, and height is an independent or explanatory variable.

Other examples of dependence relationships are exercise level (independent) and heart rate (dependent), and (on a community basis rather than an individual basis) smoking rates (independent) and death rates (dependent).

Describing a Dependence Relationship

Our first task, when setting out to explore a dependence relationship, is to find some meaningful way of describing the essence of the relationship. How exactly does age influence blood pressure, if at all? What is the relationship between weight and height, and knowing it, can we use it to estimate the weight of an individual who is 67 inches tall?

In our study of correlation we employed the scatter diagram to give us a visual impression of the type of relationship that exists in a set of measurements. When assessing that visual evidence, we mentally compared the pattern we saw to a basic straight line pattern and asked ourselves how closely the scatter of points would match a straight line drawn through their center. In studying how a dependent variable and an independent variable relate to one another, we use exactly the same approach. In fact, we take it one step further and explicitly obtain the straight line that best describes the relationship.

(A point of convention that we alluded to at the beginning of this chapter: when drawing a scatter diagram and fitting the appropriate straight line for a dependence relationship, it is conventional to plot the dependent variable (always called y) on the vertical axis and the independent variable (always called x) on the horizontal axis.)

To describe any straight line, we need to know only two values (Figure 7-10):

1. b, the slope of the line, that is, how steeply the line rises (a positive slope) or falls (a negative slope). This measures how much y increases (or decreases) for each unit change in x. This value always has a clear practical interpretation, such as the average increase in blood pressure per year increase in age.

2. a, the intercept of the line, that is, where the line starts. This is the value that y takes whenever x is zero. Sometimes this can have a practical meaning, such as the heart rate at a zero exercise level (rest); sometimes it has no real meaning, such as the weight of an individual with zero height. In both cases, however, its function is to provide a locating or starting point for the relationship line.

These come together in a familiar equation:

$$y = a + bx$$

In other words, the height of the line *(y)* at any point, *x,* is given by its initial (intercept) value plus the amount by which the line has risen (or fallen) since then.

Describing a dependence relationship, therefore, boils down to finding the values of a (the intercept) and b (the slope), which define the straight line (known as the **regression line**) that best summarizes the relationship encapsulated in the scatter diagram.

Finding the Best Description

If asked to draw the straight line that best describes a particular scatter diagram, we would instinctively draw a line through the middle of the scatter of points (as in Figure 7-11, *A*). Certainly we would not consider drawing either of the lines in Figure 7-11, *B*.

What we are doing at an intuitive level is drawing the line that comes closest to all the points in an overall sense. On a more formal level, this approach of finding the line

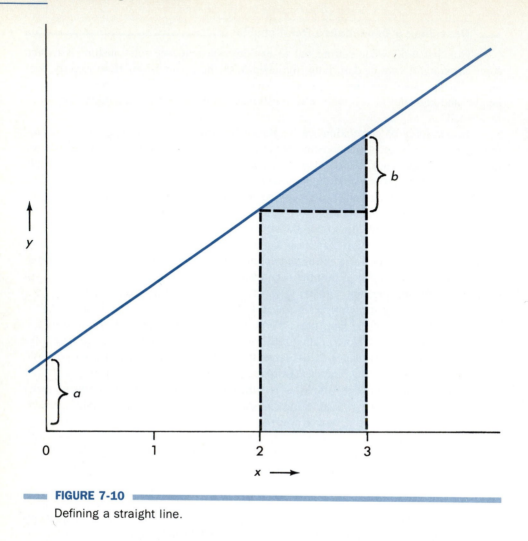

FIGURE 7-10

Defining a straight line.

that comes closest to the points overall is known as finding the line that gives the best or **least squares fit** to the points.

What exactly is a "least squares fit," and why is it the best and most reasonable way to summarize a set of points by a line? A measure of just how far any individual data point lies from a line is given very directly by the distance that point lies from the line, measured on the vertical y axis, as shown in Figure 7-12. (Remember that y, the dependent variable, is the variable whose variation we are trying to describe or explain, for example, how well can we describe or explain variation in blood pressure by the possible influence of age?) The larger this value, the less adequately the line fits or describes that particular data point, and the less impressed we are.

We might feel that an overall measure of the "closeness" of the line would be obtained by adding these distances for all the points. However, if the line goes through the middle of the points (as common sense says it must), the positive and negative distances that result will simply cancel out one another. We can easily cope with this problem,

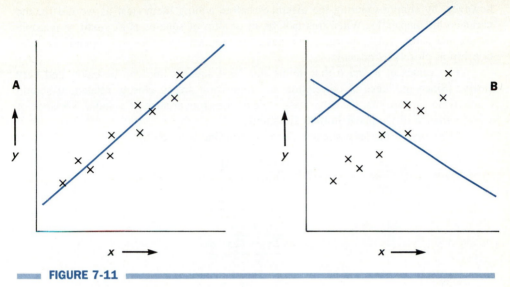

FIGURE 7-11

Fitting a line to a relationship.

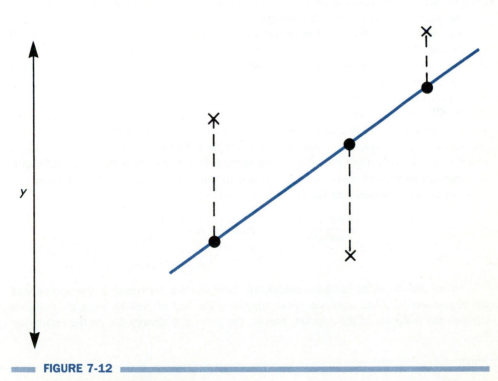

FIGURE 7-12

Measurement of lack of fit.

however, by simply squaring the distances before adding them (just as we did for the variance in Chapter 1). When this lack of fit or sum of squares is as small as possible (hence the name "least squares"), we have the line that describes the dependence relationship as closely as possible.

The values of a and b that define this least squares line can be calculated very easily. [Since the ideas of correlation and regression are so closely related, it should come as no surprise that the same basic calculations (covariance of x and y, variance of x and variance of y) appear in both situations.]

The slope of the least squares or regression line is given by

$$b = \frac{\text{covariance } x,y}{\text{variance } x}$$

$$= \frac{\dfrac{\Sigma xy - (\Sigma x)(\Sigma y)/n}{n - 1}}{\dfrac{\Sigma x^2 - (\Sigma x)^2/n}{n - 1}}$$

$$= \frac{\Sigma xy - (\Sigma x)(\Sigma y)/n}{\Sigma x^2 - (\Sigma x)^2/n}$$

Why should this provide a very reasonable measure of the dependent/independent connection, or, more precisely, a measure of just how much y changes for every unit change in x? The covariance of x and y is, of course, the basic measure of the connection between the dependent and independent variables or the pattern of change that is common to both variables. Unfortunately, as we saw in the section on The Correlation Coefficient, the covariance measures this change in composite units (mm Hg × years, in the case of the blood pressure/age relationship). The slope of a line measures change in a relative, rather than absolute, sense (mm Hg/year, if we are describing the change in blood pressure per unit increase in age). The composite measure of change provided by the covariance is therefore converted to a relative measure (the slope) by simply dividing it by a measure of variation in the independent variable (such as the variance of age, measured in years squared). Note how this simple act produces a measure of relative change in exactly the right units. Since the slope (often called the regression coefficient) is measured in relative units (such as change in mm Hg/year) and not in standardized units, it is not constrained to lie between $+1$ and -1.

$$b = \frac{\text{covariance of } x \text{ and } y}{\text{variance of } x} \text{ measured in } \frac{\text{mm Hg} \times \text{years}}{\text{years}^2}$$

or mm Hg/year

Once the slope, b, has been calculated, how can we determine a, the intercept of the regression line? As common sense suggests, the line of best fit must always pass through the midpoint of the data set, that is, the point $\bar{x}, \bar{y}$ always lies on the regression line.

Since

$$y = a + bx$$

describes the regression line, this means that

$$\bar{y} = a + b\bar{x}$$

or, equivalently,

$$a = \bar{y} - b\bar{x}$$

Provided we have calculated b, $\bar{x}$, and $\bar{y}$, we can, therefore, very easily complete our description of the regression line by calculating its intercept, a.

To see these principles in action, let's use the weight/height data of the section, Calculating the Correlation Coefficient, to obtain a relationship that might enable us to predict the weights of females using a contraceptive pill, from a knowledge of their heights.

$$b = \frac{\Sigma xy - (\Sigma x)(\Sigma y)/n}{\Sigma x^2 - (\Sigma x)^2/n}$$

$$= \frac{448.4}{59.6}$$

$$= 7.52 \text{ lbs/in}$$

that is, we expect weight to increase by 7.52 lbs for each 1-inch increase in height, which seems intuitively acceptable.

$$\text{Mean weight, } \bar{y} = 1378/10$$

$$= 137.8 \text{ lbs}$$

$$\text{Mean height, } \bar{x} = 652/10$$

$$= 65.2 \text{ in}$$

$$a = \bar{y} - b\bar{x}$$

$$= 137.8 - (448.4/59.6)\, 65.2$$

$$= -352.73 \text{ lbs}$$

A weight for a height of zero inches, of course, has no real meaning, hence the unusual value. It does, however, provide a reference or starting point for the regression line that ensures that it passes through the mean weight/mean height or "typical" individual.

The regression line for the female weight/height data set is therefore

$$y = -352.73 + 7.52\, x$$

where y is weight in pounds, and x is height in inches.

Testing the Dependence Relationship

Even if there were no real relationship between the dependent variable and the independent or explanatory variable, it is very unlikely that the value of b we calculate from our relatively small sample would work out to be exactly zero (which is what we should see if there were no connection at all between x and y). Therefore, we need to be able to test whether the value of *b* we obtain is large enough to convince us, beyond a reasonable doubt, that there is a real dependence relationship and not just some accidental deviation from a true value of zero. (In technical terms, we need to test the null hypothesis that the true population value of *b*, often referred to as β, is really zero. If we can reject this as a believable possibility, then we can reasonably conclude that the relationship is real.)

How might we tackle this problem? We derive a dependence relationship to summarize the connection between the dependent and independent variables and to help us predict or explain the behavior of the dependent variable, based on our knowledge of the independent variable. (We might, for instance, deduce that an individual has an above average weight at least partly because the individual is relatively tall. More precisely, we can predict a weight for any given height.) It would be unrealistic to expect that even a very strong relationship would enable us to predict or explain the behavior of the dependent variable perfectly. An individual will probably weigh somewhat more or less than we predict based on our summary of the weight/height relationship. In other words, some of the variation in the dependent variable will inevitably remain unexplained. In summary, the overall variation in the dependent variable (such as differences in weight) can be split into two parts, the variation that can be successfully explained by the regression relationship (our prediction of a certain weight, based on what we know of the individual's height) and the variation that cannot be explained by the relationship (the extent to which our prediction just plain gets it wrong). This split of total variation into two parts, one of which has to be just random or error variation (the unexplained variation), is very reminiscent of the analysis of variance approach of Chapter 5, and that philosophy can be applied to testing the reality of a dependence relationship.

What might we expect to happen? If the relationship is real, then much of the variation in *y* should be successfully explained by our knowledge of *x*, and the unexplained portion should be relatively small. We would therefore expect the explained variation (after taking appropriate account of the degrees of freedom involved) to be much larger than the unexplained portion, resulting in an *F* ratio greater (we hope much greater) than 1. If, however, no real relationship actually exists, then the small amount of apparently "explained" variation we might find will simply be another manifestation of the ever-present problem of random variation. A comparison of "explained" and "unexplained" variation would, in these circumstances, result in an *F* ratio very close to 1.

So much for the principles; what about the practice? How can we work out the total variation in *y* and split it into its explained and unexplained portions?

The total variation in *y* is very straightforward. It is measured by the sum of squared deviations about the mean of *y* that is, the familiar sum of squares of Chapter 1 (Figure 7-13).

$$\text{Total variation in } y = \Sigma(y - \bar{y})^2$$

For instance, in the weight/height data we referred to earlier there is an individual who is 71 inches tall and weighs 180 lbs. The overall average weight is 137.8 lbs. In other words, this individual weighs 42.2 lbs above average, and her contribution to the total variation is 42.2^2, or 1780.84 lbs squared.

The weight/height regression relationship is

$$\text{weight (lb)} = -352.73 + 7.52 \text{ height (in)}$$

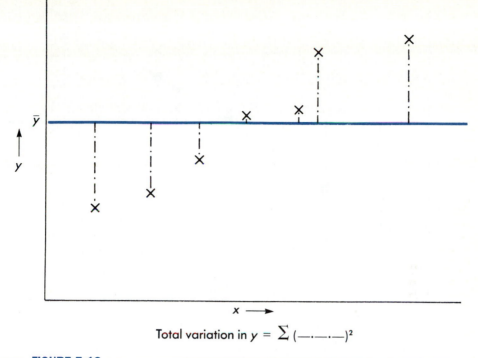

Total variation in $y = \sum (\text{—}\cdot\text{—}\cdot\text{—})^2$

FIGURE 7-13

Total variation in y.

The predicted weight (y_p) for this particular individual is therefore

$$y_p = -352.73 + 7.52 \times 71$$

$$= 181.19 \text{ lb}$$

Based on our knowledge of her height and of the relationship between weight and height, we are therefore predicting that she has a weight 43.39 lbs above average. (In other words, we feel we can explain a substantial part of her well-above-average weight by using our knowledge that she is of well-above-average height.) This individual's contribution to the explained variation is therefore 43.39^2, or 1882.69 lbs squared.

Over an entire sample (Figure 7-14), the explained variation is given by

$$\sum (y_p - \bar{y})^2$$

The dependence relationship is, of course, only a summary and does not (indeed, cannot) tell the whole story. It predicts that this particular individual should weigh 181.19 lbs, whereas she, in fact, weighs only 180 lbs, an error of 1.19 lbs. This discrepancy between relationship and reality simply cannot be explained by our knowledge of the individual's height. This individual's contribution to the unexplained variation is, therefore, -1.19^2 or 1.416 lbs squared.

Over the entire sample the unexplained variation is given by

$$\sum (y - y_p)^2$$

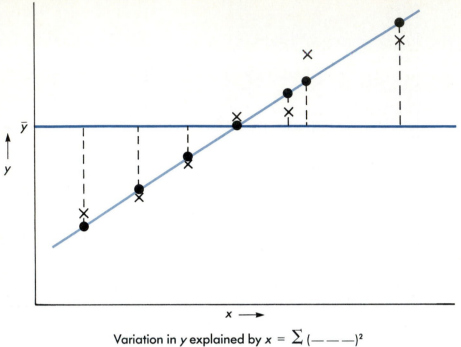

Variation in y explained by $x = \sum (\text{———})^2$

FIGURE 7-14

Variation in y explained by regression.

(Take a look at Figure 7-15 and compare it with Figure 7-12. Note that the unexplained variation is simply another name for the least squares lack of fit to the regression line).

As with the analysis of variance of Chapter 5, in practice we calculate two of these values and find the third by subtraction. The easiest two to calculate are the total variation, given by

$$\Sigma(y - \bar{y})^2 \text{ or } \Sigma y^2 - (\Sigma y)^2/n$$

and the explained variation, which is most conveniently calculated as

$$b \, \Sigma(x - \bar{x})(y - \bar{y})$$

or

$$b \, [\Sigma xy - (\Sigma x)(\Sigma y)/n]$$

The bracketed term (the "sum of cross products") is the heart of the covariance calculation and measures very directly the strength of the x, y connection. As we keep pointing out, however, it measures the connection in esoteric composite units (lbs × inches, for our weight/height study). The slope, b, calibrates the x, y connection in relative units (such as lbs/inch), and when the two are multiplied together

$$b \, [\Sigma xy - (\Sigma x)(\Sigma y)/n] \qquad \text{measured in } \frac{\text{lb}}{\text{in}} \times \text{lb} \times \text{in}$$

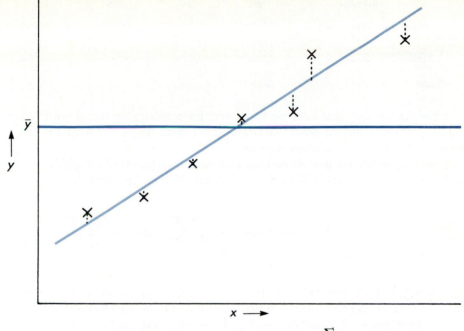

Variation in *y* not explained by $x = \sum(\text{------})^2$

FIGURE 7-15

Variation in *y* not explained by regression.

$$\text{or lbs}^2$$

the result is a measure of the *x, y* connection in terms of the variability of the dependent variable (that is, in squared *y* units). In other words, it does indeed measure the variation in the dependent variable that is successfully explained by a relationship with the independent variable.

For the weight/height data we know that the total variation in weight is

$$194{,}724 - 1378^2/10 = 4835.6 \text{ lb}^2$$

and that the variation in weight explained by a relationship with height is

$$7.52 \times 448.4 = 3371.97 \text{ lb}^2$$

The unexplained (or error) variation in weight must therefore be

$$4835.60 - 3371.97 = 1463.63 \text{ lb}^2$$

To set up an analysis of variance table in the manner described in Chapter 5 we also need to know the appropriate degrees of freedom associated with each source of variation. The total degrees of freedom are $n - 1$, as usual, since we are measuring variation in *y* about the sample mean. The explained variation has 1 degree of freedom, since it has just one explanation (the independent or explanatory variable, height in this example). This leaves $n - 2$ degrees of freedom for the unexplained or error variation (paralleling the closely related correlation situation).

VARIATION	SS	DF	MS	F
Explained by regression	3371.97	1	3371.97	18.43
Unexplained (error)	1463.63	8	182.95	
Total	4835.60	9		

For the weight/height data, we find a very large and highly significant F value (p < .01), confirming that the predictive relationship linking height and weight is very real and not just a product of random variation.

We can further say that, of the total variation in weight (4835.60), 3371.97 of this has been successfully explained by the dependence relationship with height:

that is,
$$\text{Percentage explained} = \frac{3371.97}{4835.60} \times 100$$

$$= 69.73\%$$

Note that this corresponds (except for some minor rounding error) to the r squared value we obtained for this data set earlier. (In correlation we tend to talk about percentage variation shared; in regression we talk about percentage variation explained. This percentage explained value (often referred to as r squared) is frequently quoted when a regression analysis is carried out.)

Like the correlation coefficient (or any other value derived from a sample) the slope, b, would undoubtedly fluctuate if fresh samples could be collected and analyzed. This potential fluctuation is measured by its standard error, given by

$$\text{Standard error } (b) = \sqrt{\frac{\text{Unexplained variance}}{\Sigma(x - \bar{x})^2}}$$

The unexplained random, or error, variation is, of course, exactly what would cause the value of the slope to randomly fluctuate from sample to sample, and this is best characterized by the unexplained or error variance calculated as part of the analysis of variance table (182.95, for the weight/height data). This, however, describes the random variation in terms of the dependent variable only. To measure the error variation in terms of the relative units (y/x or lbs/inch, and so forth) that characterize a slope, the error variance of y is divided by the sum of squares of x.

Once the standard error is known, it can be used to fit confidence limits around the sample slope, an important reminder that the calculated slope is simply an estimate based on a single sample and should not be taken totally seriously.

$$\text{95\% Confidence limits for the slope} = b \pm t_{.05,n-2} \, SE(b)$$

where $t_{.05,n-2}$ is the .05 critical value for t with $n - 2$ df
For the weight/height data,

$$\text{Error variance of weight} = 182.95 \text{ lb}^2$$

$$\text{Sum of squares of height} = 59.6 \text{ in}^2$$

$$\text{Standard error of } b = \sqrt{\frac{182.95 \text{ lb}^2}{59.6 \text{ in}^2}}$$

$$= 1.75 \text{ lb/in}$$

.05 critical value of t with 8 df = 2.31

95% confidence limits for change in weight per unit change in height

$$= 7.52 \pm 2.31 \times 1.75$$

$$= 7.52 \pm 4.04 \text{ lb/in}$$

In other words, based on the evidence we have available to us, the true weight/height connection for women in this situation might possibly be an increase as low as 3.48 lbs/inch or as high as 11.56 lbs/inch. (Since the null hypothesis of 0 lbs/inch is not encompassed within this range of realistic possibilities, this is quite equivalent to a rejection of the null hypothesis).

We could, quite equivalently, test the reality of the slope and hence the reality of the relationship by dividing b by its standard error (that is, calculating how many standard errors b lies away from zero) and testing this against a set of t tables with $n - 2$ degrees of freedom.

For the weight/height data

$$t = \frac{7.52}{1.75} = 4.29$$

which is highly significant ($p < .01$).

(Again note that the square of the t value is the F value we obtained earlier).

Regression and Prediction

We have already seen that we can use the regression line to predict the value of y that we would expect to correspond to some particular value of the explanatory variable, x.

For the weight/height data, the regression line

$$\text{Weight} = -352.73 + 7.52 \text{ height}$$

predicts a probable weight of 181.19 lbs for a female of 71 inches in height. Similarly, for a female of height 60 inches, the relationship predicts a probable weight of 98.47 lbs.

There is nothing (except our common sense) to stop us from attempting to predict weights for heights of 20 inches (which works out to be -202.33 lbs) or 120 inches (which works out to be 549.67 lbs). These are patently nonsensical, and no researcher would make such an obvious mistake. There is a very important point here, however. A regression line is only meaningful over the range of results on which it was based (for the weight/height example, say, between 60 and 72 inches), and there is no guarantee that the same dependence relationship will hold outside these limits. Using a regression line to predict results outside the range of evidence from which it has been constructed ("extrapolating") should never be attempted.

Even within the permissible prediction range it is important to appreciate the point that the predicted value is simply an indication of the value that the dependent variable is most likely to take in a certain specific situation. The inevitable presence of unexplained or error variation (measured by the error variance, as we discussed above) means that the true y value is likely to lie somewhat above or below its predicted location on the regression line. In addition, remember that the regression line itself may not be in exactly the right place. It must pass through the midpoint of the data set. This is determined by the sample mean and hence is itself subject to error. (The true mean of y may be somewhat higher or lower than the sample mean, and hence the sample-derived regression line may be located somewhat lower or higher than it should be.)

To further complicate the problem, the sample slope (as we saw in the last section) may somewhat underestimate or overestimate the true but unknown slope. Since the regression line must pass through the mean point, any error resulting from slope will become more and more dramatic the further one moves away from the mean of the data set. (Think of the line pivoting about the mean point and being tipped slightly further up or down.) This is another reminder that making predictions in extreme circumstances is fraught with danger.

Error resulting from unexplained or random variation is, of course, measured by the error variance, s_e^2, or the error standard deviation, s_e.

Error resulting from the use of the sample mean ($\bar{y}$) is measured by the standard error of $\bar{y}$, $s_e/\sqrt{n}$.

Error resulting from the use of the sample slope is given by the standard error of the slope times a measure of just how far the point of prediction, x_p, lies away from the mean value, $\bar{x}$.

$$(x_p - \bar{x})\,\text{SE}(b)$$

or

$$\sqrt{(x_p - \bar{x})^2 \frac{s_e^2}{\Sigma(x - \bar{x})^2}}$$

The combined effect of all these potential sources of error in prediction is given by

$$s_e \sqrt{1 + \frac{1}{n} + \frac{(x_p - \bar{x})^2}{\Sigma(x - \bar{x})^2}}$$

and we can say that the band of realistic possibilities for y when x takes the value x_p is given by

$$y_p \pm t_{.05,n-2}\, s_e \sqrt{1 + \frac{1}{n} + \frac{(x_p - \bar{x})^2}{\Sigma(x - \bar{x})^2}}$$

where y_p is the value of y predicted by the regression line when x takes the value x_p.

For instance, we have already predicted a weight of 181.19 lbs for a female of height 71 inches. The error variance we know to be 182.95 lbs^2 (see Testing the Dependence Relationship), and hence the error standard deviation is 13.53 lbs. From our previous calculations we know that the mean height is 65.2 inches (that is, we are attempting a prediction at a point on the regression line 5.8 inches above the mean height) and that the sum of squares for height is 59.6 inches^2.

The range of possibilities is therefore given by

$$181.19 \pm 2.31 \times 13.53 \times \sqrt{1 + \frac{1}{10} + \frac{33.64}{59.6}} = 181.19 \pm 40.32 \text{ lb}$$

In other words, a weight of anywhere between 140.87 and 221.51 lbs for an individual with a height of 71 inches would be quite consistent with the dependence relationship suggested by the sample evidence. A calculation like this is a graphic reminder that a predicted value is merely a statement of the dependent value we would be most likely to see in those particular circumstances and is far from a definitive statement of what must happen.

REGRESSION OR CORRELATION

As we have repeatedly seen, correlation and regression are two very closely related ideas. They use the same basic concept (the straight line pattern). They use similar calculations, and a correlation value can actually be derived from a regression approach (simply by taking the square root of the r squared value). How can we distinguish between them, and which should we use?

In correlation there is no question of a dependence relationship, and the only issue of interest is, "Are the two variables related?" Correlation assumes that both variables follow Normal distributions (more exactly, a joint Normal distribution).

In regression there is a clear dependent/explanatory relationship between the two variables (either by virtue of logic or circumstance), and explaining or describing this relationship is our key goal. In regression we only assume that the dependent variable follows a Normal distribution and that the scatter about the regression line does not start to grow or shrink as we move up or down the line. If it did change we could no longer use one value, the error variance, to characterize the entire regression line. (In fact, we should also assume that the independent or explanatory variable can be measured exactly without any error, although this assumption is treated rather liberally in practice, that is, frequently ignored. In regression there is no assumption that the independent variable has a Normal distribution.)

If there is a dependent/independent relationship, regression should be used. If not, then correlation should be used, assuming in each case that the other assumptions are not seriously violated.

Regression depends very directly on the idea of a straight line relationship. Again, examining a scatter diagram to confirm this is a very good idea. If the pattern is not a straight line (that is, "nonlinear"), simple regression (as discussed in this chapter) should not be used. When the pattern is curved, more sophisticated regression techniques ("curvilinear regression") are available to handle such data sets, and the ideas that underlie this approach will be discussed in Chapter 13.

Key Terms

Scatter diagram	Dependent variable
Co-variation	Independent variable
Covariance	Regression line
Correlation coefficient	Least squares fit

PROBLEMS

1. The following data gives weight (in lbs) and cholesterol (in mg/dl) measurements collected from 14 adult males as part of a coronary disease study. (None of the subjects were known to have experienced a coronary incident within the 10 years before the study.) Determine if there is any evidence of a relationship between the two variables.

WEIGHT	CHOLESTEROL
168	135
175	403
173	294
158	312
154	311
214	222
176	302
262	269
181	311
143	286
140	403
187	244
163	353
164	252

2. As part of a study of the microvascular complications of diabetes mellitus and their possible remediation through blood glucose control, the capillary basement-membrane width (in angstroms) and glycosylated hemoglobin level (as a percentage) of 13 diabetic patients were measured. The values are reported here. Test whether these results offer any evidence of a connection between these two patient characteristics and measure the strength of the connection, if any.

CAPILLARY BASEMENT MEMBRANE WIDTH (Å)	GLYCOSYLATED HEMOGLOBIN LEVEL (%)
2642	7.7
1180	16.7
1474	10.4
1798	10.9
2958	7.2
3480	8.5

1917	10.2
2359	12.7
1723	11.4
2414	8.1
900	11.6
2135	8.4
2347	8.3

3. In an initial trial of a newly developed anesthetic, the time from administration of the anesthetic until induction of anesthesia was recorded for 10 patients, and the effective dosage rate used (in μg/kg) was calculated from the patient's weight. Describe and test the relationship between induction time and dosage rate.

DOSE (μg/kg)	INDUCTION TIME (min)
1.3	25
1.9	13
1.5	19
1.1	23
1.0	26
1.3	21
2.1	11
0.7	31
1.6	14
0.9	24

4. Transcutaneous oxygen tension (TO) and arterial oxygen tension (AO) were measured (in mm Hg) in 15 patients undergoing thoracotomy requiring one lung ventilation. Do the results presented below give evidence of a relationship between these two measurements? If so, how strong is the relationship?

AO (mm Hg)	TO (mm Hg)
210	150
320	190
150	80
130	130
270	170
400	260
220	210

AO (mm Hg)	TO (mm Hg)
190	90
90	70
370	200
400	280
190	170
290	150
280	220
330	260

5. The fasting plasma total cholesterol levels (mg/dl) were measured for 20 male Air Force servicemen aged 20 to 24 years. After a 10-year follow-up, their fasting plasma total cholesterol levels were remeasured. Determine and test the relationship between the initial and post follow-up total cholesterol levels. Can you use this relationship to predict the most likely follow-up total cholesterol level (together with the appropriate 95% confidence limits) for a 22-year-old male Air Force member with an initial total cholesterol level of
   a. 154?
   b. 366?

INITIAL	FOLLOW-UP	INITIAL	FOLLOW-UP
150	241	215	202
186	224	122	146
130	121	159	213
116	137	179	216
168	236	143	108
205	241	154	180
101	94	172	246
135	160	147	192
192	248	164	140
160	154	157	180

6. In a study of the effects of age on the duration of anesthesia, 10 patients ranging in age from 22 to 80 years were given an epidural administration of a local anesthetic agent. Blood samples were collected at intervals after administration, and the rate of total plasma clearance (expressed in ml/min) was calculated for each patient.

PLASMA CLEARANCE (ml/min)	AGE
610	22
510	25
400	31
450	43
420	52
300	58
370	60
405	64
280	70
320	80

Describe and test the relationship between plasma clearance and age. Fit 95% confidence limits to the predicted plasma clearance rates at age 25 and at age 55.

CHAPTER 8

Designing an Experimental Study

Investigation in the health sciences, or any other area of science, is ultimately about asking, and we hope answering, questions. More precisely, it is about asking a question (or forming a hypothesis) in a way that gives the investigator a realistic chance of arriving at a meaningful and valid answer. There are many investigative situations in which the researcher may exercise a great deal of control over the research environment. In a comparison of various therapeutic approaches to a specific problem, he or she may be able to define precisely the nature of the various therapies to be compared and how they are to be administered. Also very important, he or she may have the capacity to determine the specific therapy that each of the participating patients receives. (How this is done is very important and is discussed in the section on Treatment Allocation).

When an investigator can manipulate the situation in this way, the designed experiment is the classic research device for exploring specific, detailed hypotheses. As such, an understanding of the principles of experimental design is a central part of the armory of skills needed by the health sciences investigator. (We will discuss the situation in which the investigator has little or no control over the behavior of the research subjects in Chapter 15.)

The typical experiment consists of a number of groups of subjects, each of which is exposed to a particular experimental condition. The object of the experiment is to determine whether the various experimental conditions differ in their influence on the experimental subjects. At the conclusion of the experiment, this hypothesis is tested by using the appropriate statistical technique (as discussed in Chapters 4 through 7 and 11 through 14) to analyze the experimental evidence. If the experiment has been well designed, the analysis and interpretation of the results generated should be relatively straightforward. The real skills of experimentation are invoked when the experiment is being planned, and mistakes at this stage will totally destroy the following weeks or months of effort. In this chapter we will examine the basic principles of experimental design that should be followed to ensure that an experiment will give valid results and to ensure that those results are as informative as possible.

PREVENTING BIAS

Experimentation is a comparative process, and if the experiment is to give meaningful results, those comparisons must be fair and unbiased. **Bias** is the occurrence of some systematic influence on the results that tends to make the results from one or more of the treatments or experimental conditions look consistently better or worse than they otherwise would be. If this sort of systematic influence is present in an experiment, then the comparisons are manifestly unfair, and the whole process is pointless. It is totally impossible to determine the real effects of the various treatments under comparison and hence, totally impossible to answer the question the experiment was intended to answer. Prevention of bias is absolutely vital for meaningful and valid experimentation.

The deliberate biasing of experimental results is a moral rather than biostatistical problem. However, the unintentional biasing of results can occur very easily and can present major pitfalls for the researcher who initiates an experiment without carefully considering exactly what he or she is doing. These biases, however, can be easily avoided with a little care and proper planning.

Confounding

Confounding occurs when an experimenter unintentionally constructs an experiment in such a way that there is more than one possible explanation for the results obtained. The key explanation ("the treatments are genuinely different") is said to be confounded with another irrelevant (at least from the perspective of the key research question), but possible, explanation. In this situation the experiment is totally worthless. Even if a significant final result is obtained, there is no justification for concluding that differences between the treatments have been proven since there is an alternative factor that could have produced the differences even if the treatments had been, in fact, identical in their effects.

For example, suppose we wished to compare two surgical procedures. Further suppose that in the actual experiment, procedure 1 is always carried out by surgeon A, and procedure 2 is always carried out by surgeon B. The results apparently show that procedure 1 gives superior results. However, an alternative explanation is simply that surgeon A is a better surgeon than surgeon B. The differences between the procedures have been confounded with the differences between the surgeons. It would be quite unjustified to claim that this experiment has demonstrated the superiority of procedure 1 since an alternative explanation is available. Arguing that you are convinced that the two surgeons are equally skillful does not make the comparison any more valid. This may, or may not, be true and is, after all, only your opinion. A difference in skill between the surgeons is certainly a possibility and cannot be ruled out. The mere fact that an alternative explanation exists at all is sufficient to make the conclusions meaningless.

Confounding can also easily arise in experiments in which the subjects are exposed to more than one treatment (paired or crossover experiments). Suppose that in a drug trial, 20 patients suffering from a particular condition are all treated with drug A for two weeks and their condition assessed. They are then treated with drug B for two weeks and reassessed. The results appear to indicate that drug B has given superior results. Again, such a conclusion could not be justified since an alternative explanation exists.

It is possible that the condition is such that the patients will tend to recover with time, and the apparently superior results for B simply illustrate the fact that the patients' condition is systematically improving. The difference between the drugs (the key difference) has been confounded with differences over time. Again, whether you believe that patients of this type are likely to improve (or deteriorate) over time is largely irrelevant. The presence of another possible explanation is enough to make the experiment worthless.

Confounding must be avoided at all costs. We can see from this brief discussion that confounding is very likely to be present if the investigator takes some systematic approach to deciding the order or nature of the treatment that the experimental subjects receive. How then should we approach these important questions? The techniques discussed in the next few sections offer us some practical solutions.

Treatment Allocation

One of the most obvious sources of possible bias occurs when the treatments under comparison are allocated to the experimental subjects. One possible approach has the investigator subjectively deciding which treatment should be administered to each experimental subject. This approach is potentially disastrous. There is a real possibility that the researcher may subconsciously tend to allocate the more seriously ill patients to one treatment group (say, the active treatment or a new treatment), and the less seriously ill patients to another group (say, the control group or a standard treatment group). This would obviously introduce a systematic bias into the experimental results and totally destroy the basis for a fair comparison of the treatments.

Again, protests that the allocation was scrupulously fair would be quite irrelevant. The mere fact that it might have happened (which could never be disproved) is enough to make the allocation another possible explanation for any final difference observed and to cast a major cloud over the whole experiment. Like Caesar's wife, the experimenter must be above the slightest hint of suspicion.

The only way to remove even the possibility of bias in treatment allocation is to remove allocation entirely from subjective human decision making. This can be achieved very easily through the use of a **random allocation** procedure in which the process of allocating treatments to subjects is totally outside the control of the investigator.

For an experiment involving allocating, say, 40 patients to two drug treatments, a perfectly acceptable procedure would be simply to toss a coin to decide which treatment a particular patient receives. This ensures that the decision is removed entirely from the experimenter's control and that no possible accusation of experimenter-induced bias (unintentional or otherwise) can arise. However, coin tossing has several major practical limitations. It is convenient only if two treatments are involved. Furthermore, there is no guarantee that this system will allocate the same number of subjects to each treatment group. (The 40 patients might, for instance, end up being allocated into groups of 17 and 23 subjects.) Having equal numbers in the treatment groups is highly desirable, since it generally makes an experiment more sensitive (reduces the risk of a type II error) and easier to analyze (see our discussion of multiple comparisons and specific questions in Chapter 5).

A much more convenient method of random allocation uses random number tables. These are tables of numbers (usually 2-digit) created entirely at random with no inherent order or pattern (see Table T/7 in the appendix). Their use can be illustrated by the following small-scale example.

Suppose we have 12 subjects we wish to allocate on a random basis to three treatments, while ensuring that each treatment group consists of four subjects. We simply read 12 consecutive numbers from a random number table. The approved procedure is simply to close your eyes and select a point on the random number table! From there a string of numbers may be read in any direction (horizontally, vertically, or diagonally) since the table is totally random and has no order of any sort. Should a number in the sequence repeat itself, the recurrence is ignored and a fresh number is selected.

13	52	53	94	75	45	69	30	96	73	89	26
A	B	B	C	C	A	B	A	C	B	C	A

The four smallest numbers in the sequence are located. These identify the subjects to receive treatment A. The next four largest numbers identify the subjects to receive B, and so forth. Hence, subject number 1 will receive treatment A, subject number 2 treatment B, and so on. Since the order of the numbers is entirely random, the allocation decisions are entirely random and outside the experimenter's control, and hence free of any risk of bias. In addition, the method ensures the creation of equal size treatment groups.

The other allocation technique sometimes employed is a procedure known as systematic allocation. This operates exactly as its name implies. In the above example, the first subject would receive treatment A, the second B, the third C, the fourth A, and so on. Systematic bias will arise only if there is some systematic variation among the subjects that matches the order in which the treatments are being allocated (for example, if every fourth patient tends to be particularly ill). Although this is usually highly unlikely, it does have to be viewed as a possibility.

More important, investigators are frequently required to decide if a potential subject should be allowed to participate in the experiment at all (an issue we will discuss in Chapter 10). If a systematic allocation approach is used, the investigators will know in advance which treatment a subject will receive if the subject is admitted to the experiment. This raises the possibility that an investigator might, consciously or subconsciously, tend to systematically exclude more severely ill patients from one particular treatment group, thus loading the experiment and rendering the comparisons unfair and meaningless. Again, the issue would not be whether this actually happened. The very fact that it was possible would be enough to cast a cloud of suspicion over the experiment and render it meaningless. Systematic allocation offers virtually no advantages, while presenting the real possibility of bias and hence, is not recommended as an experimental procedure.

Randomization is a flexible and easy-to-use approach to treatment allocation. It is the only approach that offers complete protection from even the possibility of bias, and it should be the standard approach to all treatment allocation decisions in the experimental situation. For instance, we pointed out earlier that systematically administering drug A first to a patient in a crossover experiment, followed by drug B several weeks later, simply confounds potential treatment differences with potential time effects. The appropriate and simple solution is to decide the order of treatment administration on a random basis, with half the patients (selected by the use of random number tables) receiving drug A first and the other half receiving drug B first.

Blinding

One easily overlooked source of systematic bias is the subjects' psychological reactions to the treatment they receive. This can produce quite dramatic differences in response. For example, patients who are not being actively treated but mistakenly believe they are being treated often respond substantially better than patients who know they are not being treated. This psychological response to the prospect of treatment is both powerful and subtle. Studies comparing sham or dummy treatments ("placebos") have confirmed that patients respond to such subtle distinctions as tablet size (the larger the tablet, the better the response), color of the medicine (the more vivid the color, the better the response), and taste of the medicine (the worse the taste, the better the response, presumably on the basis that "If it tastes this bad, it must be good for me!").

In other words, the stronger the belief in treatment, the stronger the patient's response to it will be, even if that belief has no foundation in fact. Patient response to a positive treatment environment is both real and important, which is precisely why it must be addressed in the design of an experiment. If an experiment is to be a fair comparison of the inherent effectiveness of the treatments involved and not biased by the patients' perceptions, it is vital that the experimental subjects not know which of the various experimental treatments they are receiving. Such an experiment is known as a **single-blind experiment,** since the subjects are blind to the treatments they are receiving.

For an experiment to be genuinely single blind, the subjects must be unable to distinguish the various treatments. This requires a considerable amount of planning to ensure that both the "exterior" aspects of the various treatments (such as their taste) and their methods of administration are as alike as practicable. There are, of course, situations in which this is not feasible, such as comparisons of surgical procedures. Nevertheless, every effort should be made to ensure maximum possible duplication of conditions. When an experiment involves an untreated or control subject group, it is vital that these subjects receive an apparently identical sham treatment or placebo that duplicates the "exterior" aspects of the active treatments.

We tend to think psychological response is an issue only when working with human subjects, but this is not really true. Repeated handling of animals will induce stress, with the very real possibility of a measurable effect on response. If a control group is being compared with a group whose routine is being constantly interrupted for treatment administration, then there is a real risk that systematic bias will result. Single blinding and creating apparently identical treatment environments (comparable handling of all animals, administration of dummy treatments, and so forth) should be the rule in all experimental work, if at all possible.

If the outcome of the experiment can be measured on an objective basis (such as weight gain or blood iron levels), then a randomized single-blind approach should ensure an unbiased experiment. However, many health investigations involve subjectively assessed outcomes ("Has the patient's condition improved and if so, how much?"). Such assessments present major opportunities for bias. All human beings have an extraordinary capacity for seeing what they want to see. An investigator assessing patients on a newly developed treatment for depression that he or she believes to be a major step forward, for example, may very easily "see" signs of improvement simply because he or she would like to see them.

The only sure protection against the risk of this possible bias is for the assessor to be unaware of the treatment the subject has received. An experiment in which both the subject and the assessor are blind to the treatment involved is known as a **double-blind experiment.** This requires careful planning of the total experimental enterprise with part of the research team responsible for treatment allocation and implementation, a separate section responsible for the ultimate subject assessment, and subject performance collated with the treatment received only at the completion of the study.

There is a school of thought that argues that all experiments, including those with objectively measurable outcomes, should be double blinded. This argument is based on evidence that subtle, unspoken messages can be transmitted between human beings via body language, tone of voice, facial expression, and so forth. This raises the possibility that an investigator with little faith in the treatment a patient is receiving can transmit that lack of faith to the patient, with the resultant real psychological effect on performance that we mentioned earlier. The only sure way to guard against this possibility is to ensure that the experimenters who have contact with the patient do not know which treatment the patient is receiving (that is, double blinding). Double blinding when the experimental outcome is objectively measurable is far from universal practice. Nevertheless, it is an excellent idea that offers extra protection for relatively little extra work. It is always better to be too careful rather than not careful enough.

IMPROVING SENSITIVITY

If the treatments have been allocated to the experimental subjects randomly, and the experiment has been run on a single-blind or double-blind basis, as appropriate, then the experiment can be regarded as valid. In other words, any conclusions drawn from it will be meaningful since they will be based on fair and unbiased comparisons. A good experiment is, however, more than just fair. It should offer the experimenter the best possible chance of detecting real treatment differences, if there actually are any. In other words, it should have a high power or low risk of a type II error.

As we saw when we studied Analysis of Variance (Chapter 5), the evidence for the existence of real treatment differences is assessed by comparing the variation resulting from treatment differences (the "between-treatments" variation) with the unexplained or error variation (the "within-treatments" or "random" variation). The smaller this random variation is, the more clearly any real treatment differences will show up and the more powerful or sensitive the experiment will be. For an experiment to provide maximum "value for money" and offer the best hope of detecting any real treatment differences, the error variation must be made as small as possible.

It is rarely possible to make the experimental subjects intrinsically less variable. Random variation usually cannot be reduced directly. Remember, however, that random variation is simply variation whose cause we cannot explain and that we, therefore, are forced to label "random". (In analysis of variance, recall that we split the total variation into between-treatments variation and within-treatments or random variation). Part of such variation is undoubtedly random, but at least part of it might be due to some other, so far unexplained, causes. If we could find other ways of explaining why experimental subjects differ from one another, in addition to identifying the between-treatments differences, then less of the total variation would remain unexplained. The "random" variation would, therefore, be reduced and, as a consequence, any real between-treatments differences would show up much more clearly. (We will see how to attribute variation to several causes when we meet multi-way analysis of variance in Chapter 11.)

Much of the art of experimental design involves constructing an experiment in a way that allows as much of the variation as possible to be explained and its source identified, while retaining the key feature of random treatment allocation. Time spent considering the best way to design an experiment is time well spent, since it can result in a far more effective and productive experiment.

Completely Randomized Experiments

In the simplest form of experimental design the available subjects are simply randomly allocated to the various treatments, usually so as to ensure equal numbers in the treatment groups. Because the treatment allocation is random, the comparisons involved are fair and unbiased, and a completely randomized experiment is a perfectly valid and very simple experimental procedure.

However, such a randomized experiment makes no attempt to provide any additional explanations for the variation between the experimental results. The total variation is simply split into between-treatments variation and within-treatments or unexplained ("random") variation and one compared with the other (see our discussion of analysis of variance in Chapter 5). There are no other possible explanations for the variation involved and, hence, no way of reducing the "random" or unexplained variation.

The advantage of the completely randomized experiment is that it is extremely simple to set up and carry out. Its disadvantage is the fact that it makes no attempt at all to control or reduce the random variation. As a result, completely randomized experiments are relatively insensitive and are substantially more likely to miss real treatment differences than are slightly more sophisticated experimental designs that offer more explanations for the variation observed.

Randomized Block Experiments

The completely randomized approach to comparing two treatments using 20 subjects involves randomly dividing the subjects into two groups of 10, each receiving one treatment. In Chapter 4 we pointed out that a more effective way of carrying out such an experiment would be to form the subjects into 10 pairs of subjects who were intrinsically very similar to one another, such as twins. If each twin receives one treatment (on a random basis, of course), then any treatment differences should show up very clearly since the twins are inherently very similar. Because the treatment comparisons are carried out within pairs of similar subjects rather than between groups of potentially quite dissimilar subjects, the random variation that might obscure any real differences has been carefully controlled and very considerably reduced.

A paired experiment of this type is sensitive precisely because it is designed in a controlled way that enables the experimenter to explain or account for much of the variation involved. As a result, the unexplained or random variation is much smaller than it would otherwise be. If, for instance, the 20 individuals had simply been allocated to two groups of 10 in a completely randomized design (that is, taking no account of the fact that they consisted of 10 pairs of twins), then the variation between different pairs of twins would be mixed in with the rest of the unexplained variation. The result would be a large random variation and an insensitive experiment. By carefully designing the experiment so that the twin pairs are identified, the variation between different twin pairs can be measured and isolated. The result is that the unexplained variation is greatly reduced, and any differences between the treatments show up much more clearly.

In the simple paired experiment we saw that we can achieve this reduction in random variation by concentrating on the differences between the subjects within each pair. By doing this (focusing on comparisons between similar individuals), we entirely eliminate the variation between the different pairs and obtain a much more sensitive analysis (see our discussion of the paired t test in Chapter 4).

When we discussed the comparison of more than two treatments in Chapter 5 we realized that we would have to replace the t test approach with the more flexible analysis of variance philosophy. Should we wish to compare more than two treatments in this "grouped" way (perhaps through the use of triplets) it is clear that the very direct paired t test approach simply cannot be used. However, a very simple extension of the analysis of variance (multi-way analysis of variance, which we meet in Chapter 11) can be used to control for, and eliminate, variation between the different patient groups, with the result that the treatment comparisons are made solely on the basis of the relative performance within these groups of inherently similar individuals. Just as the basic (one-way) analysis of variance can be viewed as an extension of the independent t test, so the multi-way analysis of variance of Chapter 11 can be viewed as an extension of the paired t test.

In those experiments that involve the comparison of more than two treatments, exactly the same design technique can be applied. For a four-treatment experiment, the subjects are formed into groups (or blocks) of four (or multiples of four) so that the subjects in each block are as inherently similar as possible. (Restricting the block size to multiples of the number of treatments being compared is necessary to ensure that equal numbers of subjects are allocated to each treatment). The four treatments are then randomly allocated to the subjects in each block.

Treatment differences should again show up very clearly, since they are based on the comparison of similar subjects. Variation between the different blocks can be effectively eliminated from the experiment, thus reducing the unexplained or random variation and increasing the sensitivity of the experiment. Such a design is known as a randomized block design.

BLOCK 1	T_2	T_3	T_1	T_4
BLOCK 2	T_1	T_3	T_4	T_2
	.	.	.	.
	.	.	.	.
BLOCK 10	T_2	T_4	T_3	T_1

The success of a randomized block experiment depends on how well the blocking situation succeeds in bringing together inherently similar subjects. In animal studies, blocks might well be formed using animals from the same litter. In human studies, it is usually impossible to obtain groups of siblings of the appropriate number. It is therefore usual to create "artificial" blocks based on general patient characteristics that it is believed might be related to potential variation in response. Older patients will often respond noticeably differently from younger patients, even when they receive the same treatment. Gender differences are also common. Therefore, blocks are frequently formed by bringing together patients of similar age or of the same sex.

Within each block, therefore, the treatments are being compared using individuals who are at least reasonably similar from the beginning. Variation resulting from, say, age differences, will therefore be controlled and eliminated rather than allowed to run loose and inflate the error variation, which is what would happen if the influence of age were simply ignored.

The most appropriate blocking structure will be determined by the context of the particular investigation being conducted. In an endurance study, for example, level of patient fitness (unfit, moderately fit, very fit, and so forth) might be a helpful way of blocking the patients.

More than one blocking structure may be used simultaneously in an experiment. A frequently helpful technique is to block subjects on age (say, four age categories) and gender (two categories).

GENDER BLOCKS

		Male				Female			
	Under 20	T_1	T_3	T_4	T_2	T_1	T_4	T_2	T_3
	20—34	T_2	T_1	T_3	T_4	T_3	T_4	T_1	T_2
AGE BLOCKS									
	35—49	T_3	T_1	T_4	T_2	T_1	T_2	T_4	T_3
	Over 50	T_4	T_3	T_1	T_2	T_3	T_1	T_2	T_4

The four treatments are randomly allocated to the four subjects forming each individual age/gender block, enabling the comparison of the treatments to be based on individuals of the same gender and similar ages. If gender and age differences are likely to be important in the condition under study, then controlling them in this way will result in a much more sensitive experiment.

The whole point of blocking is to bring together subjects who are inherently very similar. The ultimate similarity arises when a subject is compared with himself or herself. In this situation each patient forms an individual block, with the order of treatments decided on a random basis. A randomized block design of this type is known as a **crossover design.**

Patient 1	T_1 T_3 T_4 T_2		
PATIENT BLOCKS **Patient 2**	T_2 T_1 T_4 T_3		
.	. . .		
.	. . .		
.	. . .		
Patient 10	T_4 T_3 T_1 T_2		

When such an approach is practical, it is highly sensitive. However, there are many practical constraints that often rule out the use of crossover designs. It is essential that the treatments be temporary in their effect and that the influence of one treatment does not linger in the patient and carry over into the next treatment period. The use of inter-treatment "wash out" periods is often necessary, and the entire experimental process can become very protracted.

Randomized block experiments are more complicated to set up than completely randomized experiments, and they put more constraints on the subjects involved. (The previous age/gender blocked experiment requires four males over 50 years old, four females over 50 years old, and so on, whereas a corresponding completely randomized experiment would require the recruitment of simply 32 patients, with no gender or age constraints at all). However, randomized block experiments can offer substantial improvements in experimental sensitivity and are usually well worth the effort.

▬ Latin Squares

Using several blocking structures is a substantial help in controlling random variation, but it can require the use of large numbers of experimental subjects. The age (four categories) and gender (two categories) blocked comparison of four treatments requires a total of at least $4 \times 2 \times 4 = 32$ subjects (or some multiple of this). A six-treatment experiment with two blocking structures with 5 categories and 7 categories respectively would require a minimum of $5 \times 7 \times 6 = 210$ subjects (six in each of the 35 individual blocks). There is a way of controlling two sources of variation (that is, incorporating two blocking structures) with a much reduced number of subjects, provided that the number of categories in each blocking structure can be made equal to the number of treatments under comparison.

Such a design is called a **Latin Square.** It is most easily visualized as exactly that, a square with the first blocking structure defining the rows of the square and the second blocking structure defining the columns. The treatments are allocated randomly to the individual cells in the square, with only one treatment per cell. Each treatment must only occur once in each row and once in each column. This restriction is important since it ensures that all the treatments are assessed under comparable conditions.

The easiest way to achieve randomization while obeying these restrictions is to start with a systematic arrangement that has each treatment only once in each row and once in each column. For a four-treatment experiment (and hence, four blocking rows and four blocking columns) this would be:

		COLUMN			
		1	2	3	4
	1	A	B	C	D
	2	B	C	D	A
ROW	3	C	D	A	B
	4	D	A	B	C

where A, B, C, and D denote the four treatments.

Randomization is achieved by first randomly switching the four rows, and then randomly switching the four columns. Try it and see how the initial symmetrical pattern is completely randomized, but each treatment still occurs just once in each row and once in each column.

For example, in a study of four possible stress management strategies, it was thought that age and general physical fitness might have a substantial impact on stress levels. If these two potential blocking agents could be formed into four categories each, then it would be possible to form a Latin Square design for this experiment involving only 16 subjects rather than the 64 ($4 \times 4 \times 4$) required by a full randomized block design. After randomization in the way we have just described, the following treatment allocation plan was developed, with the very fit subject under 25 years old receiving treatment C, and so on.

	FITNESS			
	Good	**Moderate**	**Poor**	**Very Poor**
Under 25	C	D	B	A
26—39	A	B	D	C
40—54	B	C	A	D
Over 54	D	A	C	B

(**AGE** appears to the left of the rows 26—39, 40—54.)

(Again note that each treatment appears only once in each row and once in each column).

Latin Square designs offer a useful way of controlling two sources of variation while requiring relatively small numbers of subjects. However, they can be formed only when the number of blocking rows and columns each equal the number of treatments involved.

Factorial Experiments

So far in this chapter, we have talked about the comparison of different treatments. What we frequently mean, however, is the comparison of different levels of some overall treatment concept, such as the comparison of different dosage levels of a particular drug. The overall treatment concept is known as a **factor,** and the various individual treatment categories within that concept are known as the factor levels. Frequently, more than one treatment concept springs to mind when we decide to carry out a particular investigation. Rather than set up separate experiments to look at each factor, it is frequently more efficient and informative to combine the various factors in a single experiment.

An experiment that involves the simultaneous study of several factors is called a factorial experiment. For instance, in a study of dietary influences on growth, we might decide to investigate two factors: the nature of the protein source (with two levels, animal and vegetable), and the protein level (again with two levels, a high concentration and a low concentration). When combined in a single factorial experiment this combination of two factors, each at two levels, results in four unique treatment combinations: a high-concentration animal protein diet (HA), a high-concentration vegetable protein diet (HV), and so forth.

	PROTEIN SOURCE	
	Animal	**Vegetable**
High	HA	HV
Low	LA	LV

PROTEIN LEVEL (labels at left: High / Low)

In the simplest formulation of this factorial experiment the subjects would simply be randomly allocated to the four treatment combinations so as to ensure equal numbers in each combination. However, this makes no attempt at all to control the random variation in the study and is, in effect, a completely randomized approach to a somewhat more ambitious experimental objective. The variation control ideas we discussed earlier in this chapter can be very easily applied to factorial experiments. For instance, in a randomized block design the subjects would be formed into blocks of four (or multiples of four) on some suitable basis, and the individuals in each block randomly allocated to each of the four treatment combinations.

Factorial experiments are a very efficient way to make use of the available experimental subjects. If, for instance, 40 laboratory animals were available, the factorial approach would enable all 40 to be used for the evaluation of both protein levels and protein sources. The alternative would be to run two quite separate experiments, with only 20 subjects available for the evaluation of protein levels and only 20 available for the protein source evaluation.

However, the real advantage of an integrated factorial approach over the separate experiments approach is that it enables us to see not just whether the protein source or the protein level affects growth, but whether the two influence or "interact" with one another. An **interaction** occurs when the relative effectiveness of the levels of one factor is influenced by, or depends on, the presence or level of the other factor. For instance, if high concentrations produce much faster growth rates than low concentrations when the protein source is animal, but the level makes absolutely no difference when the source is vegetable, then an interaction between protein source and protein level is said to be present.

You know an interaction is present when you have to give a qualified answer to questions about your experimental conclusions. "Does protein level make a difference?" "Well, it all depends. If we're talking about protein from an animal source, yes, the level does matter. However, if the protein is from a vegetable source, the level just doesn't matter." This conditional summary of the experimental results is a clear sign that an interaction is present.

One way to get a feel for the possible presence and meaning of an interaction is to plot a very simple graphic summary of the experimental results. The experimental response (growth rate, for instance) is plotted on the vertical axis, and the levels of one of the factors (which one is usually irrelevant; in this case, protein level) are denoted on the x axis. The means of the various treatment combinations are then plotted on the graph using different symbols to distinguish the levels or identity of the second factor, in this case, protein source.

In the hypothetical situation depicted in Figure 8-1, high protein levels produce much better growth rates than low protein levels, and this is true for both animal and vegetable sources. Similarly, animal sources tend to give higher growth rates than veg-

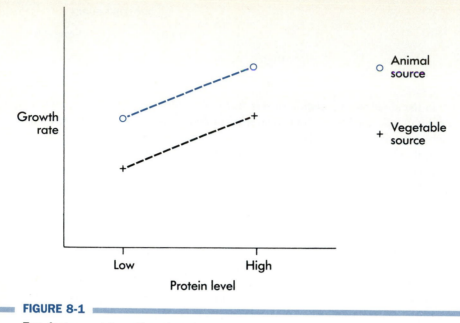

FIGURE 8-1

Two factors—interaction absent.

etable sources, a situation that holds at both high and low concentrations. The relative performance of one factor is not affected by the level of the other factor. In this situation there is no interaction between the two factors, and we can draw conclusions and make statements about one factor without having to worry about the other factor. ("Does protein level make any difference to growth rates?" "Yes it does; high protein concentrations always result in higher growth rates than low protein concentrations.") In other words, this relative performance holds, irrespective of the protein source.) The parallel lines in Figure 8-1 are visual evidence that relative performance stays constant and that no interaction is present.

In the hypothetical situation depicted in Figure 8-2 the means indicate that high protein concentrations will produce much higher growth rates than low concentrations if the protein source is animal. However, for vegetable protein sources, high and low concentrations give almost identical results, resulting in the necessity to qualify our conclusions, as we discussed earlier. Drawing conclusions about the effectiveness of high or low concentrations is meaningless without also referring to the nature of the protein source involved. In this situation the two factors interact with one another, and the effectiveness of increasing the concentration is substantially influenced by the nature of the protein source involved. Note that the lines in Figure 8-2 are clearly not parallel, indicating that relative performance changes from one situation to another.

To make real sense of a factorial experiment, it is important to determine if interactions between the factors are present. Simply asking if protein level makes a difference is a pointless exercise if an interaction is present (as in Figure 8-2), because in one situation (protein from an animal source) it does make a difference, and in another situation (protein from a vegetable source) it does not. The real power of a factorial experiment is its ability to explore the possibility of interactions between different factors.

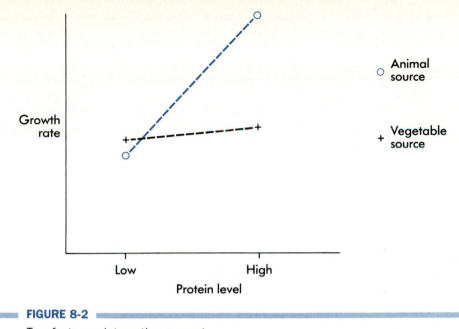

FIGURE 8-2

Two factors—interaction present .

The concept of interaction can also be applied to some of the blocking factors commonly used. In a drug comparison blocked on age, we might wish to determine whether the effectiveness of the drugs is the same in all age categories (no drug/age interaction) or whether some drugs are, for instance, clearly superior in young patients, but only marginally superior in elderly patients (a drug/age interaction).

Is the distinction between factors and blocking structures hard and fast? Factors are generally treatment concepts whose levels can be imposed on the experimental subjects at the (we hope random) whim of the investigator and that represent key issues in the experiment. Blocking structures are generally made up of inherent, unalterable characteristics of the subjects, and the role of blocking structures is primarily to control sources of variation that might obscure differences between the treatments (factor levels). However, both factors and blocks are treated in exactly the same way when the experimental results are being analyzed. In fact, the one simple technique, multi-way analysis of variance (Chapter 11), provides us with a very flexible experimental analysis tool that enables us to remove variation associated with blocks, test for factor effects, and test for the presence of interactions between factors or between a factor and a blocking structure. As a result, the distinction between factors and blocks, while valid in principle, is of much less importance in practice.

OTHER ISSUES

Another very direct way of increasing the sensitivity of an experiment is to increase the number of subjects involved, since this will decrease the standard error and hence, increase our ability to detect true treatment differences. Sample size is a very important consideration in research design, since an experiment that is too small may be so insensitive that there is a real risk that it might fail to detect even quite marked treatment differences, that is, the risk of a type II error is unacceptably high. This is potentially disastrous since the whole point of an experiment is to detect treatment differences if they genuinely exist. An experiment that has only a poor chance of detecting important treatment differences is simply a waste of time and money.

Determining the appropriate number of subjects to use in an investigation is an important research decision, and we devote a chapter (Chapter 9) to the discussion of the issues involved and how they can be translated into a practical decision-making strategy.

Another way to control the inherent variability of an experimental study is to restrict the nature (and hence the variability) of the experimental subjects (for example, study only young males). Although this will indeed reduce variability and improve sensitivity, it also unfortunately greatly reduces the ability to generalize the experimental conclusions and hence reduces their usefulness. The results of an experiment, like the results of any other statistical analysis, should never be extrapolated beyond the population from which the experimental subjects were drawn. An experiment performed solely on male subjects gives no indication of the best treatment for females. An experiment carried out on student volunteers does not allow you to draw valid conclusions about the population at large. Strictly speaking, it does not even allow us to draw conclusions about student non-volunteers.

The wider the variety of subjects included in an experiment, the wider the range of validity of the results obtained. Equally, however, the more heterogeneous the subjects, the more variable they will be, and the less sensitive the experiment will be. An experiment with more homogeneous subjects (say, all young males) will be inherently less variable and hence, more sensitive. However, the range of validity of its results will be correspondingly limited. By far the best approach is to choose the population of subjects about which you wish to draw conclusions, select the experimental subjects accordingly, and control any resulting heterogeneity through the use of an appropriate experimental design (with the caveat that it is pointless and self-defeating to use an experimental design that is more complex than research circumstances merit).

This chapter has explored the fundamental principles underlying the design of an experimental study. Chapter 9 will discuss the specific issue of study size, and Chapter 10 will discuss some of the wider issues that must be faced when planning a research project. (These issues are explored mainly in the context of planning a clinical trial. However, the issues involved are relevant to all research projects.)

Key Terms

Bias

Confounding

Random allocation

Single-blind experiment

Double-blind experiment

Crossover design

Latin square

Factor

Interaction

CHAPTER 9

What Sample Size Will I Need?

COMPARISON OR ESTIMATION?

All investigations will have one of two possible research objectives. For many research projects, the objective is to answer a question, or test a hypothesis, very often by comparing two or more groups of individuals (see Chapters 4 through 7 for a discussion of the basic ways of asking and answering research questions in a variety of investigative situations). Experimentation (see Chapter 8) is perhaps the classic example of comparison-oriented research.

In other research situations (typically surveys, see Chapter 15), the objective at the heart of the study is to estimate the level of some particular phenomenon (such as the mean systolic blood pressure, or the percentage of individuals who are diabetic) in the population. (See Chapter 3 for a discussion of the problems of estimation.)

The number of subjects you need to get the job done depends on exactly the kind of job you want to do. Comparison-oriented studies and estimation-oriented studies have different problems in this regard and have to be considered separately. Let's start by thinking about the most challenging of the two situations, deciding just how many subjects we would need to launch a valid comparison study.

COMPARISON-ORIENTED STUDIES
The Importance of Power

In the classic **comparison-oriented study** different treatments are administered to two randomly selected groups of subjects, sample evidence is collected, and finally a decision is made (based on a statistical test) as to whether the two treatments are, in fact, different. The trouble with making a decision is, unfortunately, that it is always possible to get it wrong.

As we saw in Chapter 4, these mistakes can take two forms. The first, and possibly most obvious, occurs when there is no real difference in the treatments under comparison but the sample evidence seems dramatic and appears to show the existence of a difference. In this situation we would, quite mistakenly but very understandably, conclude that the treatments were, in fact, different and thus commit a type I or α error. We tend to be well aware of this possibility, and we seek to prevent it by demanding strong evidence of a difference (a difference that is so large that it is unlikely to be accidental) before we are convinced. Normally we employ as the decision or critical value a test value that could be exceeded by pure accident no more than 5% of the time, thus ensuring that the risk of an α error is no more than .05 (a level of risk we view, somewhat arbitrarily, as acceptably small).

However, there is another, often overlooked, but very important potential error we can face in this situation. An experiment investigating treatments that are, in fact, genuinely different may, because of random variation, result in evidence that is not very convincing and that falls below the critical test value. In this situation we would quite mistakenly, but again very understandably, fail to conclude that the treatments are different and thus commit a type II or β error.

Although most researchers know the amount of risk of an α error they must face in their research, many have no idea how much risk of a β error is involved when they undertake a research project. This risk is important because the whole point of investigative work is, we hope, to find out when treatments really are different. If there is a high risk (that is, high β value) of missing genuine differences, then the whole investigative process could very easily be a waste of time and money. Would you be prepared to commit $250,000 to a project if you knew that there was a 40% risk of the study failing to detect quite genuine treatment differences?

This risk is often described by the **power** of a study, where power $= 1 - \beta$. Power is, therefore, a measure of the likelihood that the study will succeed in detecting genuine treatment differences. A powerful study is very likely to find genuine differences if they are there. A weak study could very easily miss them.

Power or β levels are, of course, very important in all research work. However, most investigations tend to be viewed a success if they demonstrate that the treatments under comparison are different (a very unscientific attitude). If we find evidence of such a difference we want to be confident that there is little chance that the evidence is merely a fluke, and we tend to give priority to controlling the α level (restricting it to .05 or less), while trying to ensure that the power is at least acceptable (a power of 80% is viewed as the absolute minimum that is acceptable in study design).

There are, however, situations in which a high level of power is absolutely vital. If the object of the investigation is to confirm that two treatments are truly the same (for example, that a new contraceptive method has no more side effects than an existing method), then we must be certain that the risk of failing to pick up any real differences (which could pose a major health risk, if left undetected) is acceptably small. In this situation the β level is considerably more important than the α level, and great care should be taken to ensure that the power of the test is acceptably high.

What Determines Power?

The power of an investigation depends on several factors, all of which must be taken into account at the planning stage. (Once the study has been launched it is far too late to remedy a lack of power, or any other problem).

Size of α.   The more severe the level of proof we demand before we accept a result as significant, the easier it becomes to fail to detect a genuine, but perhaps small, treatment difference. That is, the smaller the value of α we decide to use, the larger the value of β will inevitably be and, hence, the smaller the power, $1 - \beta$, of our study will be.

This is not usually a major planning problem. The use of a value of .05 is fairly standard for α and, hence, power is usually evaluated in relation to this. Other values of α could, of course, be used. (In the study comparing contraceptive side effects mentioned earlier, we might consider increasing the α level to .10 to increase the power of the study and decrease our chance of missing real differences.)

Sample size.   The more evidence we have available to us, the easier it becomes to detect genuine treatment differences. If a study based on 10 patients fails to show a significant difference, then we might well wonder whether the treatments under comparison are indeed equivalent (that is, the null hypothesis, H_0, is actually true) or whether the available evidence was simply so scanty that it failed to pick up the differences (that is, the alternative hypothesis, H_1, was true and we committed a β error). A non-significant result from the same study based on 1,000 patients would make us much more confident that no genuine differences were present. That is, the larger the study size, n, the smaller β will be, and the larger the power, $1 - \beta$, of the study will be.

Most power-related planning decisions involve determining the minimum sample size needed to ensure that a statistical investigation will be acceptably powerful. Making sure the sample size is adequate to do the job is a very important aspect of thorough research design. (We should remember that, since the sample size influences a statistical test by decreasing the standard error, the power of a study is actually driven by $\sqrt{n}$, rather than n itself.)

True treatment effect.   If one treatment actually has a very marked effect on patients (for example, it produces a very dramatic decrease in blood pressure), relative to another treatment or comparison group, then its effect will show up very clearly. A very dramatic treatment difference might well be detected by even a relatively small study. On the other hand, a relatively small (although quite genuine) effect on patients could easily be masked by sample variation. In other words, the larger the relative treatment effect, the smaller β, and the larger the study power, $1 - \beta$ (and correspondingly, the smaller the sample size needed to be reasonably sure of detecting the effect).

What exactly do we mean by the relative treatment effect? Well, the ultimate measure of the difference between two treatments would be the difference we would see if we could administer them to everyone in the population of eligible patients, that is,

$$\mu_1 - \mu_2$$

where μ_1 is the population mean response to treatment 1, and so forth.

When it comes to actually carrying out the study, we will, of course, have to administer the treatments to small samples of patients, and the actual difference we see will turn out to be somewhat larger or smaller than the population difference. This, however, is an unavoidable (and unfortunate) consequence of ever-present random variation, and the population or true relative effect is what we should be thinking about when trying to decide the appropriate sample size.

As we have just noted, the size of the relative treatment effect directly influences the power of a study and the sample size required. This means that, to estimate the appropriate sample size, it is necessary to state the relative effect that you believe the treatments will have. This is always easier said than done, but cannot be avoided. (We find ourselves in a classic vicious circle here. To estimate the study size needed for a project to collect evidence on potential treatment differences, we need to be able to make some statement about the likely size of these differences!)

How can we approach this tricky problem? One way might be to consult the scientific literature and examine the evidence supplied by other studies in the area. Another approach is to call on your experience and judgment to define the smallest treatment difference that you would regard as being of clinical importance to the patients involved. In a hypertension study you might feel that a treatment that reduces blood pressure by at least 10 mm Hg will offer some clinical gain for the patients, whereas a reduction of any less than this will have little real importance. By basing your sample size on a relative effect of 10, you will ensure that you run an acceptably small risk of missing a real improvement of this size or larger. Do be aware that this also means that there will be a real risk that a genuine improvement of less than 10 might not be detected. This is acceptable since you view such a difference as being of little or no clinical benefit to future patients, and failure to detect it would, therefore, not be a real blow to improved patient care.

A useful way to approach the relative treatment effect is to view it as a statement of the minimum treatment difference that you want to be able to detect with a high level of certainty (in other words, you could live with the possibility that your study might well fail to detect genuine differences that are smaller than this).

As we noted in Chapter 2, it is difficult to assess the real magnitude of a difference when it is expressed in its original units. A difference of 5 could be huge or tiny, depending on the type of phenomenon being studied, the units of measurement, and the variability of the results. The solution we devised in Chapter 2 was to express a difference in universal standardized units, that is, in standard deviations. A more meaningful and helpful indicator of relative treatment effect is therefore its standardized form, the **true treatment effect.**

$$\text{True treatment effect} = (\mu_1 - \mu_2)/\sigma$$

where σ is the true or population standard deviation of the response being measured. This means, of course, that the more variable the results, the smaller the true treatment effect will be, and the more difficult it will be to find. As we would expect, the larger the population standard deviation, the larger β will be, and the smaller the study power will be. With highly variable results, we generally need more subjects to achieve a given level of power.

In reality, of course, we will never really know what the true variability actually is. However, a researcher usually can get a good indication of how variable the results are likely to be from published work in a comparable area. If this is not available, a small preliminary or pilot study can be used to get an idea of the magnitude of variation that is likely to be involved.

It can be difficult to make the statements about relative treatment effect and variability that are necessary to define the true treatment effect and allow us to calculate the sample size we need to have a real shot at finding the true treatment effect. Jacob Cohen, in his book* devoted solely to the determination of sample size, suggests a way out

---

*Cohen JS: Statistical power analysis for the social sciences, Orlando, Florida, 1977, Academic Press.

of this. Modifying his ideas slightly, we can rather arbitrarily define a small or weak treatment effect to be one in which the treatment difference is only a quarter of the standard deviation of the measurements involved, a medium-size effect to be one in which the difference is half the standard deviation, and a large or dramatic difference to be one in which the treatment difference is at least three quarters of the standard deviation.

TRUE TREATMENT EFFECT

Small	.25
Medium	.50
Large	.75

By defining the true treatment effect in this relative way, we cunningly avoid having to make any comment on the actual size of the treatment differences and the variability of the results. All we have to do is commit ourselves to seeking a small, medium, or large treatment effect, and we have a basis for starting to calculate the sample size we will need to find that treatment effect (or, more precisely, have some acceptably high chance of finding it).

However, for all its simplicity, defining a "medium size" treatment difference to be one with a true treatment effect of .50 is totally arbitrary. This "generic" approach is acceptable as a last resort, but it is much better to arrive at the true treatment effect through a more precise definition of the minimum difference being sought in the study and the anticipated variability of the results.

The Power Index

When we venture into any investigative situation we face two possibilities: Either there are no real treatment differences, H_0, or there are real treatment differences, H_1. Our task as researchers is to decide between these alternatives. We must use the evidence available to us to reach the most reasonable conclusion. No matter which conclusion we draw, there is, of course, always the possibility that we were misled by an untypical set of results. We may reject H_0 when there are no real differences, that is, commit a type I error, α. Alternatively, we may fail to reject H_0 when there are genuine differences, that is, commit a type II error, β. Our challenge, as responsible researchers, is to find some way of controlling both these sources of error and ensuring that they are both acceptably small.

The α error is usually directly controlled when we carry out a test. By adopting a critical value of 1.96 when testing means, we ensure that the risk of exceeding this accidentally if there is no real difference is less than 5%, that is, $\alpha < .05$ (Figure 9-1). This risk is conventionally regarded as being so small as to be quite acceptable.

(Why are we using the z or normal table value of 1.96 in this discussion, rather than the .05 value from a set of t tables? Well, the vicious circle rears its head again. We would need to know the sample size to determine the degrees of freedom at which to access the t tables, and the whole point of the discussion is to find some way of determining sample size. Using the z value is an effective solution. Less pragmatically, sample size discussions are based on statements about true treatment effects, and no modification for sample variability is necessary. For most sample sizes, the difference between the normal and t values will be very small in any case.)

However, how can we control the β error? We know that the power of a study

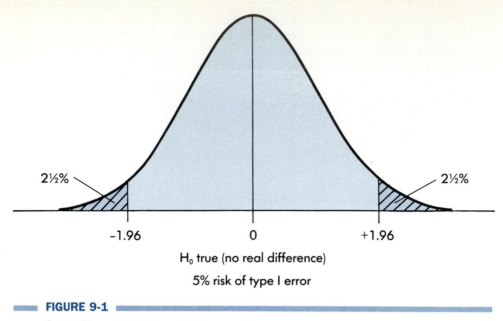

2½%   2½%

-1.96   0   +1.96

H_0 true (no real difference)

5% risk of type I error

FIGURE 9-1

Controlling the type I error ($\alpha = .05$).

depends on the true treatment effect and the sample size (we will see how these are brought together in the next section). Since we are using 1.96 as our critical value, should we pick our sample size so that we would obtain a test value of 1.96, if the true treatment effect is actually present? (After all, this would give us a significant result and point to the conclusion that the treatments are truly different. If the true treatment effect is present then the treatments are indeed different and this conclusion is absolutely appropriate.)

Aiming for a test value of 1.96 is not a good idea. Remember we will have to make the call based on imperfect sample evidence and hence the test value we actually achieve in practice will vary about this anticipated figure. If the sample data "overperforms" (that is, the sample treatment differences turn out to be somewhat more dramatic than we had anticipated) then all will be well. The sample test value will be somewhat greater than 1.96 (a nice bonus), we will still have a significant result, and we will still draw the appropriate conclusion.

If, however, the sample data "underperforms" (even slightly) and demonstrates a study treatment effect somewhat less dramatic than the true treatment effect, we will have a major problem. The sample test value will fall below 1.96, the result will not be significant, and we will have to draw the quite inappropriate conclusion of "no real difference." Since a little bit of good luck (overperforming) and a little bit of bad luck (underperforming) are equally likely, there is a 50% chance that, in this situation, we would fail to get a significant result, despite the fact that the treatments are genuinely different (Figure 9-2). In other words, the β error would be .50, and the power of the study would be only .50, which is far too small to be acceptable.

To put it bluntly, planning to come up with a test value of 1.96 is "sailing far too

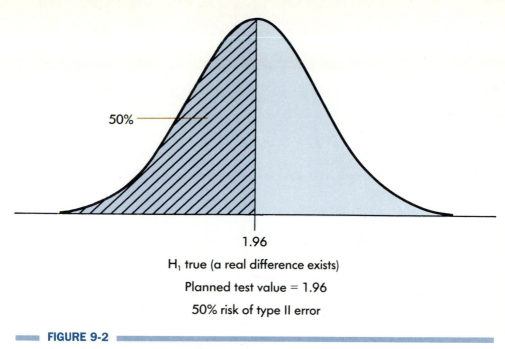

50%

1.96

H₁ true (a real difference exists)

Planned test value = 1.96

50% risk of type II error

FIGURE 9-2

Controlling the type II error ($\beta = .50$).

close to the wind" and allows absolutely no margin for error. If it is 2500 miles from New York to Los Angeles, and your private jet burns fuel at the rate of one gallon every 10 miles, would you load 250 gallons of fuel in New York? With a tail wind, you'll be fine. With even the slightest head wind, you won't make Los Angeles, and there are no prizes for getting close. Your survival becomes a 50/50 proposition.

The sensible thing to do, of course, is to aim for a larger test value than we actually need (load some extra fuel) so that, even if we get some bad luck or underperformance in our study sample, we will still come up with a test value of at least 1.96. To be realistic, we must acknowledge that there will still be some situations in which extreme bad luck (hurricane force head winds) will result in a test value under the critical 1.96 level. In other words, a β error is still possible, although we hope the risk has been much reduced.

Just how high should we aim? Well, it depends on just how much protection we want to give ourselves. Let's assume that we believed we could live with a 5% risk of failing to detect the true treatment effect (that is, a study power of 95%). Take a look at Figure 9-3.

If we aim for a test value of 3.60 there will be only a 5% chance that the sample evidence will so dramatically underperform that the sample test value will fall below 1.96. (Recall that, in the Normal distribution, only 10% of the results will fall more than 1.64 standard units away from the center. However, we are dealing here with a one tailed situation. We will commit a β error only if the sample data underachieves. If it overachieves, we certainly can't complain. As a result, if we aim for a test value of 3.60 (1.64 above 1.96), there will be only a 5% risk of slipping below the critical 1.96 value, failing to detect the genuine difference, and committing a β error).

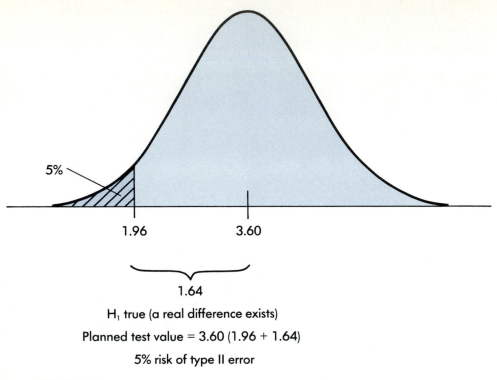

5%

1.96   3.60

1.64

H₁ true (a real difference exists)

Planned test value = 3.60 (1.96 + 1.64)

5% risk of type II error

FIGURE 9-3

Controlling the type II error ($\beta = .05$).

This target test value (3.60 for .05 α and .05 β) is called the **power index** and gives us the opportunity to control both the α and β errors in our study. Insisting on a test value of at least 1.96 will protect us against an α error, if there is no real difference. Aiming for a test value of 3.60 will protect us against a β error, if the true treatment effect is present.

Other α and β error levels are, of course, possible. An α value of .05 is almost universally employed. The β value, on the other hand, is sometimes permitted to rise as high as .20. This will occur when raising the power above .80 would require an impractically large sample. The decision to relax the β level rather than the α level is based on the argument that β errors (that is, missing a genuinely better new treatment and, hence, continuing with an established and reasonably effective treatment) usually have less serious consequences than α errors (that is, switching from an established, effective treatment to a new, relatively unknown treatment that is not actually any better). In general, α should always be .05 or less, and β should never be greater than .20. A β value of .10 is much preferable, and a β value of .05 is ideal, if practicable. (It very often isn't practicable because it often requires an impractically large sample size.) In situations where the objective is to establish the similarity of two treatments rather than their difference, then the β error becomes of paramount importance, and the α error should be relaxed if necessary.

α (two-tailed)	β (one-tailed)	POWER INDEX
.05 (1.96)	.05 (1.64)	3.60
.05 (1.96)	.10 (1.28)	3.24
.05 (1.96)	.20 (.84)	2.80

If a one-tailed test is appropriate, then this should be reflected in the power index by using a 1.64 rather than 1.96 value for the α contribution, (for example, the .05 α (one-tailed), .05 β power index is 1.64 + 1.64 = 3.28.

Studies Involving Two Means

How do all these ideas come together? Consider again the classic research situation involving the comparison of the means of two different groups. We discussed this situation in Chapter 4 and developed the t test as a sensible and meaningful way of assessing the strength of the sample evidence. To refresh our memories, the t test (see Chapter 4) is given by

$$\frac{\bar{x}_1 - \bar{x}_2}{s_p \sqrt{\frac{1}{n} + \frac{1}{n}}} \quad \text{or} \quad \frac{\bar{x}_1 - \bar{x}_2}{s_p \sqrt{(2/n)}}$$

where $\bar{x}_1$ and $\bar{x}_2$ are the means of the two groups,

s_p is the pooled within-groups standard deviation,

and n is the number of individuals in each group

(we assume that the two groups are of equal size).

At the planning stage, before any sample evidence is available, we have to state this test statistic in terms of the true treatment effect:

$$\frac{\mu_1 - \mu_2}{\sigma \sqrt{(2/n)}}$$

As we saw in the previous section, to ensure that we run only a 5% risk of failing to detect this true treatment effect, we should plan (in effect, pick our sample size, n) so that this should give us a test value of 3.60.

Somewhat more generally, we should plan our sample size so that

$$\frac{\mu_1 - \mu_2}{\sigma \sqrt{(2/n)}} = PI$$

where PI is the power index incorporating the α and β levels we wish to employ in our study.

Rearranging the above relationship leads us to

$$n = 2 \left(PI \frac{\sigma}{\mu_1 - \mu_2} \right)^2$$

Please remember that n is the number of subjects we will require in each of the two groups (that is, a total study size of $2n$).

Note that, just as we had anticipated, the more variable the data, the more subjects we will need; the more dramatic the difference, the fewer subjects we will need; and the higher the desired power (the larger the power index), the more subjects we will need.

To see these ideas in action, let's imagine a situation in which we are comparing two hypertension treatments, and we wish to have a 90% power of detecting a difference of 5 mm Hg or greater, when we believe the standard deviation of the results is likely to be around 12 mm Hg (with testing to take place at the usual .05 level).

$$PI = 1.96 \ (.05 \ \alpha, \ \text{two-tailed}) + 1.28 \ (.10 \ \beta, \ \text{one-tailed})$$

$$= 3.24$$

$$n = 2 \left(3.24 \times \frac{12}{5} \right)^2 = 120.93$$

that is, a minimum study size of 242 individuals, split into two groups of 121 subjects each.

If we think we cannot specify the true treatment effect but can, at least, anticipate that it is a "medium size" effect, then we can substitute .50 (see the section, What Determines Power?) for the true treatment effect.

For instance, to have an 80% power of detecting a "medium size" difference, when carrying out a two-tailed test at the usual 5% level,

$$PI = 1.96 \ (.05 \ \alpha, \ \text{two-tailed}) + .84 \ (.20 \ \beta, \ \text{one-tailed})$$

$$= 2.80$$

would require a sample size of

$$2 \left(2.80 \times \frac{1}{.50} \right)^2 = 62.72$$

or a minimum of 63 subjects in each of the two study groups.

Studies Involving a Single Mean

A single sample study of this type usually involves the comparison of a sample mean with some reference or null hypothesis value (usually zero). It occurs, for instance, in the analysis of paired results (such as the paired t test) when we effectively have a single set of results (the difference between the pairs), and we wish to establish whether the mean difference is significantly different from zero.

At the planning stage the test value is therefore

$$\frac{\mu_d}{\sigma/\sqrt{n}}$$

where μ_d is the true mean difference, and σ is the true standard deviation of the differences, and, as in the previous section, we should plan so that

$$\frac{\mu_d}{\sigma/\sqrt{n}} = PI$$

or

$$n = \left(PI \ \frac{\sigma}{\mu_d} \right)^2$$

In a crossover experiment, patients suffering from anxiety will receive two different tranquilizers (in random order, naturally) and, after each, will be rated on an anxiety scale. We believe that there will be a relative effect of 10 (that is, $\mu_d = 10$) and that the true standard deviation for results of this type will be 15. To have only a 5% risk of missing an effect of this size (that is, a power of .95), while maintaining the usual α of 5%, we require a sample size of

$$n = \left(3.60 \times \frac{15}{10}\right)^2$$

$$= 29.16$$

that is, a sample of at least 30 individuals.

The appropriate sample size should always be calculated in this way before an experiment is conducted. Sometimes, however, the sample size may be determined by factors beyond the investigator's control. In this situation it is prudent to determine the likely power of such an experiment. This can be readily done by simply reversing the calculation and obtaining the power index:

$$n = \left(PI \frac{\sigma}{\mu_d}\right)^2$$

and hence

$$PI = \sqrt{n} \, (\mu_d/\sigma) \quad \text{one mean}$$

[or

$$PI = \sqrt{n} \, (\mu_1 - \mu_2)/(\sigma\sqrt{2}) \quad \text{two means}]$$

The value of β (and hence of the power, $1 - \beta$) corresponding to this power index for a particular desired α (usually .05) can then be obtained from our knowledge of how the power index is put together.

If only 20 patients were available for the anxiety study, the power index of the proposed investigation would be

$$PI = \sqrt{20} \times (10/15)$$

$$= 2.98$$

$$\alpha = .05, \text{ corresponds to } 1.96$$

The β portion is therefore

$$2.98 - 1.96 = 1.02.$$

Inspection of normal tables (table T/1 in the Appendix) shows that normal values will fall outside 1.02 15.39% of the time (one-tailed). A study based on only 20 patients would therefore have a β value of .1539 and a power of .8461.

▬ Studies Involving Two Proportions ▬

Many health studies involve the comparison of qualitative variables (usually through the use of Chi squared tests; see Chapter 6). A typical example would be the comparison of two cancer treatments, the outcome variable being either survival past some defined time or death. Here the object of the exercise is to compare the two survival proportions for evidence of a significant difference. There are clear parallels, of course, with the comparison of two treatment means that we discussed earlier, and we might expect to see these parallels reflected in the calculation of sample size.

Survival or death is a binomial situation, and we know from our discussion of the Binomial distribution (Chapter 2) that the mean number of survivors (out of a total of n) $= np$, with a standard deviation of $\sqrt{np(1-p)}$ (where p is the probability of survival). Expressing these same results as a proportion of n, we see that

$$\text{the mean proportion of survivors} = \frac{1}{n} np$$

$$= p$$

$$\text{with a standard deviation of} \quad \frac{1}{n}\sqrt{np(1-p)}$$

$$= \sqrt{p(1-p)/n}$$

(As we pointed out in Chapter 3, this is also a standard error, as we might well guess from its form.)

The two measures p and $\sqrt{p(1-p)/n}$ in the categorical context clearly correspond to μ and $\sqrt{\sigma^2/n}$ in the continuous context, and it seems reasonable to come up with a sample size formula for the comparison of two proportions by simply appropriately modifying the very similar situation of comparing two means, in other words, replacing μ by p and σ^2 by $p(1-p)$.

There is one complication. In the comparison of means, the value of the variance is unaffected by the value of the mean (that is, μ has absolutely nothing to do with σ^2). However, for proportions, the value of the variance is very directly controlled by the value of the proportion, that is, p clearly determines the size of $p(1-p)$. Since we are comparing two proportions, p_1 and p_2, what value of p should we use to define the study variance? A reasonable solution seems to be to simply average p_1 and p_2,

that is, to use

$$\bar{p} = (p_1 + p_2)/2$$

The sample size calculation for the comparison of two proportions therefore becomes (by analogy with the comparison of two means)

$$n = 2\,\text{PI}^2\,\frac{\bar{p}(1-\bar{p})}{(p_1 - p_2)^2}$$

Again, this gives the desired sample size for each of the two groups.

For example, in a clinical trial of two arthritis treatments, the standard treatment seems, on the basis of previous evidence, to produce a successful outcome 20% of the time. What size study would we need to be 95% certain of detecting a true success rate of 40% in a new drug, while maintaining an α level of .05?

This is a one-tailed situation (since we are anticipating that the new drug will be superior), and the appropriate power index is

$$1.64\ (.05\ \alpha,\ \text{one-tailed}) + 1.64\ (.05\ \beta,\ \text{one-tailed}) = 3.28$$

$$\bar{p} = (.20 + .40)/2$$

$$= .30$$

$$n = 2 \times 3.28^2 \times \frac{.30 \times .70}{(.2 - .4)^2}$$

$$= 112.96$$

or a minimum of 113 patients per drug.

Other Comparisons

In this chapter we have discussed the issues that must be addressed when determining the sample sizes needed to carry out a valid comparison study, and we have developed specific strategies to determine sample sizes for the two-mean (independent *t* test), one-mean (paired *t* test), and two-proportion (Chi squared) situations. There are, of course, other comparison or hypothesis-testing situations that the investigator can encounter, such as the comparison of several means (analysis of variance), or relationships between continuous variables (correlation/regression). The general principles (anticipating the magnitude of the effect you are exploring and, based on that, aiming for a test value comfortably greater than the critical value) are exactly the same, although the calculations may become slightly more intricate. In these situations, one is best advised to consult a specialist text on the subject, such as Jacob Cohen's *Statistical Power Analysis for the Social Sciences* (published by Academic Press, Orlando, 1977).[1]

ESTIMATION-ORIENTED STUDIES

As we noted in Chapter 3, the mean of a sample is only an estimate of the true or population mean value. Fresh samples would undoubtedly produce somewhat different sample means. To make an honest and realistic statement about the phenomenon under investigation, we quote our sample mean (the best estimate we have of the true mean), together with confidence limits (an acknowledgment of the limitations of our study). The narrower these confidence limits, the more precise the statement we can make about the true mean of the population. The width of the confidence limits depends on the variability of the phenomenon being measured and on the number of measurements collected. The role of sample size in estimation-oriented studies is very direct. The more data we collect, the narrower the resulting confidence limits will be, since the standard error of the mean will decrease in direct proportion to the square root of the sample size. The number of subjects you need depends very directly on how precise you want your sample estimate to be.

Estimating a Mean

The initial step in determining sample size is, therefore, to define just how precisely we wish our study to estimate the value in question. In a blood pressure study, we might wish to estimate the mean systolic blood pressure of a population of elderly male smokers to within ±4 mm Hg. We also need to specify the level of confidence to be associated with these limits. This will usually, although not necessarily, be 95%.

$$95\% \text{ confidence limits} = \pm 1.96 \text{ SD}/\sqrt{n}$$

Calculating confidence limits requires a knowledge of the standard deviation involved. Just as we must do when calculating the sample size for comparative studies, we must, by necessity, come up with a statement about a fundamental property of the final set of results before a single sample has been collected. In practice, published work in a related area can often be used to provide an estimate of the likely standard deviation. Failing this, a small-scale pilot survey can be conducted simply to get some idea of the probable variability. Suppose that previous studies have shown there is a likely standard deviation of about 20 mm Hg. We, therefore, now have:

$$4 = 1.96 \times 20/\sqrt{n}$$

(Note that we are again using the z or normal value of 1.96, rather than a t value, for the same reasons we discussed earlier in this chapter.)

$$\sqrt{n} = 1.96 \frac{SD}{(\text{Confidence limit})}$$

or

$$n = \left(\frac{1.96 \times SD^2}{\text{Confidence limit}} \right)^2$$

In our example,

$$n = (1.96 \times 20/4)^2$$

$$= 96.04$$

Therefore a minimum sample size of 97 subjects is required. A sample size of 100 or 110 (to allow for losses) would be more prudent.

Estimating a Proportion

The standard deviation of a proportion depends on the proportion's size (see the section, Studies Involving Two Proportions) and, therefore, specifying the standard deviation we will face requires us to state the anticipated proportion in the population under study. (This is, of course, the very thing we are trying to estimate and places us in yet another classic vicious circle.)

For example, suppose we wish to estimate the percentage of sexually active individuals in a particular high school population to within $\pm 10\%$, with 95% confidence. We believe that the percentage is likely to be near 40%.

Since the standard deviation of a proportion is

$$\sqrt{p(1 - p)/n}$$

and 95% confidence limits for a proportion are given by

$$\text{Interval} = 1.96 \sqrt{p(1 - p)/n}$$

(as we pointed out earlier, a proportion is a group measurement, and its standard deviation is already a standard error)

then

$$n = 1.96^2 \frac{p(1 - p)}{\text{Interval}^2}$$

For our sexual activity survey,

$$n = \frac{1.96^2 \times .40 \times .60}{.01}$$

$$= 92.20$$

we would need to survey at least 93 students.

Do be aware that, the closer p is to .50 or 50%, the greater the standard deviation, and the greater the required study size for any given level of precision. When in doubt, it is therefore wise to err towards the 50% region, thus ensuring a sample size that may be too large rather than possibly too small.

Key Terms

Comparison-oriented study   **True treatment effect**

Power   **Power index**

PROBLEMS

1. A health education study proposes to look at possible personality differences between individuals at high risk of AIDS who have positive and negative attitudes toward condom use. The participating individuals will be assessed on a scale that measures self-confidence. The literature suggests a standard deviation for the scale would be about 5.5, and you believe that a mean difference of 3.0 or greater is of psychological importance.

   What size study (assuming an α level of .05) would be required to
   a. Have a 95% power of detecting this difference?
   b. Have an 80% power of detecting this difference?

   If you felt that a mean difference of 1.0 was of psychological importance, what size study would be needed to have an 80% chance of detecting such a difference (again assuming testing at $\alpha = .05$)?

2. In a study of the role of the administration site on the efficacy of an anticlotting agent, intravenous administration will be compared with intracardial administration. Pilot studies indicate probable success rates of 40% and 65% respectively.

   What size study would be needed to detect a difference of this size with
   a. An α of .05 and a β of .10?
   b. An α of .05 and a β of .20?

3. As part of a study of the physiology of acclimation to cold, rectal temperature in adult males is to be measured before and after 90 minutes' immersion in cold water. You anticipate that the change in rectal temperature will have a standard deviation of 0.4° C and that a decrease in rectal temperature of 0.3° C would be of interest.

   What size study would be needed to ensure a 90% power of detecting this difference or greater, when testing at the usual 5% level?

4. In a study of the effects of a physical exercise program, patients' systolic blood pressure is to be measured before and after such a program. You believe that a decrease in systolic blood pressure of 5 mm Hg would benefit the patients. Previous studies suggest that the standard deviation of such a change is likely to be about 6 mm Hg. What size sample would be needed to detect a decrease of 5 mm Hg with:
   a. An α of .05 and a β of .05?
   b. An α of .05 and a β of .20?

5. In a related study two different exercise regimes are to be compared. The mean systolic blood pressure of each patient group will be compared after the patients

complete the exercise program. Again a difference of 5 mm Hg is viewed as important. The standard deviation of the results is thought likely to be about 10 mm Hg. What size sample would be needed to detect a difference of 5 mm Hg with:

a. An α of .05 and a β of .05?

b. An α of .05 and a β of .20?

6. The proportions of patients successfully completing the two exercise programs mentioned in problem 5 are also to be compared. You believe that the overall average completion rate should be around 80% and that a difference in rates between the two programs of 10% would be important. What size sample would be needed to detect a 10% difference between rates with an α of .05 and a β of .10?

7. You are involved in planning a study to compare the ages at which alcoholic men and women were first admitted to a treatment facility. Previous studies suggest a variance for age of 73 years^2. You feel that a mean age difference of 5 or more years is of demographic and sociological importance.

What size study (assuming an α level of .05) would you need to have

a. A 95% power of detecting such a difference?

b. A 90% power of detecting such a difference?

c. An 80% power of detecting such a difference?

d. If you believed that an age difference of 2 years was important, what size study would you need to have a 90% power of detecting such a difference when testing at 5%?

e. If you believed that an age difference of 1 year was important, what size study would you need to have a 90% power of detecting such a difference when testing at 5%?

8. As part of the same study, you are interested in determining whether male and female alcoholics differ in their willingness to admit that alcohol is their main problem. The literature suggests that about 50% of alcoholics (males and females combined) will admit that alcohol is the main problem in their lives. You believe that, of the percentages prepared to admit this, a difference of 30% or greater between the sexes is of real importance.

What size study, assuming the use of an α level of .05, would you need to have

a. A 95% power of detecting such a difference?

b. A 90% power of detecting such a difference?

c. An 80% power of detecting such a difference?

CHAPTER 10

Planning a Clinical Trial

WHAT IS A CLINICAL TRIAL?

The **clinical trial** (the planned comparison of alternative treatments) has a deserved reputation as the classic investigative technique in health research and has come to epitomize the need to bring objective evaluation techniques (in other words, the ideas and tools of biostatistics) to the health sciences. Indeed, it is far from coincidental that the explosion of clinical trials since the first truly meaningful trial in 1950 (one that used a randomly selected control group) has mirrored the explosion in the use of biostatistics over the same period.

Clinical trials have attained such major importance in health research because they offer the only truly fair and objective way of comparing the value of a new treatment to that of a currently employed treatment. Therefore, clinical trials offer the only reasonable, and indeed the only truly ethical, basis for choosing the best possible future health care for patients. For this reason alone, clinical trials and their underlying issues would merit discussion. However, the problems that have to be faced and the planning process that one must follow to implement a clinical trial are applicable to just about any health-oriented investigation. Developing a framework for planning a clinical trial will also give us a framework we can apply to the planning of virtually any other health sciences research project.

Before we start to worry about how one plans a clinical trial, it might be prudent to take a moment to reflect on what exactly a clinical trial actually is. Much of the experimentation in health research requires human subjects (human-based experiments draw on the general experimental principles we discussed in Chapter 8, but pose additional ethical and practical problems of their own), and such studies are generally called clinical trials. One more detailed definition of a clinical trial is a prospective study (that is, one that follows the same subjects from treatment to outcome) that examines the effectiveness of one or more interventions on human subjects by comparing the intervention groups with a suitable control group.

Reflecting this definition, we can say that the three key features of clinical trials (in addition to the involvement of human subjects) are as follows:

1. They involve some form of active intervention applied in a clearly defined, standardized fashion to the subjects scheduled to receive it.
2. The value of the intervention is assessed by comparing it with a meaningful control or reference patient group. This will usually be a group receiving the current standard therapy for the condition involved, but might also be a group receiving no intervention at all. The various patient groups should be similar to one another at the start of the trial.
3. The performances of the various interventions are observed by following the progress of the patients from initial intervention to some clearly defined end point.

TRIAL PROTOCOLS

"Look before you leap" is one of those pieces of advice with a message so sensible and obvious that we often fail to heed it and inevitably end up paying the price. In the research context, the same advice might read, "Think before you do," or, better still, "Think, and then write it down." Planning, preferably documented planning, is absolutely vital for the success of any enterprise.

In many ways most of the work in a clinical trial is, or should be, done long before the trial actually commences. This is the stage at which you determine the precise objectives of the trial, the best way to achieve those objectives (and why), the problems that might arise during the trial, and how those problems might be solved. It is much easier to respond to a problem with a game plan already in place than to have to rely on ad hoc solutions developed in haste. This planning process culminates in the writing of the trial **protocol**, essentially a master game plan that rigorously documents the purpose of the trial and how this will actually be achieved in practice.

A trial protocol breaks down into a series of basic sections:
1. Background and general aims (Why is the study worth doing?)
2. Specific objectives (What exactly do we want to learn from the trial?)
3. Study population (Who precisely will we study?)
4. Treatments (What exactly will we do to the subjects?)
5. Trial design (How will the trial be organized?)
6. Procedures (How will the trial be administered?)
7. Statistical analysis (How will the results be analyzed and conclusions drawn?)
8. Facilities and budget (How will we pay for it all?)

These sections reflect the key issues that must be faced when defining the *why*'s, *what*'s, and *how*'s of the project.

The various components of the trial protocol do not, of course, exist in isolation. Decisions made about one aspect of the trial will affect other aspects, and the final integrated protocol will invariably result from a repeated cycle of revision and modification (Figure 10-1).

Writing, like facing a hanging, has a way of concentrating the mind, and the real advantage of the trial protocol is that it does require writing. It forces investigators to consider very clearly exactly what they are trying to do and precisely how they can achieve it, and to commit themselves to this in print. This requires detailed planning of every aspect of the trial, from the exact question being asked, to the precise way that the trial outcome will be measured, to the most appropriate statistical analysis to be used.

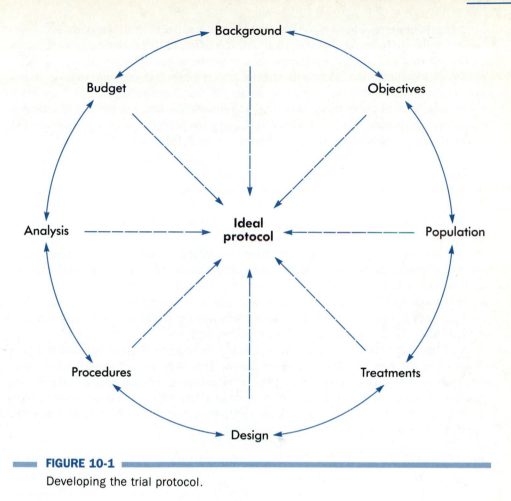

FIGURE 10-1
Developing the trial protocol.

The time to ask and answer questions such as these is before the trial (or any other investigation) starts, not after it is underway.

Stating the Research Question

The very first question to be addressed in any project is, "Why bother doing this at all?" It is a valid and important question and deserves (indeed, demands) an answer. The trial protocol should start by giving its readers a framework within which they can understand the relevance and importance of the study. This background statement should present an overview of the problem, outline why the trial is worthwhile and should be carried out, and establish the aims of the study in a general way.

Developing a general statement of objectives is an appropriate place to start. However, every study has (or should have, if it is to be implementable and productive) some very specific objectives, each definable as a precisely worded hypothesis or question. These specific objectives will consist of a primary question and, probably, a series of secondary questions.

The **primary question** is the one that lies right at the heart of the study and without which the study becomes pointless. It will most frequently be concerned with establishing whether one particular intervention is superior to one or more alternative interventions. Sometimes the study will attempt to establish that one intervention is not worse than an alternative (in a situation where the first intervention has other ancillary advantages, such as convenience or cost). The primary question will be used to establish the appropriate sample size (see Chapter 9). Stating the primary question is equivalent to stating the fundamental hypothesis that the trial is intended to test.

In addition to the single primary question, the researchers may have a number of **secondary** (or supplementary) **questions.** These questions might look at alternative measures of treatment success or examine the effectiveness of the various treatments in subgroups of the population (such as young males or severely ill patients). Again, these secondary questions should be specified in detail before the trial begins and not formulated after an inspection of the trial results.

When a number of secondary questions are posed, it should be remembered that the risk of obtaining apparently significant results accidentally (a type I error) may be well above the usual .05 level. In this case, the researcher should consider testing the secondary questions at a more severe significance level. (See the section on Statistical Analysis later in this chapter for a discussion of the same problem in a slightly different context. The advice given there is also applicable here.)

Stating the specific questions to be asked also requires that the researcher define exactly how the trial outcome will be measured. This may involve a quantitative measure such as, say, blood pressure reduction, or a qualitative measure such as success (for example, a blood pressure reduction of more than a specified amount) or failure or survival/death. In either case, the appropriate outcome variable must be defined precisely, and it must be measurable in all the trial subjects.

Defining the Study Population

Once the study objectives have been defined, careful thought should be given to defining the precise type of patient who will be studied in the trial. This is important because it will determine the extent to which the trial conclusions can be generalized to future patients. Remember that results can only truly be generalized to the population from which the study sample was drawn, see Chapter 8. The patient definition also will determine how easy it will be to recruit subjects for the trial.

Defining the study population involves specifying in detail and unambiguously the eligibility and exclusion criteria that will be used to govern the admission of patients to the trial. If, for instance, it is decided to exclude patients with severe hypertension, then a precise definition of severe hypertension must be given, together with details of when and how blood pressure should be measured. It is also preferable to state why the inclusion or exclusion criteria have been imposed.

Eligibility or inclusion criteria are usually developed in accordance with the investigator's view of the type of patient the investigator believes can benefit from the treatment under test and to whom they would therefore wish to be able to generalize the trial conclusions. Correspondingly, the principal aim of the exclusion criteria is to protect patients who may be at substantial risk of adverse side effects of the various treatments. (For example, pregnant women are often considered subject to unacceptable risk levels from any adverse consequence of intervention.)

The ultimate objective of any clinical trial is to produce guidelines for the best possible treatment of future patients with a specific condition. If admission to the trial is tightly restricted to, say, males under 30, then the trial may well be more sensitive since the subjects will be inherently less variable. However, the extent to which the results of the trial can be generalized to future patients will be correspondingly restricted. This may seriously compromise the usefulness of the trial and should be considered when defining the inclusion/exclusion criteria.

Since restrictive inclusion criteria tend to make the study sample less variable, they will also tend to reduce the sample size needed to achieve a sensitive or powerful trial. This may seem attractive. However, remember that restrictive eligibility criteria will also reduce the number of patients who meet the criteria and may well make it substantially more difficult to recruit the required number of patients.

Specifying the Interventions

Once the research questions and study population have been defined, the various treatments or interventions to be employed in the trial must be specified in detail. This is essential to ensure absolute consistency in the way the interventions are administered. This is always important, but is doubly so when the trial is to be carried out at more than one location (a multicenter trial), as is often the case.

Once again, the specifications must be detailed and totally unambiguous. It is probably wise to consider every possible misunderstanding or potential treatment shortcut, and to draft the specifications accordingly. For a drug study, for instance, these should include:

1. The drug formulation
2. The amount and frequency of dosage, the duration of the treatment, and the method of administration

A similar degree of detail should be applied to the definition of any intervention. In addition, this section of the protocol should include potential side effects and complications, together with the treatment practices to be implemented in these situations. (Remember our earlier advice to anticipate problems and plan their solution before they happen.)

The Trial Design

The trial design section of the protocol will detail the experimental strategies (see Chapter 8) to be employed to ensure that the trial comparisons are fair and meaningful and that the study will have acceptable power. It will outline the sample size to be used, the justification for the particular sample size proposed (see Chapter 9), the method of randomization to be used for treatment allocation, a definition of the blinding procedures to be employed, and a description of any blocking structure or other technique that will be employed to reduce unexplained variation.

Randomization. A clinical trial should always be a **randomized clinical trial** (RCT). As we discussed in Chapter 8, this is the only approach to treatment allocation that ensures that there is no risk of bias. It also tends to ensure that the various treatment groups consist of generally similar subjects, since there will be no systematic difference between the patients allocated to each treatment.

Randomization is invariably based on random number tables (given in Appendix T/7), usually employed to ensure equal subject numbers in each treatment group. This then leads to an allocation list that specifies that, for instance, the first patient entering the trial will receive treatment C, the second treatment A, and so forth. It is important that the investigator responsible for the inclusion/exclusion decision not know which treatment a potential trial subject is scheduled to receive if admitted to the trial, since this may bias the admission decision. The chief objection to a systematic rather than random method of allocation is that the treatment a patient is scheduled to receive, if admitted to the trial, is always known. (This risk can be reduced by using very specific admission criteria.) Consequently, the random allocation list should be held by another investigator who refers to it only after the admission decision has been made. Alternatively, the list may be held in a series of sealed envelopes, to be opened after a patient is admitted to the trial.

Allocation should generally proceed on the basis of equal sample sizes per treatment, since this ensures that the study will have the maximum power for a given total study size. If a trial is scheduled to be relatively large, say, 200 patients to be admitted to four treatments, there are advantages in ensuring that group size is kept equal in a series of steps. The first 20 patients might be randomly allocated with five to each of the four treatments. This is repeated (with fresh randomization) for each succeeding group of 20 patients. This ensures that, if the allocation is actually terminated at 100 patients, 120 patients, or some other multiple of 20, the group sizes will still be equal.

If the trial is to incorporate a blocked structure (say, blocked on age and gender), then separate random allocation lists must be held for each block (also sometimes called a stratum), such as a list for males under 20, a list for females between 20 and 29, and so forth. Clearly, the block sizes must be multiples of the number of treatments, if group size is to be kept equal. As well as controlling variation, blocking or stratification offers the opportunity to look at the influence of the blocking criteria (such as age) on treatment effectiveness (that is, an age/treatment interaction) and might increase the usefulness of the trial.

Controls. The effectiveness of a treatment can only be meaningfully assessed by evaluating its performance relative to some alternative or control treatment. The only truly fair way to do this is to evaluate the performance of randomly allocated treatment patients and control patients concurrently under otherwise identical medical and social conditions. The inherent reasonableness of such a comparison makes the randomized controlled clinical trial the classic mode of treatment evaluation.

One alternative evaluation procedure sometimes employed is to compare a treatment group to a set of historical controls. One argument in favor of this says it is more ethical, since all current patients can receive the new treatment. This argument is of very doubtful validity, however, since there is no evidence that the new treatment is actually more efficient or safer. This is the whole point of initiating the clinical trial in the first place. A second argument says that the use of historical controls is more efficient since the evidence about the controls already exists and does not require further investigation. In addition, all new patients can be directed into the new treatment group, resulting in a larger new treatment group.

The pitfalls, however, far outweigh these advantages. There is no guarantee at all that the historical control groups are truly similar to the new treatment group. The records concerning them may be inaccurate and incomplete, and it is highly unlikely that the admission criteria and treatment conditions are consistent with the new conditions. In addition, the impact of the disease in question may well have changed over time with changing social attitudes. (Altered dietary, exercise, and smoking patterns, for instance, have had a marked influence on observed patterns of coronary heart disease.) Without the built-in guarantee of similarity offered by randomized concurrent controls, any conclusions reached are of very dubious validity. (Indeed, there is ample evidence that using historical controls tends to exaggerate the advantages of new therapies.) In short, the use of historical controls offers no basis at all for a fair and meaningful comparison of interventions and should never be preferred to a randomized clinical trial.

Blinding. The key to the importance of the clinical trial in health research is its guarantee of an inherently fair comparison of treatments. If it is to deliver on this promise of fairness, a clinical trial must be run on a single-blind basis (unless this is physically impossible). It is probably better for a clinical trial to be double blind since the patient/physician bond can play a strong role in a patient's response to treatment, and this bond may be compromised by the physician's attitude toward the various treatments. Blinding a trial does make it more complex to set up and administer (a price well worth paying). In a single-blind drug trial, for instance, the alternative medications ideally should be identical to the new drug in their taste, color, odor, texture, tablet size, and so forth. This is particularly vital in crossover studies, but is also very important in trials involving separate patient groups, since different responses may simply reflect the patients' psychological responses to the perceived "potency" of the medication.

Double-blind trials also inevitably make administration more complex, and this must be incorporated in the planning. The (we hope) indistinguishable treatments must be identified to the participating health professionals by some coding system, to preserve their blindness. Consequently, an investigator not involved in direct patient contact (possibly a pharmacist) must maintain a master treatment code list and ensure that the correct treatment is packaged under the correct code. Should an emergency arise with a patient, ethics, of course, dictate that the blinding be broken and the exact treatment being administered to that patient ascertained.

Sample size. The sample size to be employed is central to the likely power of the whole trial. A trial carried out on an inadequate number of patients runs a real risk of failing to detect even quite marked benefits from the use of a particular treatment. This is extremely bad experimental and ethical practice. It is a waste of the investigators' time, a waste of limited research resources, and an unwarranted and unacceptable misuse of the patients involved. The sample must be large enough to ensure that there is an acceptably small risk of failing to detect significant treatment differences (see Chapter 9 for a discussion of the principles and practice of sample size determination).

Once the overall sample size has been determined, the recruitment of trial subjects must be planned. Investigators usually overestimate (often by a factor of two or more) the rate at which they will be able to admit patients to the study. (Overoptimism is a very human failing, especially when we want to believe that a pet project really is feasible.) Recruitment rates should be considered very carefully and realistically since they determine how long the trial will have to run. It is always advisable to have fall-back plans available in case the trial has to be terminated before the anticipated study size is reached (for example, use the "stepped" randomization process that ensures group equality at various intermediate study sizes).

The use of blocking structures, discussed in Chapter 8, is an excellent way of controlling and reducing the error variation in a study. However, there is as always a down side; blocking may substantially complicate and extend the recruitment process. In a four-treatment study blocked on age and gender with eight patients per block, it will be necessary to continue recruiting until, say, eight females under 25 years old have been recruited, even if patients of this type are relatively rare. By this time 40 males over 55 may have presented themselves, even though only the first eight can actually be admitted to the trial. Like a convoy proceeding at the pace of the slowest ship, a blocked trial is a prisoner of the rarest block.

Specifying the Procedures

The day-to-day running of a trial depends on the successful (and appropriate) administration of a whole host of procedures. It is important that everyone involved in the study, in whatever capacity, know exactly what they should do in any given situation that arises within their sphere of responsibility. All procedures to be employed in the trial must be specified in detail, whether they involve the methods of administering the treatments, methods of evaluating the outcome (or outcomes, if secondary questions are asked that relate to various alternative measures of success), the routine clinical management of the patient, or the treatment of side effects or intercurrent illness. Clear and unambiguous procedures should be laid down for every eventuality. (Not every eventuality can be foreseen but, with a little thought, most problems can be anticipated.) The objective should be to ensure consistent treatment (both the specific intervention and general clinical management) and evaluation of all the patients in the trial.

Remember that the patients, not the researchers, are the most important people in any clinical trial. When the investigator is defining procedures, it is a good idea to view the trial from the patients' perspective as well as the investigators'. Procedures that are pleasant to follow and that do not impose unnecessary inconvenience (or just plain indignities) on the patients are likely to encourage patient compliance with the intervention regimes and limit trial dropouts.

Statistical Analysis

The method of statistical analysis to be employed is often determined after the results have been collected. Nothing could be more inappropriate and unnecessary. Like any other research planning choice, the method of analysis should be defined in the trial protocol, long before the first patient is recruited. The decision is usually clear cut since the appropriate statistical analysis will be determined by the research question posed (is it a one-tailed or two-tailed hypothesis, are two or more groups being compared, and so forth?) and the nature of the trial outcome variable (is it continuous or categorical; if continuous, does it follow a Normal distribution, and so forth?). Much of the rest of this book is devoted to the methods of statistical analysis, and we will not discuss them any further here.

There are, however, two situations that arise quite frequently in clinical trial work and that have direct implications for the final statistical analysis.

The first is the problem of patient withdrawals. It is not uncommon for some patients in a trial to fail to complete their scheduled treatments and hence, to be effectively lost from the trial. Should these patients be discounted from the final analysis of the trial results? Remember that failure to complete a course of treatment is, in a real sense, a failure of the treatment itself. No matter how successful a treatment is, its usefulness is severely limited if it is difficult or unpleasant to complete it. Trial withdrawals are predictive of future potential treatment failures of this type. Therefore, withdrawal patients should always be included in the final treatment assessment, where possible.

When treatment assessment is made on a qualitative basis (such as success or failure), this is quite practicable. The withdrawal patients are included with the full treatment failures as patients who have not achieved success on this intervention (that is, any patient who is not a success is a failure). The percentage of successes among the various interventions can then be compared.

When treatment assessment is made on a quantitative basis (such as reduction in blood pressure), then unfortunately no results will be available for patients who have not completed the treatment course. The final analysis will have to be based on only the patients completing the trial. However, the number of withdrawals should always be reported and preferably tested for differences among the various interventions. (Commenting that treatment X is more successful than treatment Y if the full treatment course is followed, but treatment X also has a significantly higher withdrawal rate, is both more honest and more useful than simply reporting that treatment X is superior.)

The second problem that has implications for the final statistical analysis is the question of repeated testing of the trial results. It is understandable that the investigator wishes to prove the superiority of a new treatment as soon as possible (if it is indeed superior), since this will enable as many future patients as possible to receive the best possible treatment. Some long-term trials seek to achieve this by analyzing the results on an ongoing basis. In a 200-patient trial the treatments might be compared after 50 patients, after 100, and so forth. It is important to remember that repeated testing of this type substantially increases the risk of a type I error occurring at some point. This is important in clinical trial work, since concluding that one treatment is superior to another when it is not could have serious effects on future patients.

If repeated testing is to be employed, then the individual tests should use a more severe critical value to maintain the overall risk of a type I error at .05. This can be achieved by carrying out each test at $.05/c$, where c is the number of tests to be carried out. (As always, repeated testing should be planned and not undertaken after the trial has begun.) Using this rule can result in significance levels not detailed in most tables! As a convenient rule of thumb, using a .025 critical value for two or three tests, and a .01 critical value for four to 10 tests, will maintain the overall risk of a type I error acceptably close to .05. (Carrying out more than 10 tests during the course of a trial is usually quite excessive and unnecessary.)

Facilities and Budget

The most imaginative and elaborate trial is totally pointless if it is simply unaffordable, or if the equipment or staff required to implement it are just not there. This portion of the protocol should present a realistic assessment of the facilities that will be needed to complete the trial, the facilities that will actually be available, and the projected trial budget. The key word here is *realistic*. Proposing a complex, large-scale trial is a complete waste of time if funds will be insufficient to implement it.

If the estimated budget for the proposed trial turns out to be excessive, then key elements in most of the other protocol sections may well have to be modified to reconstruct the trial to make it more practical and realistic.

━ ETHICS OF CLINICAL TRIALS ━

Make no mistake about it, clinical trials involve experimentation on human beings. Clearly that raises major ethical issues, issues that must be faced head on. Is it ethical at all to experiment on human beings? Can a physician in conscience decide a patient's therapy on the toss of a coin (literally or figuratively)?

It is a basic and unalterable principle of medical ethics that each patient must receive the best possible health care currently available. How can this be reconciled with the apparently haphazard approach to care inherent in the clinical trial? In the case of a newly developed therapy, there is no current knowledge as to whether it is actually superior (or indeed inferior) to the equivalent existing therapy. The question of which treatment in fact offers the best possible health care is unresolved. A clinical trial, with its random treatment allocation, in this situation can be viewed as ethical from the perspective of the participating patient (since there is no basis for favoring either alternative), and ethical from the perspective of the future patient (since, in the future, treatment decisions can be made from a basis of knowledge).

The ethical perspective demands that the control patients receive the best possible currently available treatment for their condition. There are few situations in which it is ethically acceptable to leave the control patients effectively untreated. Either the condition must be relatively minor and not a serious threat to the patient's health (such as in a headache trial), or a placebo must be genuinely viewed as potentially offering as much hope as any of the other alternative interventions. Once real evidence becomes available that one treatment is indeed superior to the other, it is clearly unethical to continue treating patients with the inferior treatment.

The general requirements for an ethically acceptable clinical trial are laid out in the **Declaration of Helsinki** (1975 revision). Anyone involved in clinical trial work (and indeed any research work involving human subjects) should ensure that he or she is familiar with this important and thoughtful document.* In summary it states that:

1. Clinical trials should be carried out only by properly qualified personnel who are fully conversant with all available relevant evidence.
2. Any risks involved are believed to be predictable, and the level of the potential risk is compatible with the foreseeable benefits of the trial. (In other words, seriously ill patients would generally be prepared to accept a greater risk than mildly ill patients, since the potential gain for them is correspondingly greater. However, the level of acceptable risk is ultimately a decision for the participants and not the physicians, a point we will explore when we talk about informed consent).
3. The dignity, integrity, and privacy of the trial subjects are crucial and must be respected at all times.
4. The informed consent of each subject must be obtained.

This last point is important. It is a fundamental violation of a person's rights and dignity to experiment on the person (no matter how well intentioned) without his or her explicit and fully informed consent. Informed consent must be an integral aspect of any clinical trial. The Helsinki Declaration's comments on informed consent are obligatory reading:

────

*See Pocock SJ: Clinical trials: a practical approach, Chichester, 1983, John Wiley & Sons, pp. 100-102.

In any research on human beings, each potential subject must be adequately informed of the aims, methods, anticipated benefits, and potential hazards of the study and the discomfort it may entail. He or she should be informed that he or she is at liberty to abstain from participation at any time. The doctor should then obtain the subject's freely-given informed consent, preferably in writing.

When obtaining informed consent for the research project, the doctor should be particularly cautious if the subject is in a dependent relationship to him or her or may consent under duress. In that case, the informed consent should be obtained by a doctor who is not engaged in the investigation and who is completely independent of this official relationship.

In case of legal incompetence, informed consent should be obtained from the legal guardian in accordance with national legislation. Where physical or mental incapacity makes it impossible to obtain informed consent, or when the subject is a minor, permission from the responsible relative replaces that of the subject in accordance with national legislation.

PLANNING OTHER RESEARCH PROJECTS

In this chapter we have concentrated on the issues that must be addressed when planning a clinical trial. These issues are, however, also central to almost every other research endeavor in the health sciences. Blinding may be less crucial in animal-based experimentation, and random allocation is not an issue in observational studies where the study groupings (such as smoker/nonsmoker) are self selected. Nevertheless, the questions "What exactly do we want to do, why is it important and worthwhile, how exactly will we do it, how many subjects will we need, what individuals should be eligible to participate in the study, how will we analyze the results, what will we do if something goes wrong?" and so forth, are universally applicable.

Developing a comprehensive game plan that addresses the issues of why, whom, and how will repay the budding researcher 100-fold. It will bring vague concepts into sharp focus, ensure that everyone involved knows exactly what they should be doing and why, defuse problems before they arise, and even give you a head start on the final project write-up.

To end on perhaps a rather gloomy perspective, with a well-written research protocol in place, you could be struck by a car and die secure in the knowledge that the project could be carried on to successful completion in exactly the way you had planned. In a very real sense, the development of the research protocol represents the creative portion of the research experience. Everything after that is merely operational.

Key Terms

Clinical trial

Protocol

Primary question

Secondary question

Randomized clinical trial

Declaration of Helsinki

CHAPTER 11

Multi-Way Analysis of Variance

ANALYZING SEVERAL SOURCES OF VARIATION

The simplest form of experiment, the completely randomized experiment, involves forming the subjects into a number of groups (ideally of equal size) and posing the question "Do the groups differ from one another in their mean performance?" The analysis of variance, which we met in Chapter 5, gives us an elegant way of answering exactly this question. Reviewing the basic ideas behind the analysis of variance is worthwhile at this point (reread Chapter 5 if you need a refresher). In the analysis of variance, we calculate the total variation present in the study and the variation resulting from the between-group differences. The difference between these two values must be within-group or purely random variation.

(From now on, we will use the term **error variation** to describe such variation that has no known cause and that must be assumed to be totally random. The expression, "within-groups" becomes ambiguous when there are several ways of grouping our data, a situation we will discuss shortly.)

The between-groups variation is compared to the error variation (after converting both to mean squares or variances), and if the between-groups differences are much larger than can be explained by error variation (the ratio of variances is much larger than 1), then we conclude that real differences do exist between the groups. This type of analysis of variance, in which only one possible cause of variation ("differences between the groups") is tested, is often called a **one-way analysis of variance.**

So much for analyzing a very simple experiment. In Chapter 8 we saw that there are considerable advantages in organizing the available experimental material in a more structured way. These more complex experiments have a number of influences or potential sources of variation incorporated within them. The experimental material might be arranged so that similar individuals are grouped together into blocks (a randomized block experiment), two different treatment factors might be incorporated into a single experiment (a factorial experiment), or both approaches might be employed simultaneously.

Provided that such an experiment has been properly balanced (an important point that we will consider in more detail in the next section), we can use the basic ideas of analysis of variance to measure the amount of variation that is due to each of the influences or causes that have been built into the experiment. An analysis of variance that is used to examine more than one source of variation is known as a **multi-way analysis of variance.** Untangling the variation in this comprehensive way enables us to substantially reduce the amount of variation that we simply cannot explain and that, therefore, we are forced to regard as genuinely random or "error" variation. This is, of course, the benchmark against which we measure the size of the treatment effect (and any other known effect). Reducing the error variation will, therefore, make it considerably easier to detect genuine treatment differences, and thus will substantially increase the sensitivity (power) of our experiment. In addition, information on the size or importance of the other sources of variation will usually be of considerable interest in its own right.

How can we measure the variation resulting from a variety of causes? Remember that the variation between groups of individuals is obtained very simply by using the totals for each group *(T)* and the number of individuals in each group *(n)*:

$$\frac{T_1^2}{n_1} + \frac{T_2^2}{n_2} + \ldots + \frac{T_k^2}{n_k} - \frac{(\Sigma x)^2}{N}$$

It is obvious what we mean by groups in a simple completely randomized experiment. However, what do we mean by the term "groups" in, say, a randomized block experiment intended to compare several drug treatments? We could clearly group the subjects on the basis of the treatment they had received. Just as logically, we could take groups to mean the groups of individuals in each block of the randomized block experiment (or, indeed, any other grouping of individuals that has been built into our experimental design). Both are valid and meaningful groupings of the experimental material, and they correspond to the two sources of variation that may be present in the experiment. In both situations, the variation resulting from a particular grouping ("between treatments" or "between blocks") is obtained using the appropriate group totals and number of individuals per group in the usual manner.

For instance, in a simple randomized block experiment using 12 subjects arranged in four blocks of three to compare three treatments the between-treatments variation (that is, grouping the subjects in terms of treatment), is given by

$$\frac{T_1^2}{n_t} + \frac{T_2^2}{n_t} + \frac{T_3^2}{n_t} - \frac{(\Sigma x)^2}{N}$$

where T_1 is the total for the four individuals receiving treatment 1, and so on, and n_t is the number of individuals per treatment (that is, four).

On the other hand, we could just as validly think of the experiment as one in which the 12 individuals have been split into four groups (the blocks) of three. Correspondingly, the between-blocks variation is given by

$$\frac{B_1^2}{n_b} + \frac{B_2^2}{n_b} + \frac{B_3^2}{n_b} + \frac{B_4^2}{n_b} - \frac{(\Sigma x)^2}{N}$$

where B_1 is the total for the three individuals constituting block 1, and so on, and n_b is the number of individuals per block (that is, three).

The variation resulting from any particular grouping of individuals can be calculated in this way, provided the various groupings have been properly balanced.

The degrees of freedom associated with any particular source of variation are, as usual, the appropriate number of categories (such as number of treatments or number of blocks) − 1. The unexplained or error variation (the variation that still cannot be attributed to any known source or cause) is simply the total variation less all the explained variation, and the error degrees of freedom are correspondingly the total degrees of freedom less all the "explained" degrees of freedom.

THE IMPORTANCE OF BALANCE

The whole point of a more complex experimental design (and of multi-way analysis of variance) is that it enables us to isolate variation from several causes and study it. However, we can only truly isolate variation from a number of causes if those causes (as defined by the groupings of the experimental material) are truly distinct and do not have any built-in connection with one another. This will only happen if the various groupings in the experiment are constructed in a fair and balanced way.

To understand what we mean by **balance** and why it is important, consider a simple experiment in which we wish to compare two drugs, A and B. The available patients vary in the severity of their illness, so we decide to block our experiment into severely ill and moderately ill patients, using a total of 16 patients.

		SEVERELY ILL	MODERATELY ILL
DRUG	A	4	4
	B	4	4

The above allocation of patients, with four in each treatment situation, is clearly balanced and manifestly fair. Each drug is administered to corresponding numbers of severely ill and moderately ill patients and thus has an equivalent opportunity to demonstrate its worth. A comparison between severely ill and moderately ill patients would be equally fair and unbiased since both groups have half their patients treated with drug A and half treated with drug B. The between-drugs and between-severity variation are truly separate or independent.

		SEVERELY ILL	MODERATELY ILL
DRUG	A	8	0
	B	0	8

By contrast, in the above extreme situation the two sources of variation are hopelessly tangled. A comparison between the two drugs is also a comparison between the two severity levels. (Is it drug A vs. drug B or severely ill vs. moderately ill?) In other words, the two effects are fully confounded. Any comparison between the drugs is patently unfair, and separating the two effects is quite impossible. No sane researcher would structure an experiment in such a totally unbalanced way. However, what would the effect be if we allocated the subjects in the less dramatic manner shown as follows?

		SEVERELY ILL	MODERATELY ILL
DRUG	A	6	2
	B	2	6

The situation is almost as bad as before. The drug comparison is clearly unfair, since A is being administered to a much higher proportion of severely ill patients than B is. Although the two effects are not fully confounded, they are still heavily intertwined, and hence between-drugs and between-severity variation cannot be regarded as independent. The two effects cannot be separated, and a comprehensive multi-way analysis of variance is impossible.

Proper balance is assured whenever the available experimental material is split evenly into the various possible groupings. (This is implied in the notation of the previous section where we assumed that, say, the number of individuals per treatment will be the same for all treatments.) However, an experiment will still be balanced if the proportions, if not the exact number, of the various individuals in each category remain the same.

		SEVERELY ILL	MODERATELY ILL
DRUG	**A**	3	9
	B	1	3

A moment's reflection will confirm that the comparisons in the above experiment are still fair (25% of the patients are severely ill in both drug groups) and that treatment and severity are still independent. Multi-way analysis of variance can be used quite appropriately here, provided we remember to use the correct number of individuals for each of the totals involved in the calculation of the sums of squares associated with the various sources of variation. (The drug A total is based on 12 individuals, for instance, while the drug B total is based on only 4). However, it is much more advisable to balance an experiment by distributing the available subjects in an even way. First, it is easier to do. Second and more important, it ensures that the experiment will be as powerful as possible.

——— MULTI-WAY ANALYSIS OF VARIANCE IN ACTION ———————

To see just how easy it is to take the basic analysis of variance approach and apply it to the analysis of an experiment with multiple sources of variation, let's analyze a hypothetical experiment in which the reaction times of rats responding to a stimulus are compared under three different drug treatments. The experiment is constructed as a randomized block experiment with four blocks of six rats (litter mates) each. The reaction times obtained (in hundredths of a second) are given below.

	DRUG					
	1		**2**		**3**	
Block 1	12	10	14	15	18	21
Block 2	17	14	17	13	23	24
Block 3	26	21	28	29	31	35
Block 4	13	16	13	12	16	18

Summing all 24 results gives

$$\Sigma x = 456 \qquad \Sigma x^2 = 9724$$

$$\text{Total variation} = 9724 - 456^2/24 = 1060.0$$

$$\text{Total df} = 24 - 1 = 23$$

The first step in calculating the variation resulting from the two known causes (drugs and blocks) is to obtain the appropriate drug and block totals.

$$D_1 = 129 \qquad D_2 = 141 \qquad D_3 = 186 \qquad n_d = 8$$

where D_1 = Total for rats on drug 1, and so forth,

and n_d = number of rats on each drug.

$$B_1 = 90 \quad B_2 = 108 \quad B_3 = 170 \quad B_4 = 88 \quad n_b = 6$$

where B_1 = total for rats in block 1, and so forth

and n_b = number of rats in each block

$$\text{Between-drugs variation} = \frac{D_1^2}{n_d} + \frac{D_2^2}{n_d} + \frac{D_3^2}{n_d} - \frac{(\Sigma x)^2}{N}$$

$$= \frac{129^2}{8} + \frac{141^2}{8} + \frac{186^2}{8} - \frac{(456)^2}{24}$$

$$= 225.75$$

$$\text{Between-drugs df} = 3 - 1 = 2$$

$$\text{Between-blocks variation} = \frac{B_1^2}{n_b} + \frac{B_2^2}{n_b} + \frac{B_3^2}{n_b} + \frac{B_4^2}{n_b} - \frac{(\Sigma x)^2}{N}$$

$$= \frac{90^2}{6} + \frac{108^2}{6} + \frac{170^2}{6} + \frac{88^2}{6} - \frac{(456)^2}{24}$$

$$= 737.33$$

$$\text{Between-blocks df} = 4 - 1 = 3$$

The amount of variation left unexplained in the experiment is therefore

$$\text{Error variation} = \text{Total variation}$$

$$- \text{ Between-drugs variation}$$

$$- \text{ Between-blocks variation}$$

$$= 1060.0 - 225.75 - 737.33$$

$$= 96.92$$

$$\text{Error df} = 23 - 2 - 3 = 18$$

The appropriate analysis of variance table for this experiment is

SOURCE	SS	df	MS	F
Between drugs	225.75	2	112.88	20.98
Between blocks	737.33	3	245.78	45.68

SOURCE	SS	df	MS	F
Error	96.92	18	5.38	
Total	1060.00	23		

The F values confirm that highly significant differences exist between the drugs and between the blocks. (The purpose of the blocks here is solely to control variation, and it is not really necessary to carry out a formal test for differences. However, there will be many situations in which the nature of the blocks will make testing for block differences of genuine interest, for example, when there are possible age or gender differences.)

The wisdom of using a randomized block design is apparent if we consider the analysis of variance table that would result if the blocking had been ignored and the experiment analyzed as a completely randomized experiment.

SOURCE	SS	df	MS	F
Between drugs	225.75	2	112.88	2.84
Error	834.25	21	39.73	
Total	1060.00	23		

Since the substantial differences that exist between animals from different litters is no longer explicitly isolated in the analysis, the inevitable result is that this variation is lumped in with the error variation (the variation for which we can find no explanation or cause). This key benchmark becomes greatly increased, and the previously significant drug differences now disappear, overwhelmed by the huge rise in error variation. The advantages of blocking are not always as dramatic as this. However, the amount of variation controlled by blocking (providing it is done on a sensible basis) will normally more than offset the loss in error degrees of freedom and result in a much reduced error variance and a much more sensitive and powerful analysis.

Isolating the variation between blocks means that the drug differences are being compared within blocks (groups of inherently similar individuals), and this enables us to generalize the pairing principle we first met when we discussed the paired t test back in Chapter 4. Note that the drug differences (particularly the longer reaction times on drug 3) show up clearly within each block. This is confirmed by the multi-way analysis. However, these relatively small differences would be completely obscured by the much larger between-blocks differences (compare blocks 3 and 4, for instance) and would be missed in a simple one-way analysis, as we saw in the previous table.

Once the existence of drug differences has been established by the multi-way analysis, a multiple comparison technique such as Tukey's test (Chapter 5) can be used to determine precisely which of the drugs differ from the others. Just as before, Tukey's test employs the error mean square (from the full multi-way analysis of variance table, since this is the only variation without any known cause), the number of individuals in each of the groups being compared, and the appropriate q value (determined by the number of groups being compared and the error df) to determine the minimum difference between the group means that we must observe before the groups are to be considered different.

For the drug comparison in this example

Number of drug groups = 3

Number of individuals/group = 8

Error mean square = 5.38

q (3 drugs, 18 error df) = 3.61

Tukey's value for α = .05 = 3.61 $\sqrt{5.38/8}$ = 2.96

MEANS	DRUG 1	DRUG 2	DRUG 3
	16.13	17.63	23.25

The drug differences are therefore due to the fact that drug 3 produces a significantly longer reaction time than drugs 1 or 2, which do not differ significantly from one another.

This particular example involves two sources of variation (drugs and litters or blocks) but the principle generalizes very simply to the analysis of experiments with three or more sources of variation built into them. Consider an experiment to compare three diets that is blocked on gender and obesity (with four defined obesity categories). Six subjects are recruited per obesity level/gender combination (resulting in an overall study size of 48) and two randomly allocated to each of the three diets under test. Since the experiment is fully balanced, the between-diets, between-genders and between-obesity levels sums of squares can be calculated in the way we have just discussed, and the error or unexplained variation found by subtracting these from the total sum of squares. Questions relating to possible diets, gender or obesity differences can then be tested by comparing the appropriate "between" mean square to the error mean square. The skeleton of the appropriate analysis of variance table is given below. (Writing down this outline is often a very helpful way to start a multi-way analysis of variance.)

SOURCE	SS	df	MS	F
Between diets		2		
Between genders		1		
Between obesity levels		3		
Error		41		
TOTAL		47		

TESTING FOR INTERACTION

Measuring and testing the significance of the various sources of variation (often called the **"main effects"**) built into a designed experiment is the most obvious objective of the analysis of variance. However, in our discussion of factorial experiments in Chapter 8 we introduced the concept of an interaction, the idea that the relative performance of the various levels of one factor (or main effect) might change dramatically, depending on the level of the other factor that was present (for example, one drug might be clearly superior to another in young people but be slightly inferior in older people).

To obtain a truly comprehensive and honest understanding of the main effects under study, we need to be able to determine if they do, in fact, interact with one another.

To see how we can use our analysis of variance tools to do this, let us suppose we have a randomized block experiment in which two drugs are to be compared. The experiment is blocked into three age categories with eight patients per block. As well as testing for drug and age differences, (the two main effects in the experiment), we would like (and it would be important) to determine if there is an age/drug interaction, that is, if the relative effectiveness of the two drugs varies substantially from one age category to another.

How might we approach the problem of testing whether an interaction is present in a study? Well, if the relative effectiveness of a drug depends on the age of the patient receiving it (an age/drug interaction), then the key groupings in the experiment are not the various drug groups or age blocks, but the various age/drug combinations (the under-30-year-olds who receive drug A, and so forth). The variation between these various age/drug combinations should be large, we believe, if an interaction between age and drug is indeed present (such as, if drug A is producing excellent results in young patients but is much less effective in older patients).

We can calculate this "between-age/drug combinations" variation in the usual way by

$$\frac{T_{a1}{}^2}{n_c} + \frac{T_{a2}{}^2}{n_c} + \ldots + \frac{T_{b3}{}^2}{n_c} - \frac{(\Sigma x)^2}{N}$$

where T_{a1} is the total for those patients who received drug A and were in age category 1, and so forth, and n_c is the number of patients in each individual age/drug combination.

The corresponding degrees of freedom are given, as always, by the number of combinations (or groupings) involved less 1.

That is,

$$ad - 1$$

where a is the number of age categories and d is the number of drug categories.

Before we accept this "between-combinations" variation as the answer to our interaction problems, we need to appreciate one very important point. By isolating all the various age/drug combinations, we have also isolated the age and drug categories.

For example

	T_{a1}	T_{a2}	T_{a3}		T_{b1}	T_{b2}	T_{b3}
Drug		A				B	

In other words, both the age groupings and the drug groupings are contained in, or can be recreated from, the age/drug groupings. The consequence of this is that both the age and drug main effects are already incorporated within the between-combinations variation we have just calculated. To obtain the variation resulting solely from the presence of a drug/age interaction, the between-drugs and between-ages variation must be calculated in the usual way and simply subtracted from the between-combinations variation.

The variation resulting from an age/drug interaction (usually simply called age × drug) is, therefore, given by

Between-age/drug combinations variation − Between-ages variation − Between-drugs variation

The degrees of freedom associated with age × drug variation are correspondingly given by

$$\text{Age} \times \text{drug df} = ad - 1 - (a - 1) - (d - 1)$$

This, in fact, works out very simply to give the relationship

$$\text{Age} \times \text{drug df} = (a - 1) \times (d - 1)$$

Very appealingly, the degrees of freedom associated with any interaction are therefore simply the multiplication of the degrees of freedom associated with the corresponding main effects.

When $a = 3$ and $d = 2$, for instance,

$$\text{Age} \times \text{drug df} = (2 \times 3 - 1) - (3 - 1) - (2 - 1)$$

$$= 2$$

Or equivalently $= (3 - 1) \times (2 - 1)$

$$= 2$$

Once the sum of squares and degrees of freedom are established, the interaction is then tested against the error variance for significance in exactly the same way as the main effects. Please remember that if a significant interaction exists, the two main effects involved are influenced by one another, and it is therefore pointless and potentially quite misleading to discuss, or attempt to interpret, either of the main effects in isolation. Interpretation of the results should center on the interpretation of the interaction itself. The preparation and inspection of a graphic description of the interaction (as discussed in Chapter 8) can be very helpful. When interpreting a significant main effect, we use a multiple-comparison test such as Tukey's test to determine precisely where the significant differences lie. When an interaction is significant, it is the various combinations involved that are of key importance in understanding how the factors involved affect one another. Just as before, Tukey's test can be used to compare mean performance between the relevant combinations, to determine which situations differ and which do not.

To illustrate the calculations involved, let's expand the analysis of the drug/litter experiment in the previous section to include a test for the existence of a drug × litter interaction, that is, let's examine the possibility that the relative effectiveness of the drugs differs from one litter to another.

$$\begin{array}{l} \text{Between-drug/litter} \\ \text{combinations} \end{array} = \frac{22^2}{2} + \frac{29^2}{2} + \ldots + \frac{34^2}{2} - \frac{456^2}{24}$$

$$= 1012.0$$

where 22 is the total for the 2 rats in litter 1 receiving drug A, and so forth.

$$\text{Drug} \times \text{litter SS} = 1012.0 - 225.75 - 737.33$$

$$= 48.92$$

where 225.75 is the between-drugs SS

and 737.33 is the between-litters SS

The full analysis of variance table for this experiment is therefore

SOURCE	SS	df	MS	F
Drugs	225.75	2	112.88	28.22
Litters	737.33	3	245.78	61.45
Drugs × Litters	48.92	6	8.15	2.04
Error	48.00	12	4.00	
Total	1060.00	23		

and this confirms that no significant interaction is present in this study. The drug and litter main effects can therefore be meaningfully interpreted as separate entities.

The calculation of the appropriate between-combinations variation, followed by removal of the main effects variation involved, can, of course, be applied to the study of any desired interaction. However, it is important to realize that there must be two or more results available for each combination for the calculation to be feasible. If only one result is available within each combination, then we have no measure of how variable each, say, age/drug combination is and, hence, no way of assessing how impressive any differences between them are. This need to ensure that two or more results will be available for each combination should be considered at the experimental design stage, if you intend to test for interactions in your final analysis.

In a study with three or more main effects, interactions can, of course, exist between all pairs of main effects, and are tested in the way we have described. These are sometimes called **first-order interactions.** It is also possible to have higher order interactions. If the hypothetical age/drug study discussed at the start of this section had also been blocked on gender then the existence of an age × drug × gender (or second-order) interaction would be possible. Such an interaction would indicate that the nature of the age × drug interaction changes significantly from one gender to the other. In other words, to really paint a fair and honest picture of our results, we would need to discuss the implications of a subject's drug treatment in the context of the subject's age and gender.

A comprehensive analysis of variance table for such a study would have the following general structure.

SOURCE	SS	df	MS	F
Drugs		1		
Age		2		
Gender		1		
Age × drugs		2		
Gender × drugs		1		
Age × gender		2		
Age × drugs × gender		2		
Error		12		
Total		23		

The age $\times$ drugs $\times$ gender sum of squares would be calculated by calculating the "between-age/drugs/gender combinations" variation, removing the age, drugs, and gender main effects and then also removing the age $\times$ drugs, gender $\times$ drugs, and age $\times$ gender interaction sums of squares. Remember that all the two-effect combinations can be re-formed from the even finer age/drugs/gender combinations, and hence the various first-order interactions are hidden inside the combination's sum of squares. As before, the relevant degrees of freedom are obtained by simply multiplying together the degrees of freedom of the various main effects involved. Higher order interactions can rapidly become very difficult to interpret in any meaningful way, and it is rarely helpful to explore interactions involving more than three main effects.

— SPLIT-UNIT EXPERIMENTS ———————————————

Crossover designs, in which various treatments are administered to the same individual, are highly sensitive ways to compare treatment effectiveness. Since the comparisons are made solely within individuals, the potentially large variations that can exist between individuals and that can obscure the treatment differences are totally irrelevant. In practice we use multi-way analysis of variance to calculate and isolate the between-individuals variation (obtained by treating each individual as a block). The remaining variation (which must, of course, be purely within-individuals variation) is then similarly split into between-treatments variation (which is within individuals because of the nature of the crossover design) and the remaining unexplained within-individuals variation. Testing is accomplished by comparing the between-treatments variation to the within-individuals error variation. The actual test of significance is, therefore, based solely on within-individuals variation.

The following skeleton analysis of variance table illustrates the analysis of a crossover design in which three drugs were administered (in random order) to each of 12 individuals, yielding a total of 36 results.

	SOURCE	SS	df	MS	F
	Between individuals		11		
Within Individuals {	**Between drugs**		2		
	Error		22		
	Total		35		

Although they are very attractive in principle, crossover comparisons are only practicable if the effects involved are temporary and of relatively short duration. Responses to certain drugs or to stimuli might well meet these criteria. However, very many comparisons involve treatments that have a permanent (or at least long-term) effect on an individual (such as surgical techniques) or involve inherent, unalterable characteristics of the individual (such as the individual's gender or age). Comparisons of this type must be carried out using distinct groups of individuals. In this situation, analysis of variance involves simply splitting the total variation (which is all between individuals) into between-treatments and the unexplained (error) variation, and comparing the two. Testing treatments administered to different individuals, therefore, involves solely between-individuals variation. (It can hardly be otherwise, since, in an experiment based entirely on different individuals, all the variation must be between individuals). The comparison of 6 males and 6 females (yielding 12 results, all from different individuals) would, for example, result in the following skeleton analysis of variance table.

	SOURCE	SS	df	MS	F
Between Individuals	Between genders		1		
	Error		<u>10</u>		
	Total		11		

Factorial experiments are experiments in which more than one treatment concept is applied to the same set of experimental subjects (see Chapter 8). When one of these factors is suitable for administration within subjects (that is, in a crossover manner) while the other must be given to separate subjects, we have a special type of factorial experiment variously known as a split-plot, repeated measures, or **split-unit experiment.**

The following is an example of a split-unit experiment in which one of the factors (a comparison of three drugs) is suitable for crossover administration. Each patient will, therefore, receive all three drugs (in a random order and with suitable intervals between them) and be tested after each drug. The other factor involves a comparison of possible gender differences. This, of course, must involve distinct individuals. The 12 experimental subjects are therefore split into groups of 6 males and 6 females.

		DRUG	
	Subject 1	2 1 3	
		. . .	. . .
Males		. . .	. . .
		. . .	. . .
	Subject 6	1 3 2	
	Subject 7	1 2 3	
		. . .	. . .
Females		. . .	. . .
		. . .	. . .
	Subject 12	3 1 2	

(The name "split-unit" reflects the fact that the basic unit for one of the comparisons (the subject) is split into sub-units for the other comparison.)

The analysis of a split-unit experiment requires extra care. Recall from our earlier discussion that one of the comparisons (between genders, in this example) involves only between-subjects variation, while the other comparison (between drugs) involves only within-subjects variation. In a sense, therefore, two separate analyses are required, one (testing the between-genders comparison) based solely on the between-subjects variation and the other (testing the between-drugs comparison) based solely on the within-subjects variation.

The first step, therefore, is to split the total variation (and degrees of freedom) into between-subjects and within-subjects portions. The various sums of squares are calculated in the usual way.

$$\text{Total SS} = \Sigma x^2 - \frac{(\Sigma x)^2}{N}$$

$$\text{Between-subjects SS} = \frac{S_1^2}{n_s} + \frac{S_2^2}{n_s} + \ldots + \frac{S_{12}^2}{n_s} - \frac{(\Sigma x)^2}{N}$$

where

$$S_1 = \text{Total for subject 1, and so forth,}$$

and

$$n_s = \text{Number of results/subject (3 in this example).}$$

Since variation must be either between subjects or within subjects, the within-subjects SS is simply given by:

$$\text{Within-subjects SS} = \text{Total SS} - \text{Between-subjects SS}$$

The within-subjects df are obtained by subtraction in the same way, where

$$\text{Total df} = \text{Total number of results} - 1$$

and

$$\text{Between-subjects df} = \text{number of subjects} - 1$$

Once this basic split has been made, the two separate analyses can be carried out.

Let us consider the between-subjects situation first. The between-genders variation is given by

$$\frac{M^2}{n_g} + \frac{F^2}{n_g} - \frac{(\Sigma x)^2}{N}$$

where M is the total for males, and so forth, and n_g is the number of results for each gender (18 in this example).

The between-subjects unexplained or error variation is therefore

$$\text{Between-subjects SS} - \text{Between-genders SS}$$

The final skeleton analysis of variance table for the between-subjects situation is, therefore:

SOURCE	SS	df	MS	F
Between genders		1		
Error		10		
Between subjects		11		

Now let's consider the within-subjects situation. The between-drugs variation is given by

$$\frac{D_1^2}{n_d} + \frac{D_2^2}{n_d} + \frac{D_3^2}{n_d} - \frac{(\Sigma x)^2}{N}$$

where D_1 is the total for drug 1, and so forth, and n_d is the number of results for each drug (12 in this example).

There is, however, one other effect that is of interest in a split-unit experiment, and this is the possible existence of an interaction between the two factors involved (such as a drug $\times$ gender interaction). Since this involves a between-subjects factor (gender) and a within-subjects factor (drugs), it is initially unclear whether we should place this in the between-subjects or within-subjects portion of our overall analysis. Remember that the key issue that we will be testing here is whether the relative efficiency of the drugs (that is, a within-subjects effect) varies from one situation to another. Therefore, this should be tested in a within-subjects context. An interaction between a between-subjects factor and a within-subjects factor is always placed in the within-subjects portion of the analysis.

Variation between gender/drug combinations is given by

$$\frac{C_{m1}^{2}}{n_c} + \frac{C_{m2}^{2}}{n_c} + \ldots + \frac{C_{f3}^{2}}{n_c} - \frac{(\Sigma x)^2}{N}$$

where C_{m1} is the total for males on drug 1, and so on, and n_c is the number of results in each combination (six in this example).

The variation resulting from gender $\times$ drug interaction is obtained (as we saw in the previous section) by removing the gender and drug main effects variation.

Gender $\times$ drug SS = Between-combinations SS $-$ Between-drugs SS $-$ Between-gender SS

With the two sources of within-subjects variation measured, the within-subjects unexplained or error variation is obtained by subtraction (as are its degrees of freedom).

Within-subjects error SS = Within subjects SS $-$ Between-drugs SS $-$ Drugs $\times$ gender SS

The final within-subjects skeleton analysis of variance table for this example is, therefore

SOURCE	SS	df	MS	F
Between drugs		2		
Drugs × gender		2		
Error		20		
Within subjects		24		

Although a split-unit analysis of variance is most logically considered as two distinct analyses of variance, the results are usually presented, for convenience, in one combined analysis of variance table. (The fictitious sums of squares and other columns are included as follows purely to illustrate the layout of the final table.)

Once the existence of significant differences has been established, then multiple-comparison tests can be used to determine the precise nature of the differences, as with any other multi-way analysis of variance. Do remember, however, that tests for between-subjects differences should be based on the between-subjects error mean square, while tests for within-subjects differences should be based on the within-subjects error mean square. (It really is best to think of a split-unit analysis of variance as two quite separate analyses being carried out in parallel.)

SOURCE	SS	df	MS	F
Between subjects	400	11		
Between genders	100	1	100	3.33
Error	300	10	30	
Within subjects	200	24		
Between drugs	100	2	50	20.00
Drugs × gender	50	2	25	10.00
Error	50	20	2.5	
Total	600	35		

MISSING VALUES

We have already commented in this chapter on the importance of assigning subjects to experimental treatments in a balanced way. Even with the most careful planning, however, results can be lost from an initially balanced experiment. Animals can die, patients can fail to complete their course of treatment, and so forth. When this happens, the final set of results will be unbalanced, with all the resultant complications we have already discussed.

With a completely randomized experiment this creates no problems. There is no need for an experiment with only one source of planned variation to have equal numbers per group, since there is no other source of variation to entangle with the planned variation. A one-way analysis of variance can, therefore, still be applied in the normal way.

If, however, there is more than one source of variation built into the experiment, then loss of a result and loss of balance create major complications. One way out of this dilemma is to replace the missing value with an "artificial" result that restores the balance of the experiment. What value could one reasonably use in this situation? Since the missing value replacement is not a genuine experimental result, it is vital that it not distort the true results in any way. To arbitrarily alter the various treatment and other effects (as characterized by their means) would simply make a mockery of the experimental process.

In a replicated experiment (one in which more than one result is available for each combination of conditions), this can be done very easily. One or more companion results (that is, results obtained under fully identical conditions) will exist for the missing value. Therefore, the missing value can simply be replaced by the mean of its companion values. This restores balance while leaving the mean of that group of results completely unaltered and still a fair reflection of the genuine experimental evidence.

	FACTOR B	
	1	**2**
	12	16
1	17	18
	19	20

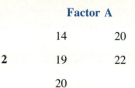

In the above hypothetical two-factor experiment with two levels per factor and three results (that is, three replicates) at each combination of levels, one of the 12 results has been lost. The appropriate replacement value is 21, since this leaves the mean of the particular treatment combination unaltered and, therefore, in no way alters the genuine experimental evidence. The analysis would then proceed using 12 results. It is, however, important to realize that there are still only 11 genuine results involved. (The 12th exists only to restore balance while not distorting the facts. It is, therefore, merely a restatement of the other results.) This means that the total degrees of freedom involved is 10 (11 − 1) and not 11 (12 − 1). As a result, the error df will also be reduced by 1. In general the total df are reduced by the number of missing value replacements used.

In an experiment with several effects and with no replication, deciding which value we can insert to replace a missing value without distorting the results is much more difficult. After all, there are no truly similar results available to serve as guidelines. It is necessary, therefore, to develop some indirect way of generating a replacement value that is totally consistent with the available experimental evidence. Over the next few paragraphs, we will explore just how we can reasonably do this. If you prefer to simply accept the formula that results, feel free to skip ahead. (It's always better, however, to understand the ideas behind the formula and know why it results in something very sensible.)

Let's start by considering a simple randomized block experiment with two blocks, three treatments per block and one missing result.

	T_1	T_2	T_3
B_1	6		10
B_2	5	7	9

There are obviously two influences in this experiment: possible differences between the treatments and possible differences between the blocks. Inspection of the available results indicates, for instance, that block 1 results tend to be higher than block 2 results. Put another way, block 1 tends to be above average, and block 2 tends to be below average. It also indicates that treatment 2 results tend to be intermediate between treatments 1 and 3, and probably close to the overall average.

If the replacement value we come up with is to be fair, it must be totally consistent with this available evidence. It must reflect the fact that block 1 results tend to be above the overall average. It must simultaneously reflect the position of treatment 2 relative to the other treatments and the overall mean. We can summarize these two requirements by saying that

Result for block 1, treatment 2 = Overall average value
+ Amount that B_1 is better (or worse) than average
+ Amount that T_2 is better (or worse) than average

Unfortunately, we cannot work out the overall average or how much T_2 tends to differ from it, and so forth, because some of the evidence is missing. We can sidestep this vicious circle by calling the missing value x and proceeding as if its value were known.

	T_1	T_2	T_3
B_1	6	x	10
B_2	5	7	9

$$T^* = 7 \qquad B^* = 6 + 10 \qquad G^* = 6 + 10 + 5 + 7 + 9$$
$$= 16 \qquad\qquad = 37$$

To allow us to generalize our results, we will call the incomplete total (that is, the total based on only the available results) for the treatment with the missing value T^*, the incomplete total for the block with the missing value B^*, and the overall incomplete total G^*. Let b denote the number of blocks and t the number of treatments.

The overall average (if we knew x) would, therefore, be:

$$\frac{G^* + x}{bt} \quad \text{that is,} \quad \frac{37 + x}{6}$$

The average for T_2 would be

$$\frac{T^* + x}{b} \quad \text{that is,} \quad \frac{7 + x}{2}$$

and the average for B_1 would be

$$\frac{B^* + x}{t} \quad \text{that is,} \quad \frac{16 + x}{3}$$

Combining the block and treatment effects (that is, the amount by which they tend to differ from the overall mean) with the overall average shows that the result for treatment 2, block 1 that fits all the available evidence (that is, x) should be

$$x = \left(\frac{G^* + x}{bt}\right) \quad \text{Overall average}$$

$$+ \left(\frac{T^* + x}{b} - \frac{G^* + x}{bt}\right) \quad \text{Amount by which } T_2 \text{ is better (or worse) than average}$$

$$+ \left(\frac{B^* + x}{t} - \frac{G^* + x}{bt}\right) \quad \text{Amount by which } B_1 \text{ is better (or worse) than average}$$

$$= \frac{(T^* + x)}{b} + \frac{(B^* + x)}{t} - \frac{(G^* + x)}{bt}$$

$$= \frac{t(T^* + x)}{bt} + \frac{b(B^* + x)}{bt} - \frac{(G^* + x)}{bt}$$

that is,

$$x = \frac{tT^* + bB^* - G^* + (t + b - 1)x}{bt}$$

$$btx - (t + b - 1)x = tT^* + bB^* - G^*$$

$$(bt - t - b + 1)x = tT^* + bB^* - G^*$$

$$(b - 1)(t - 1)x = tT^* + bB^* - G^*$$

$$x = \frac{tT^* + bB^* - G^*}{(b - 1)(t - 1)}$$

This is the basic formula for calculating the appropriate replacement for a missing value in a randomized block experiment without replication. All that it does is give us a value that is totally consistent with all the available evidence (as incorporated in the various incomplete totals). It enables the usual analysis of variance procedures to be carried out, while not distorting the genuine experimental evidence in any way.

Again, the only modification to the analysis of variance table that is required is the reduction of the total (and, hence, error) df by 1, reflecting the fact that this missing value replacement is simply a restatement of the other results and not, in any sense, a new observation.

For the sample experiment quoted earlier:

$$B^* = 16 \qquad T^* = 7 \qquad G^* = 37$$

$$b = 2 \qquad t = 3$$

$$x = \frac{3 \times 7 + 2 \times 16 - 37}{2} = 8$$

	T_1	T_2	T_3
B_1	6	8	10
B_2	5	7	9

Note how the missing value replacement reflects the fact that, on average, block 1 results tend to be one higher than block 2 results, and that treatment 2 results tend to be halfway between treatment 1 and treatment 3 results. Therefore, the replacement value simply mimics the trends already present in the available results and introduces no distortion or bias to the final analysis.

A missing value for a latin square experiment (see Chapter 8) is calculated using exactly the same principles.

$$x = \frac{k (R^* + C^* + T^*) - 2G^*}{(k - 1)(k - 2)}$$

where R^*, C^*, T^* and G^* are the incomplete totals for the row, column, and treatment with the missing value and the incomplete overall total, respectively, and k is the number of rows, columns and treatments.

If more than one value is missing, then the situation is much more difficult, since the value we calculate for one missing value is needed to calculate the other missing value, and vice versa. The only practical solution to this very vicious circle is to use a procedure known as iteration. In this technique we guess a value for one of the missing values (by inspecting the evidence and making a common-sense judgment call) and then use this value with the usual formula to calculate the other missing value. The value that was initially guessed is then recalculated using our calculated result for the second missing value, and so on. This cyclic process is continued until neither of the missing values being calculated changes very much (say, by less than .1%).

MISSING VALUES

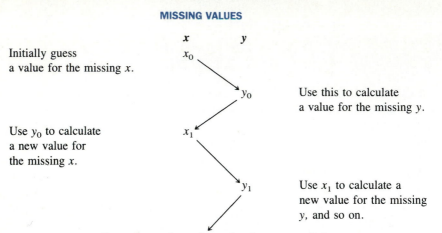

Initially guess a value for the missing x.

Use this to calculate a value for the missing y.

Use y_0 to calculate a new value for the missing x.

Use x_1 to calculate a new value for the missing y, and so on.

Stop when values of x *and* y change very little.

When more than two values are missing, this same approach can be used. It is, however, very tedious to do by hand, although it can easily be done by computer.

We have been discussing the use of missing value replacements to restore balance to an initially balanced experiment that has suffered some unplanned loss. When an experiment is not properly balanced to begin with, methods of analysis (based on the ideas discussed in Chapter 13) are available that will isolate the intertwined effects as much as possible. However, these unbalanced analyses of variance are more complex to implement and interpret and are no substitute for a properly balanced experimental design.

___ Key Terms ___

Error variation   **First-order interaction**
One-way analysis of variance   **Split-unit experiment**
Multi-way analysis of variance

___ PROBLEMS ___

1. The effect of alcohol abuse on concentration and coordination was studied by comparing the times (in seconds) taken to complete a task involving hand/eye coordination skills of 12 patients undergoing treatment for alcohol abuse and 12 non-abusers. To control any possible influence of age and gender, the 24 subjects were blocked into 3 age categories and by gender. Analyze the results given below and outline your conclusions.

AGE (yr)	GENDER	ABUSE? NO		YES	
15-30	M	23	16	28	22
	F	18	26	24	29

			ABUSE?		
AGE (yr)	**GENDER**	**NO**		**YES**	
31-45	**M**	21	19	30	25
	F	24	21	32	22
46+	**M**	28	23	40	43
	F	26	31	41	46

2. As part of an examination of the dose-response relationship of bupivacaine when used for lumbar sympathetic blockade, 36 patients were blocked on the basis of gender and age (<35, 35-50, >50) into 6 blocks of 6 patients each. Within each age/gender block 3 bupivacaine doses (5, 10, and 15 ml respectively) were randomly allocated to 2 patients each.

   The response measured was the increase in the temperature (in degrees Celsius) of the dorsum of the foot 1 hour after injection. (The greater the increase, the more successful the blockage.) The results are presented below. Carry out the most appropriate analysis and outline your conclusions.

		DOSE OF BUPIVACAINE (ml)					
Age	**Sex**	**5 ml**		**10 ml**		**15 ml**	
<35	**M**	0.5	1.4	1.6	0.8	1.9	1.1
	F	0.4	1.1	1.7	1.0	1.0	1.9
35-50	**M**	0.3	1.7	0.7	1.4	1.3	1.8
	F	1.1	1.3	1.6	1.3	0.8	1.6
<50	**M**	0.2	0.1	1.0	0.3	0.6	1.3
	F	0.6	0.3	0.7	0.8	0.9	1.0

3. In an attempt to study the influence of the drug Pentoxifylline on exercise tolerance in patients suffering from intermittent claudication, a crossover experiment was designed in which the patients were treated with the drug and a placebo, in random order. Walking distance on a treadmill was measured after each treatment. To test for possible gender differences in exercise tolerance, the group was split into 5 males and 5 females. The results (in meters) obtained are given below. Test for the presence of gender and drug effects.

			PLACEBO	**ACTIVE**
Males	**Subject**	1	52	58
		2	43	40
		3	31	37
		4	56	46
		5	42	45

			PLACEBO	ACTIVE
Females	Subject	6	48	54
		7	36	30
		8	39	42
		9	29	34
		10	38	36

4. Cloridine has been shown to provide a high level of analgesia in animals and to be a good adjunct to narcotic analgesia in humans. As part of a study of the drug's potency, 5 dogs each received 5 doses of cloridine (tetracaine 4 mg as control, tetracaine 4 mg plus cloridine 50 mg, tetracaine 4 mg plus cloridine 100 mg, tetracaine 4 mg plus cloridine 150 mg, tetracaine 4 mg plus cloridine 200 mg) in random order over a 4-week period. Effectiveness of anesthesia was measured by time to arousal (time from administration to spontaneous movement, in minutes).

Unfortunately two of the dogs died before they could receive all five treatments. The available results are given below. Test for the presence of a dose/response relationship.

CLORIDINE DOSE (mg)

	0	50	100	150	200
Dog 1	21	28	38	39	37
Dog 2		18	32	30	31
Dog 3	29	40	44	48	46
Dog 4	27	28	39		40
Dog 5	17	31	32	37	39

5. To determine if the metabolism of halothane is changed in the presence of cirrhosis, 24 male Wistar rats were randomly allocated to two groups of 12. Cirrhosis was generated in one group by a standard procedure while the other group received a harmless placebo treatment. Serum fluoride levels were measured in each rat before exposure to halothane and at 4 hours.

Analyze the results given below using the appropriate analysis of variance procedure.

		SERUM FLUORIDE	
		0 hr	4 hr
Rats without cirrhosis	1	2.0	3.1
	2	2.2	3.3
	3	2.6	3.1
	4	2.4	3.7
	5	2.6	3.9
	6	3.1	3.8

		SERUM FLUORIDE	
		0 hr	**4 hr**
	7	2.9	4.0
	8	2.7	3.6
	9	1.7	2.6
	10	2.1	3.4
	11	2.3	2.9
	12	2.1	2.7
Rats with cirrhosis	13	2.1	1.9
	14	2.5	3.0
	15	2.3	2.3
	16	1.8	2.0
	17	2.0	2.4
	18	2.2	2.8
	19	1.4	1.9
	20	1.7	1.8
	21	2.9	3.1
	22	2.5	2.9
	23	2.3	2.5
	24	1.8	2.5

6. It is common surgical practice to drain the subhepatic region after cholecystectomy. In a study to compare the effectiveness of three different drainage procedures, 24 cholecystectomy patients were blocked on the basis of gender and age (55 and under, over 55) into 4 blocks of 6 patients each. Within each age/gender block the three treatments under comparison (LP, low pressure drainage; HP, high-pressure drainage; PD, passive drainage) were randomly allocated to two patients each. The success of the drainage procedures was determined by measuring the serum gentamicin levels (mg/l) 1 hour after gentamicin was instilled into the peritoneal cavity (low levels therefore corresponding to more efficient drainage). Analyze the data presented below and outline your conclusions.

		LP	HP	PF
	≤55	1.6	2.4	3.0
		2.1	2.7	2.2
Male				
	>55	2.2	3.4	2.9
		2.4	3.8	2.4

		LP	HP	PF
	≤55	2.3	2.5	2.7
		1.7	2.1	2.2
Female				
	>55	2.5	3.5	3.1
		2.4	3.7	2.8

7. In a study of the effects of wheat bran and barley husk on nutrient utilization, 12 rats were structured into four blocks (using litter mates as blocks). Within each block, one rat received a control diet (without either wheat or barley), one a diet rich in wheat bran, and one a diet rich in barley husk. After 9 days the net protein utilization of each rat was measured and is given below as a proportion. Two results were lost because of incorrect diet administration. Calculate replacement values for these missing values and complete the analysis of variance.

		DIET		
		C	W	B
Block	1	.852	.881	.863
	2	.811		.802
	3	.824	.844	.830
	4	.860	.888	

Analysis of Covariance

VARIATION FROM A CONTINUOUS SOURCE

One of the primary objectives of the experimental design techniques we discussed in Chapter 8 was to structure the experimental material so that as much of the variation as possible could be measured and explained, thus limiting the unexplained or error variation that could obscure the treatment or group differences. If, for instance, we believed that the sexes might show differing levels of response to treatment, we might block the experiment by gender, thus enabling us to measure and remove any gender-related variation from the final analysis.

Variation can, of course, be caused by continuous as well as categorical influences. It is well known that blood pressure, for instance, tends to increase with age. In a hypertension study, differences in the ages of the subjects will result in variation in the subjects' blood pressures, which might obscure possible differences between the various hypertension treatments. Using experimental design techniques, we could control this source of variation by forming blocks of subjects of similar ages and measuring the between-blocks variation. However, this is, at best, only a partial solution. A division into, say, three age blocks (such as under 35, 35-55, over 55) is, after all, completely arbitrary. Even worse, within each block, there will still be age differences (a 39-year-old, a 52-year-old, and so on), and hence age-related variation, although greatly reduced, will not be totally eliminated.

There is another, more direct way to tackle this problem. Both age and blood pressure are continuous variables. We can also logically view blood pressure as a dependent variable and age as an independent or explanatory variable. The relationship between the two could, therefore, be examined using the regression analysis approach of Chapter 7. The regression equation not only describes the relationship:

$$y = a + bx$$

where

$$y = \text{Blood pressure (dependent or outcome variable)}$$

$$x = \text{Age (independent, explanatory, or control variable)}$$

$$a = \text{Intercept}$$

$$b = \text{Slope}$$

but also very important, the regression equation enables us to measure the amount of variation in blood pressure resulting from variation in age.

Recall from Chapter 7 that the variation in y explained by x

$$= b[\Sigma xy - (\Sigma x)(\Sigma y)/n]$$

$$= \frac{[\Sigma xy - (\Sigma x)(\Sigma y)/n]^2}{[\Sigma x - (\Sigma x)^2/n]}$$

$$= \frac{SCP^2}{SSX}$$

where *SCP* stands for the sum of cross-products (of x and y, of course) and *SSX* stands for the sum of squares of x.

(We will use this less cumbersome shorthand notation for the remainder of our discussion of the analysis of covariance.)

Used in this way, regression serves exactly the same purpose as measuring the between-blocks variation in a blocked experiment. It measures, and enables us to isolate, variation that is due to a particular cause. Therefore, it enables us to reduce the amount of unexplained variation and renders the experiment more sensitive. In addition, it expands our understanding of the influences acting on the variable of primary interest (such as blood pressure).

An analysis of variance that isolates and measures variation resulting from some source described on a continuous scale, as well as variation resulting from differences between various groups or categories, is known as an **analysis of covariance.** In effect it combines regression analysis and conventional analysis of variance into a single analysis.

The explanatory variable used (the **covariate,** also known as the control variable or concomitant variable) can be any continuous variable that we believe might influence the outcome variable and thus result in variation that might obscure between-treatment differences or group differences. Age is a frequently used covariate. Duration of illness might be a covariate in a clinical trial, or initial weight could well serve as a covariate in a study of dietary regimes in which weight loss was the outcome of interest. (We would, for example, instinctively be happier comparing groups of individuals who are all the same age or all the same weight. This is rarely, if ever, practical, and these differences in age or weight might well be reflected in patient behavior. Eliminating this extraneous variation will improve our capacity to detect between-group differences.)

CONTROLLING CONTINUOUS VARIATION

Let us start by considering a single set of results in which we have measurements for systolic blood pressure and age (Figure 12-1). To understand and describe the nature and strength of the relationship between these variables we carry out a regression analysis, relating blood pressure to age. (For convenience, for the rest of this chapter we will refer to systolic blood pressure simply as blood pressure or BP).

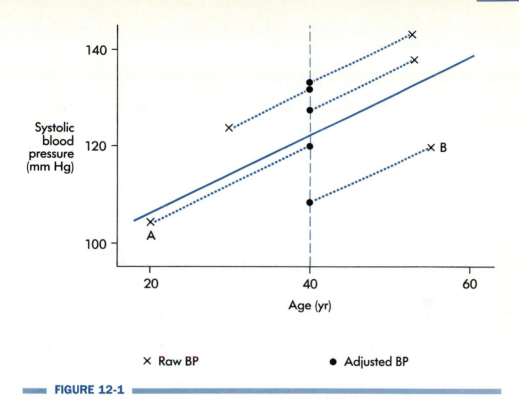

FIGURE 12-1

The regression line in Figure 12-1 describes the nature of the blood pressure/age relationship. Clearly, blood pressure increases with age. The fact that the individual points do not all lie perfectly on the regression line shows that not all the variation in blood pressure can be explained by variation in age. The amount by which the individual points lie above or below the regression line shows the extent of the variation in blood pressure that is not age-related. The variation in blood pressure remaining after the variation explained by age has been removed,
that is,

Total sum of squares for BP

− Sums of squares for BP explained by age

or

$$SSY - SCP^2/SSX$$

is simply the sum of squares of these deviations from the regression line.

To get a feel for how covariate-related variation can seriously distort our perceptions of a group of individuals (and therefore how important it is to control and eliminate it) let's consider two of the individuals depicted in Figure 12-1. Individual A, for instance, has a very low blood pressure (relative to the mean BP for the study). However, Individual A is also very young (again, relative to the mean age for the study). When we take account of the age/blood pressure relationship, as summarized by the regression line, it becomes clear that this very low blood pressure is almost totally attributable to A's very young age.

Suppose that, for this group of individuals,

$$\text{Mean BP} = 122 \qquad \text{Mean age} = 40$$

and that the regression equation that describes the age/blood pressure relationship is

$$\text{BP} = 90 + .8 \text{ age}$$

and that individual A has an age of 20 and an actual blood pressure of 104.

The age/blood pressure relationship predicts a blood pressure of

$$90 + (.8 \times 20) = 106$$

for an individual of this age. The actual blood pressure for A is 104, giving this individual a deviation from the age-related pattern (and, therefore, a deviation unexplained by age) of -2. In other words, individual A is, in fact, a very typical 20-year-old with a blood pressure only slightly below what we would expect for a person of this age. Once A's age has been taken into account we realize that A does not have a dramatically low blood pressure at all.

Individual B, on the other hand, has a markedly higher blood pressure (120) than individual A. However, this is largely explained by the fact that B is also considerably older (he is 55) than A. In fact, the age/blood pressure relationship predicts a blood pressure of

$$90 + (.8 \times 55) = 134$$

for someone of age 55. The actual observed value for B lies 14 below this. This means that if the age effect and the age-related variation were removed, B would be considered to have a much greater tendency toward below-average blood pressure than A. Individual B, in fact, gives a much more impressive performance than individual A. If the purpose of the study were to evaluate the effect of lifestyles on blood pressure, ignoring the effect of the covariate (age) could seriously distort our understanding of the evidence. We would naively laud A as having a lifestyle superior to B, when the truth is the exact reverse. When we compare the blood pressure status of A and B, the ability to eliminate an age difference and its effects will both decrease the variability of the data and improve our ability to interpret it.

Removing age-related variation from a set of data is equivalent to making everyone the same age. We can, in fact, do this very easily by converting each individual result to its equivalent result at some reference age. The most obvious and sensible reference age is the sample mean age. In our hypothetical example the mean age is 40, and this corresponds to the mean blood pressure of 122.

$$122 = 90 + (.8 \times 40)$$

(Recall that a regression line must pass through the sample means.) Individual A has a blood pressure only 2 units below the anticipated result for his age. If he were 40 years old, therefore, the equivalent result would be 120 (122 − 2). Similarly, individual B's deviation of −14 from his age standard is equivalent to a blood pressure of 108 at age 40. Moving individuals A and B to a common age reinforces the point that B's blood pressure is truly more impressive than A's. The patterns of variation that are not age-related are faithfully retained, other relationships are clarified, and the amount of variation involved is reduced. Take another look at Figure 12-1 and see how making everyone the same age (that is, eliminating age-related variation) has greatly reduced the blood pressure variability, while giving a much fairer reflection of relative performance.

COVARIATES AND GROUPS

So far we have considered the measurement and control of variation resulting from a continuous source in a single set of results (a classic regression situation). In the analysis of variance context, however, the data will form two or more groups. How will this internal structure affect the way we can control such variation? In other words, how can we convert a conventional analysis of variance into an analysis of covariance? (For discussion purposes, let's stick to systolic blood pressure as the characteristic we wish to compare between groups, and age as the continuous variable that we suspect is related to blood pressure).

One possibility would be to simply ignore the group structure and, just as before, use the overall age/blood pressure relationship to eliminate age-related variation. As we have seen, this is equivalent to standardizing all the subjects to a common age (Figure 12-2). The regression line and the amount of variation explained are calculated from the

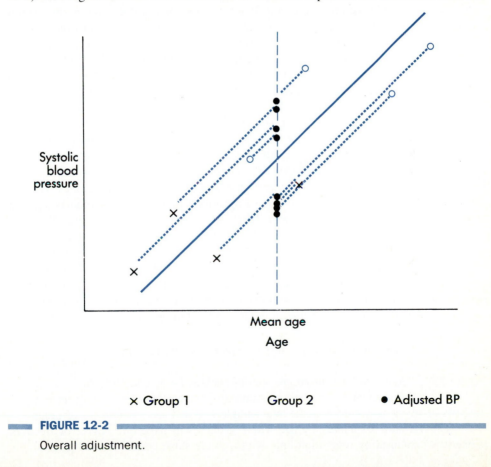

× Group 1   Group 2   ● Adjusted BP

FIGURE 12-2

Overall adjustment.

overall sum of age/blood pressure cross-products (SCP) and the overall age sum of squares (SSX). The overall variability in blood pressure after variation resulting from age has been removed is then simply the overall sum of squares for blood pressure (SSY) less this calculated amount.
That is,

$$\text{Overall variation in } y \text{ after removal of age effects} = SSY - \frac{SCP^2}{SSX}$$

SSY, the original total sum of squares for blood pressure, will have $N - 1$ degrees of freedom associated with it (where N is the total sample size).

The sum of squares explained by age will have one degree of freedom associated with it, effectively corresponding to the slope of the age/blood pressure regression line.

The blood pressure variation that is left after removal of the age-related variation (the adjusted overall variation) will, therefore, have $N - 2$ degrees of freedom associated with it.

While this approach certainly reduces the variability in blood pressure, it takes no account at all of the existence of different patient groups. In addition, a single overall regression line will not necessarily fit the age/blood pressure relationship in the various portions of the data set very well. The most logical way to do this is to fit an age/blood pressure relationship within each group. If we can assume that the same basic relationship holds within each group (that is, that the regression slope is constant from group to group), then the relationships can be described efficiently and very conveniently by fitting parallel regression lines inside each group, as depicted in Figure 12-3. (This simplifying assumption parallels the assumption we make in analysis of variance that the groups behave in the same way, that is, have comparable variances, thus permitting us to utilize pooled within-groups variation.)

Fitting regression lines (with a common slope) within each group of results enables us to eliminate age-related blood pressure variation from the within-groups variation (again, by effectively converting all subjects to a common age). The within-groups blood pressure variation is, of course, the error variation in a conventional one-way analysis of variance. The amount of variation explained by age, within groups, is calculated exactly as before (that is, the sum of age/blood pressure cross-products squared, divided by the sum of squares of age). However, instead of using the overall sum of cross-products and sum of squares of age, we use the corresponding within-groups results. (Calculation of these is considered in detail in the next section. However, we can say that, just as analysis of variance is used to break the total variation in blood pressure into within-groups and between-groups components, so the same basic approach can be used to break up variation in age, and even variation in the age/blood pressure cross-products, into their respective between-groups and within-groups components.)

$$\text{Within-groups variation in } y \text{ after removal of age effects} = SSY_w - \frac{SCP_w^2}{SSX_w}$$

where SSY_w denotes the within-groups sum of squares for y, and so forth.

SSY_w, the original or unadjusted within-groups sum of squares, will have $N - k$ degrees of freedom, where N is the total sample size and k is the number of groups.

The variation explained by age within groups will have one degree of freedom, since it is explained by regression lines with a single slope common to all groups.

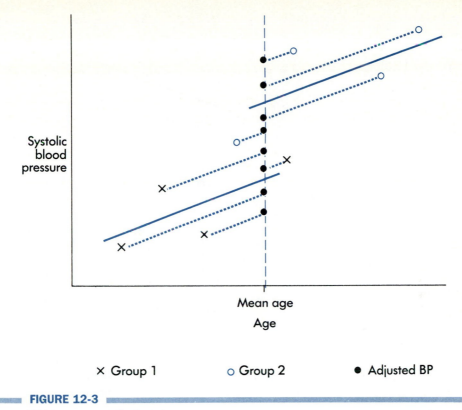

Systolic blood pressure

Mean age

Age

× Group 1   o Group 2   ● Adjusted BP

FIGURE 12-3

Within-groups adjustment.

The within-groups blood pressure variation left after removal of age-related variation (the adjusted within-groups variation) will, therefore, have

$$N - k - 1 \text{ degrees of freedom}$$

Removing age-related variation will considerably reduce the within-groups (error) variation. (Take a look back at Figure 12-3 and see how adjusting every individual in the study to the same age, such as the overall mean age, has substantially reduced the range of blood pressures within each of the groups.) This capacity to substantially reduce the unexplained or error variation, and thus increase the power of an analysis, is one of the principal reasons for using the analysis of covariance when an explanatory variable or covariate is present.

So far we have talked about the effect removing covariate-related variation has on the total variation in the outcome variable and on the within-groups variation. Knowing what we know about analysis of variance, we might well suspect that it also will affect the between-groups variation.

In both the overall and within-groups situations, removing the influence of age is, as we have seen, equivalent to placing all the subjects at a common age. This is exactly the way it operates in the between-groups (that is, between-means) situation as well (Figure 12-4). The actual means of the groups are effectively replaced by the group

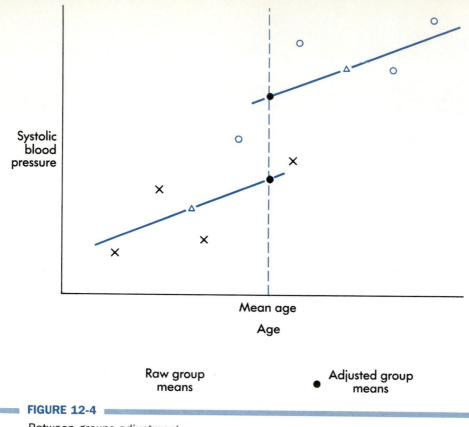

Systolic blood pressure

Mean age

Age

Raw group means

Adjusted group means

FIGURE 12-4

Between-groups adjustment.

means we would have seen if everybody in the study had a common age (say, the over-all mean age for the total data set). These are known as the **adjusted means.**

Suppose, for instance, that the common (pooled) slope is .8.

(This would be calculated just as for any regression line, except that it is based on the within (pooled) values

$$SCP_w/SSX_w$$

If you need to refresh your memory on calculating the slope of a regression line, reread Chapter 7, the section on Finding the Best Description).

Let's also suppose that the two groups involved have respective means:

Group 1	Mean age	35
	Mean blood pressure	118

(that is, the regression line for group 1 must be

$$BP = 90 + .8\ age$$

since the line must pass through the group mean, and the intercept is calculated accordingly).

Group 2	Mean age	45
	Mean blood pressure	136

which means that the regression line for this group must be

$$BP = 100 + .8 \text{ age}$$

If the overall mean age is 40, then the respective adjusted means (remember that the mean must always lie on the regression line) would be

Group 1 $90 + (.8 \times 40) = 122$

Group 2 $100 + (.8 \times 40) = 132$

In this instance, the removal of age-related variation has the following effect:

	TRUE MEAN BP	**ADJUSTED MEAN BP**
Group 1	118 (at mean age 35)	122 (at mean age 40)
Group 2	136 (at mean age 45)	132 (at mean age 40)

It decreases the difference between the groups and would, therefore, reduce the between-groups variation. This is very fair, since the difference between the groups was being exaggerated by the fact that the group with the higher blood pressure also happened to consist of generally older individuals. After we eliminate this purely age effect, the two groups can be compared on a fair and unbiased basis. Equally, if age differences had been masking inherent group blood pressure differences, removal of age-related variation would increase the difference between the groups (that is, restore the true inherent difference) and increase the between-groups variation. This ability to make groups truly similar (in terms of the covariate) and thus make comparisons of them fair and unbiased is the other major motivation for using the analysis of covariance.

The analysis of covariance will always decrease the total and within-groups sums of squares. However, as we see in Figure 12-5, it can increase or decrease the between-groups sums of squares. This makes calculating the adjusted between-groups sums of squares a more tricky proposition than calculating either the adjusted overall or within-groups sums of squares. In fact, there is absolutely no need to carry out a separate calculation for it. We have already seen, in principle, how to calculate

$$\text{Adjusted overall } SSY = SSY - \frac{SCP^2}{SSX}, \text{ with } N - 2 \text{ df}$$

$$\text{Adjusted within-groups } SSY = SSY_w - \frac{SCP_w^{\,2}}{SSX_w}, \text{ with } N - k - 1 \text{ df}$$

The adjusted between-groups sums of squares for *y* are simply given by

$$\text{Adjusted overall } SSY - \text{adjusted within-groups } SSY$$

$$\text{with degrees of freedom} = N - 2 - (N - k - 1)$$

$$= k - 1$$

The between-groups degrees of freedom are, therefore, unaltered by the adjustment. This is reasonable since the number of means being compared (*k*) is not affected.

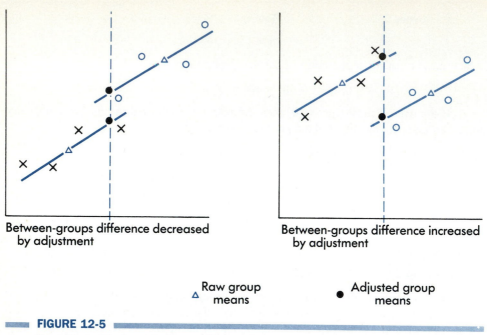

Between-groups difference decreased by adjustment

Between-groups difference increased by adjustment

△ Raw group means

● Adjusted group means

FIGURE 12-5

The effect of between-groups adjustment.

Once the adjusted between-groups and within-groups sums of squares have been obtained, they are simply compared using the standard analysis of variance approach.

The analysis of covariance is, therefore, nothing more or less than a conventional analysis of variance in which the influence of a continuous source of variation (the covariate) has been removed by effectively adjusting all the subjects to a common covariate value. This adjustment both reduces the unexplained variation and eliminates possible bias in the between-groups comparisons.

Conventional analysis of variance − variation related to covariate = Analysis of covariance (adjusted analysis of variance)

SOURCE	df		SOURCE	df		SOURCE	df
Between	$k - 1$					Between	$k - 1$
Within	$N - k$	−	Within	1	=	Within	$N - k - 1$
Total	$N - 1$	−	Total	1	=	Total	$N - 2$

CARRYING OUT AN ANALYSIS OF COVARIANCE

Although it has advantages, the analysis of covariance requires more calculation steps than just about any other commonly used statistical method. Since it is an amalgamation of analysis of variance and regression analysis, its complexity hardly is surprising. However, there is no need for panic. The calculations involved should be very familiar by now, and approaching the analysis in a systematic, organized way will defuse any potential problems.

We will, as before, denote the outcome variable (such as blood pressure) by y and the control variable or covariate (such as age) by x.

To implement an analysis of covariance, we first need to obtain the following totals.

Overall data set:

Σx	Σy	Σxy
Σx^2	Σy^2	N

where N is the total sample size.

Group 1

Σx_1	Σy_1	$\Sigma x_1 y_1$
Σx_1^2	Σy_1^2	n_1

where Σx_1 denotes the total for the x values in group 1, and so forth, and n_1 is the number of individuals in group 1,

and similarly for groups 2 through k, where k is the number of groups involved.

Once these basic totals are obtained, the next step is to calculate the overall sums of squares for x (that is, SSX) and for y (that is, SSY), and the sum of cross-products of x and y (that is, SCP), and then split these into between-groups and within-groups components. This is done, as you might expect, by using the analysis of variance approach.

Outcome variable, y

Total variation,

$$SSY = \Sigma y^2 - (\Sigma y)^2 / N$$

Between-groups variation,

$$SSY_b = \frac{(\Sigma y_1)^2}{n_1} + \ldots + \frac{(\Sigma y_k)^2}{n_k} - \frac{(\Sigma y)^2}{N}$$

Within-groups variation,

$$SSY_w = SSY - SSY_b$$

SSY_b and SSY_w form the basis of the original unadjusted analysis of variance used to test for possible differences between groups in the outcome variable:

SOURCE	SS	df
Between groups	SSY_b	$k - 1$
Within groups	SSY_w	$N - k$
Total	SSY	$N - 1$

The total (SSX), between-groups (SSX_b) and within-groups (SSX_w) sums of squares for the covariate, x, are calculated in exactly the same way, using the corresponding x totals.

Calculation of the sum of cross-products or covariation has clear parallels with the calculation of a sum of squares. Similarly, its partition into between-groups and within-groups components closely parallels the analysis of variance approach used to split up a sum of squares.

Total covariation,

$$SCP = \Sigma xy - (\Sigma x)(\Sigma y)/N$$

Between-groups covariation,

$$SCP_b = \frac{(\Sigma x_1)(\Sigma y_1)}{n_1} + \ldots + \frac{(\Sigma x_k)(\Sigma y_k)}{n_k} - \frac{(\Sigma x)(\Sigma y)}{N}$$

Within-groups covariation,

$$SCP_w = SCP - SCP_b$$

Once the total and within-groups variation in x and y, and their covariation have been obtained, the total and within-groups covariate-related variation can be calculated in the manner outlined in the discussion of Covariates and Groups and removed from the appropriate outcome variation. (Remember that the required between-groups adjustment is not calculated directly.)

Adjusted overall variation in y,

$$\text{Adjusted } SSY = SSY - \frac{SCP^2}{SSX}$$

Adjusted within-groups variation in y,

$$\text{Adjusted } SSY_w = SSY_w - \frac{SCP_w^2}{SSX_w}$$

The final step in the calculation sequence, the adjusted between-groups sum of squares, is obtained by subtracting the two values above. The adjusted between-groups and within-groups values are then compared in a classic variance ratio test for group differences in the outcome variable, y.

SOURCE	SS	df	MS	F
Between groups	adjusted SSY_b	$k - 1$		
Within groups	adjusted SSY_w	$N - k - 1$		
Total	adjusted SSY	$N - 2$		

Let's put these ideas into practice and carry out an analysis of covariance to test for evidence of differences in systolic blood pressure (the outcome, y) between two hypertension treatments, while adjusting for the patients' ages (the covariate, x). The results obtained for the two eight-patient treatment groups are given below.

TREATMENT 1

BP	120	114	132	130	146	122	136	118
Age	26	37	31	48	55	35	40	29

TREATMENT 2

BP	109	145	131	129	101	115	133	105
Age	33	62	54	44	31	39	60	38

To save you some rather tedious calculation, the totals we will need for the various groupings are given below.

GROUP 1

$\Sigma x_1 = 301$	$\Sigma x_1^2 = 12,001$	$\Sigma x_1 y_1 = 38,832$
$\Sigma y_1 = 1018$	$\Sigma y_1^2 = 130,340$	$n_1 = 8$

GROUP 2

$\Sigma x_2 = 361$	$\Sigma x_2^2 = 17,311$	$\Sigma x_2 y_2 = 44,923$
$\Sigma y_2 = 968$	$\Sigma y_2^2 = 118,848$	$n_2 = 8$

OVERALL

$\Sigma x = 662$	$\Sigma x^2 = 29,312$	$\Sigma xy = 83,755$
$\Sigma y = 1986$	$\Sigma y^2 = 249,188$	$N = 16$

Since blood pressure is the focus of the study, let's start by obtaining the total variation in systolic blood pressure present in the study and splitting it into its between-treatments and within-treatments parts.

$$\text{Total SS } (SSY) = 249,188 - (1986)^2/16$$

$$= 2675.75$$

$$\text{Between groups SS } (SSY_b) = \frac{1018^2}{8} + \frac{968^2}{8} - \frac{1986^2}{16}$$

$$= 156.25$$

$$\text{Within-groups SS } (SSY_w) = SSY - SSY_b$$

$$= 2675.75 - 156.25$$

$$= 2519.50$$

We could now, if we wanted, carry out a conventional analysis of variance that would compare the mean systolic blood pressures in the two treatment groups (and that, of course, would take absolutely no account of the patients' ages).

SOURCE	SS	df	MS	F
Treatments	156.25	1	156.25	.868
Error	2519.50	14	179.96	
Total	2675.75	15		

Based on this we would have to conclude that, if the variation attributable to age is left uncontrolled, there is no evidence of any significant difference in systolic blood pressure between the two treatments.

Taking the age-related variation into account is, of course, what the analysis of covariance is all about. To do this, we first need to split the variation in age into its between-treatment and within-treatment components.

$$\text{Total SS } (SSX) = 29{,}312 - (662)^2/16$$

$$= 1921.75$$

$$\text{Between-groups SS } (SSX_b) = \frac{301^2}{8} + \frac{361^2}{8} - \frac{662^2}{16}$$

$$= 225.00$$

$$\text{Within-groups SS } (SSX_w) = SSX - SSX_b$$

$$= 1921.75 - 225.00$$

$$= 1696.75$$

Finally, we need to split the blood pressure/age covariation into its between-treatments and within-treatments components. (Do remember that, unlike a sum of squares, a sum of cross-products can be negative.)

$$\text{Total } SCP \text{ } (SCP) = 83{,}755 - \frac{662 \times 1986}{16}$$

$$= 1584.25$$

$$\text{Between-groups } SCP \text{ } (SCP_b) = \frac{301 \times 1018}{8} + \frac{361 \times 968}{8} - \frac{662 \times 1986}{16}$$

$$= -187.50$$

$$\text{Within-groups } SCP \text{ } (SCP_w) = SCP - SCP_b$$

$$= 1584.25 - (-187.50)$$

$$= 1771.75$$

With these facts at our fingertips, we are finally in a position to calculate and eliminate the variation in blood pressure that is simply a reflection of the differing ages of the study patients.

Let's do this for the overall blood pressure variation first, using the total SCP and SSX values.

$$\text{Variation in } y \text{ due to } x = \frac{1584.25^2}{1921.75}$$

$$= 1306.02 \text{ (with 1 df)}$$

When this explained variation is removed, the total blood pressure variation (and df) is reduced from

SS	df
2675.75	15
to 1369.73	14

Now let's make the adjustment for the within-treatments blood pressure variation, using the within-treatments *SCP* and *SSX*.

$$\text{Variation explained by age} = \frac{1771.75^2}{1696.75}$$

$$= 1850.07 \text{ (with 1 df)}$$

When this explained variation is removed, the within-treatments or within-groups (error) blood pressure variation (and df) is reduced from

SS	df
2519.50	14
to 669.43	13

After the influence of age has been eliminated from the overall and within-treatments blood pressure variation, we have

	SS	df
Adjusted total variation in BP	1369.73	14
Adjusted error variation in BP	669.43	13

The adjusted between-treatments variation must therefore be

	700.30	1

With all the adjustments made and age-related variation completely eliminated, we can now put together the final analysis of covariance table.

SOURCE	SS	df	MS	F
Between groups	700.30	1	700.30	13.60
Error	669.43	13	51.49	
Total	1369.73	14		

Our age-adjusted analysis indicates that a very highly significant difference in blood pressure exists between the two treatments. Note, by comparing this table with the unadjusted analysis of variance, how eliminating age-related variation has greatly reduced the unexplained variation. Note also how the adjusted between-groups variation is much higher than the corresponding unadjusted variation, reflecting the fact that the treatment differences were being masked by the fact that the group 1 mean age (37.625) was substantially lower than the group 2 mean age (45.125).

To appreciate exactly why the adjusted between-treatments variation is higher, let's finish by calculating the adjusted mean systolic blood pressures for the two treatment groups.

The common pooled slope is given by:

$$b = \frac{SCP_w}{SSX_w} = \frac{1771.75}{1696.75} = 1.044$$

For Group 1:

$$\bar{x}_1 = 37.625 \text{ years} \qquad \bar{y}_1 = 127.25 \text{ mm Hg}$$

and the regression relationship for group 1 is

$$y = 87.97 + 1.044\,x$$

(using $\bar{x}_1$ and $\bar{y}_1$ to calculate the intercept, a).

The mean age for the total data set is 41.375 years.

The group 1 mean blood pressure, adjusted to this overall mean age, would, therefore, still lie on this regression line but at the overall mean age:

$$87.97 + (1.044 \times 41.375)$$

$$= 131.17 \text{ mm Hg}$$

For Group 2

$$\bar{x}_2 = 45.125 \text{ years} \qquad \bar{y}_2 = 121.00 \text{ mm Hg}$$

and the regression relationship for group 2 is, therefore:

$$y = 73.89 + 1.044\,x$$

The group 2 mean blood pressure adjusted to the overall mean age (41.375) is, therefore:

$$73.78 + (1.044 \times 41.375)$$

$$= 117.09 \text{ mm Hg}$$

We can summarize the change in mean behavior that has resulted from adjusting everyone in the study to the same age, as follows.

	UNADJUSTED MEANS		ADJUSTED MEANS	
	BP	Age	BP	Age
Group 1	127.25	37.625	131.17	41.375
Group 2	121.00	45.125	117.09	41.375
Difference	6.25		14.08	

Note how the extent of the true treatment difference is masked in the unadjusted situation. The below-average age in group 1 has depressed the group 1 mean blood pressure, while the mean blood pressure in group 2 is elevated by the group's above-average mean age, obscuring the substantial differences that do exist between the two treatments.

With only two groups being compared, the conclusion that they differ from one another doesn't leave any unanswered questions. A significant result from an analysis of covariance involving three or more groups, however, begs the question, "Exactly which of the groups are different?" This is exactly the same problem we encountered when we first met the analysis of variance, and it is solved, as before, by conducting a follow-up analysis with a multiple comparison test such as Tukey's test (see Chapter 5). Since an analysis of covariance is simply an analysis of variance that has been adjusted to allow for the influence of a covariate, the means being compared in such a test are the adjusted means, and the error variance (and associated df) involved is the adjusted error variance.

COMPARING COVARIATE RELATIONSHIPS

The analysis of covariance, as we have met it to date, depends on the assumption that the same outcome/covariate relationship holds within each of the groups (Figure 12-6, *A*). This is equivalent to assuming that the slopes of the individual regression lines we could fit in each group are all exactly the same. If this is true, then it makes good sense simply to calculate one common slope for all the groups, using the pooled within-groups sums of squares and cross-products. This considerably simplifies the situation (particularly the comparison of the group means, as we shall see in the next section) and is the standard approach to the analysis of covariance.

However, this need not always be the case, (Figure 12-6, *B*) and it would be reassuring to be able to test whether the idea of using a single common slope is reasonable. The basic analysis of covariance does not include a test for whether the slopes actually do differ from one another. In fact, it does not even include a test of whether any real relationship exists at all between the outcome variable and the covariate (that is, a test of the null hypothesis that the true slope is actually zero). Both these questions are usually of considerable interest, and it is easy to develop a test for both questions that can be used in conjunction with the basic analysis of covariance.

The outcome/covariate (such as blood pressure/age) relationships are concerned solely with describing the variability that exists within the various data groups. This might be one common relationship (that is, a single common slope) or different relationships (that is, distinct slopes) in each group. Either way, describing the relationship(s) is equivalent to trying to explain at least a part of the within-groups variation in the outcome variable (that is, the within-groups sums of squares, SSY_w, with $N - k$ df).

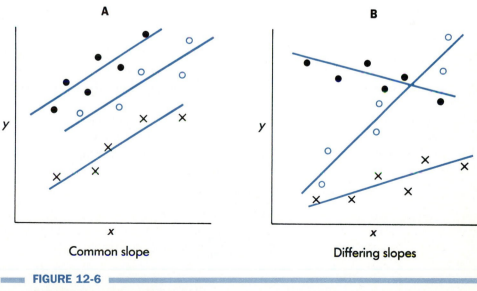

A

y

x

Common slope

B

y

x

Differing slopes

FIGURE 12-6

Outcome/covariate relationships.

Let us initially stick to our assumption of a common relationship or slope. We have seen in the previous section that the within-groups variation in y explained by a common relationship with x is given by:

$$\frac{SCP_w^2}{SSX_w} \text{ with 1 df (resulting from the single common slope)}$$

This is the amount by which we adjusted the within-groups sum of squares in the analysis of covariance (such as the amount of within-groups variation in blood pressure that is actually attributable to variation in age). The adjustment is based on pooled within-group values, which is very reasonable as long as a common relationship exists within all the groups.

Now suppose that we believe (or fear) that different relationships exist within each group. Since differing outcome/covariate relationships may exist inside each study group, the obvious approach is to calculate separate regression lines for each groups (that is, analyze each group separately).

For example, the regression slope that would be appropriate for group 1, and the variation in outcome that is explained by a relationship with the covariate within that particular group, are given by

$$\text{Slope} = \frac{SCP \text{ for group 1}}{SSX \text{ for group 1}}$$

In other words, a regression line defined by

$$b_1 = \frac{\Sigma x_1 y_1 - (\Sigma x_1)(\Sigma y_1)/n_1}{\Sigma x_1^2 - (\Sigma x_1)^2/n_1}$$

$$a_1 = \bar{y}_1 - b_1 \bar{x}_1$$

where $\bar{x}_1$, $\bar{y}_1$ are the respective means for group 1.

In group 1 the variation in y explained by variation in x

$$= \frac{(SCP \text{ for group 1})^2}{SSX \text{ for group 1}}$$

This variation has 1 df associated with it, corresponding to the regression slope for group 1.

Similar calculations would be carried out for each of the other groups.

The variation in y explained by variation in x over all k groups is simply given by the sum of the amounts explained by each of the k individual relationships. That is,

$$\frac{(SCP \text{ for group 1})^2}{SSX \text{ for group 1}}$$

$$+ \frac{(SCP \text{ for group 2})^2}{SSX \text{ for group 2}}$$

$$\cdots$$

$$\cdots$$

$$\cdots$$

$$+ \frac{(SCP \text{ for group } k)^2}{SSX \text{ for group } k}$$

with k df (one for each of the k individual slopes).

This very flexible and totally unrestricted approach enables us to explain as much of the variation in y as we possibly can. The question is, "Does it enable us to explain substantially more of the outcome variation?" Since it is more flexible than the "one common slope" approach, it must explain more of the outcome variation. Is this additional explanatory power large enough, however, to take seriously?

The additional variation explained by using k different slopes, rather than one common slope, is given by the difference between these two situations:

$$\frac{SCP_1^2}{SSX_1} + \frac{SCP_2^2}{SSX_2} + \ldots + \frac{SCP_k^2}{SSX_k} - \frac{SCP_w^2}{SSW_w}$$

with $k - 1$ degrees of freedom,
where SCP_1 is the sum of xy cross-products for group 1 only, SSX_1 is the sum of squares of x for group 1 only, and so forth, and SCP_w, SSX_w are the pooled within-groups sums of cross-products and squares, as defined in the previous section on Carrying Out an Analysis of Covariance.

This additional variation represents the additional descriptive power we have gained by allowing the individual slopes to deviate from a single common slope, and this additional variation is usually described as "variation resulting from deviations from common slope." A test of this variation will show if there is any benefit in using individual slopes and if there is any evidence of genuine deviations from a single common slope.

The approach here is very similar in concept to the measurement of interaction effects in analysis of variance (see Chapter 11). Calculating the interaction variation in that situation involves calculating the variation resulting from various possible combinations (analogous to the individual slopes) and then removing the variation resulting from the main effects (analogous to the common slope). (We do not have to eliminate between-groups variation, the other main effect, since we are working solely with within-groups variation.)

The "deviations" degrees of freedom can also be worked out in a way exactly analogous to the analysis of variance interaction approach.

"Deviations from common slope" df = Common slope df

$\times$ between-groups df

$= 1 \times (k - 1)$

$= k - 1$

In a real sense the "deviations from common slope" term does correspond to a groups-by-slope interaction, since, if it exists, it means that the size of the slope depends on the group involved. As in any interaction situation, the two influences or effects must be considered simultaneously if they are to make sense.

The within-groups variation in y, the outcome variation, can, therefore, be split into three portions.

1. Variation resulting from a common slope

$$\frac{SCP_w^{\,2}}{SSX_w} \text{ with 1 df}$$

2. Variation resulting from deviations from a common slope

$$\frac{SCP_1^{\,2}}{SSX_1} + \ldots + \frac{SCP_k^{\,2}}{SSX_k} - \frac{SCP_w^{\,2}}{SSX_w} \text{ with } k - 1 \text{ df}$$

and

3. The balance of the within-groups variation, which has no explanation, and that we, therefore, view as residual or error variation

$$SSY_w - \left(\frac{SCP_1^{\,2}}{SSX_1} + \ldots + \frac{SCP_k^{\,2}}{SSX_k}\right) \text{ with } N - k - (k - 1) - 1$$

$$= N - 2k \text{ df}$$

This splitting of the within-groups variation into "common slope," "deviations from common slope," and unexplained portions now enables us to test the various regression questions we posed earlier ("Is analysis of covariance worthwhile?" "Is there any evidence that the group slopes differ from one another?") using the standard analysis of variance approach.

SOURCE	SS	df	MS	F
Common slope		1		
Deviations from common slope		$k - 1$		
Residual		$N - 2k$		
Within groups		$N - k$		

Having already carried out an analysis of covariance, it might be a wise (if somewhat belated) idea to test whether our use of a common blood pressure/age slope of 1.044 mm Hg/yr in both groups is justified, and to establish whether any real connection between blood pressure and age exists at all. (If it doesn't, then carrying out an analysis of covariance using age as the covariate is rather a waste of time.)

From our analysis of the two group blood pressure/age data presented in the section on Carrying out an Analysis of Covariance we know that the unadjusted within-groups variation in blood pressure was:

2519.50 with 14 df

We also know that the within-groups variation explained by age (assuming a common slope) was:

1850.07 with 1 df

Using the totals given, we can calculate that the variation in blood pressure explained by age within group 1 alone (allowing group 1 to have its own unique slope) is:

$$\frac{SCP_1^2}{SSX_1} = \frac{[\Sigma x_1 y_1 - (\Sigma x_1)(\Sigma y_1)/n_1]^2}{\Sigma x_1^2 - (\Sigma x_1)^2/n_1}$$

$$= \frac{[38,832 - (301 \times 1018)/8]^2}{12,001 - 301^2/8}$$

$$= 415.217 \text{ with } 1 \text{ df}$$

Similarly the variation explained within group 2 is:

$$\frac{SCP_2^2}{SSX_2} = \frac{[44,923 - (361 \times 968)/8]^2}{17,311 - 361^2/8}$$

$$= 1511.021 \text{ with } 1 \text{ df}$$

The variation in blood pressure explained by age when we allow for different slopes is, therefore:

$$(415.217 + 1511.021) \text{ with } (1 + 1) \text{ df}$$

$$= 1926.238 \text{ with } 2 \text{ df}$$

The variation explained by possible deviations from a common slope is therefore

Variation resulting from different slopes − Variation resulting from a common slope

$$= 1926.238 - 1850.07$$

$$= 76.168 \text{ with } (2 - 1) = 1 \text{ df}$$

The analysis of variance table for testing the various hypotheses about the blood pressure/age relationship is, therefore:

SOURCE	SS	df	MS	F
Common slope	1850.070	1	1850.07	37.42
Deviations from common slope	76.168	1	76.17	1.54
Unexplained	593.262	12	49.44	
Within groups	2519.500	14		

A very significant (F = 37.42 with 1, 12 df, p < .001) relationship is, therefore, confirmed between age and blood pressure. In other words, using age as a covariate has explained a substantial portion of the within-groups variation, and our decision to employ age as a covariate is fully vindicated. Reassuringly, there is no evidence of any differences between the group slopes, and our use of a common pooled slope in the analysis of covariance appears totally reasonable.

ANALYSIS OF COVARIANCE WITH MULTIPLE SLOPES

If the analysis of the previous section confirms that the assumption of a single common slope is reasonable, then we can use the ideas we have developed in this chapter with a clear conscience. What do we do, however, if we discover that the outcome/covariate slope really does differ from group to group?

To appreciate just how helpful the assumption of a common slope is, and just how much more complicated life becomes when we have to use different slopes in each group, take a look at Figure 12-7.

When the regression lines are parallel (Figure 12-7, *A*), the difference between the groups is constant, and the covariate value to which we decide to adjust the subjects is quite immaterial. The overall mean covariate value is usually used (it seems a reasonable value to use), but adjusting everyone to, say, age 20 or age 60, would result in exactly the same between-groups difference. In other words, we can unequivocally ask the straightforward question "Do the groups differ after we allow for the influence of the covariate?"

The situation is much more complicated when the slopes differ (Figure 12-7, *B*). Here the difference between the groups changes as the value of the covariate changes, and the universal question, "Do the groups differ after we adjust to a common covariate value?" has no meaning. To ask a question of this type, we would have to state the specific covariate value we are talking about, such as "Would the treatments differ if all the subjects had been age 60?"

Such a question really involves comparing the difference between two regression lines at a specified value of x. As we saw in Chapter 7, a regression line

$$y = a + bx$$

or

$$y = \bar{y} + b(x - \bar{x}) \quad \text{(since } a = \bar{y} - b\bar{x})$$

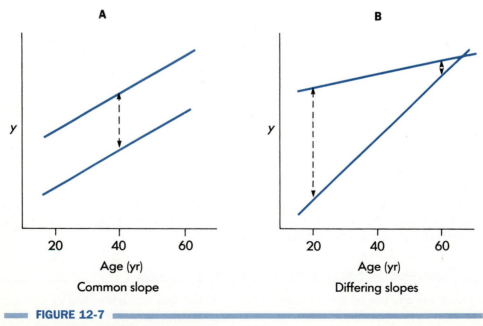

FIGURE 12-7

The problem of differing slopes.

cannot be taken totally seriously because the parameters that define it, $\bar{y}$ and b, are simply estimates and would change somewhat if a fresh sample were studied.

The variance of y_1, the point on the group 1 regression line corresponding to a particular covariate value, x, is given by

$$\text{Variance } (y_1) = s_e^2 \left(\frac{1}{n_1} + \frac{(x - \bar{x}_1)^2}{SSX_1} \right)$$

where n_1 is the number of individuals in group 1

$\bar{x}_1$ is the covariate mean for group 1

SSX_1 is the covariate sum of squares for group 1

and s_e^2 is the unexplained or error variance used to test for deviations from a common slope in the previous section.

(If necessary, reread the Regression and Prediction section of Chapter 7 to reacquaint yourself with the variability of a regression line.)

The difference between two regression lines at a particular covariate value, x, is

$$d = \bar{y}_1 - b_1(x - \bar{x}_1) - [\bar{y}_2 - b_2(x - \bar{x}_2)]$$

and the variance of the difference is given by

$$\text{Variance } (d) = s_e^2 \left(\frac{1}{n_1} + \frac{1}{n_2} + \frac{(x - \bar{x}_1)^2}{SSX_1} + \frac{(x - \bar{x}_2)^2}{SSX_2} \right)$$

The question of whether the lines really do differ at this point (that is, would the groups differ if the subjects were all adjusted to covariate value x?) can now be answered by a simple t test.

$$\frac{d}{\sqrt{\text{variance } (d)}} = t \text{ with } N - 2k \text{ df}$$

If there are more than two groups in the study, then the problem of multiple comparisons arises. The solution, as always, is to strengthen the critical value used for testing (with 3 or 4 groups, with 3 or 6 possible comparisons respectively, using a .01 critical value will ensure that the overall risk of a type I error is close to .05). In addition, it is important to have a clear reason for adjusting the groups to a specific covariate value. Simply picking a value of x because the groups seem to differ there is not ethical hypothesis testing!

This more complex version of the analysis of covariance should, of course, only be used if significant deviations between the group slopes actually exist. If only the common slope is significant, then the basic analysis of covariance outlined earlier in this chapter is quite appropriate. If neither the common slope nor the deviations from the common slope are significant, then there is no evidence of any relationship between the outcome variable and the covariate. In this situation, there is no merit in employing an analysis of covariance at all, and a simple unadjusted analysis of variance is perfectly adequate.

— ANALYSIS OF COVARIANCE AND COMPLEX EXPERIMENTAL DESIGNS —

Up to now we have discussed analysis of covariance solely in the context of the comparison of several groups, that is, an adjusted one-way analysis of variance. In experimental design terms, this is equivalent to a completely randomized design with no additional effects (such as blocks) involved. (This is why we have permitted the possibility of different sample sizes in the various groups.) This is the simplest way to demonstrate the concept of analysis of covariance. This situation also is frequently encountered in practice. Since use of a covariate controls variation and reduces the unexplained variation, even this simple design can be a sensitive analysis technique.

However, there will be situations in which we might wish to control variation from a continuous source as well as from one or more categorical sources (such as blocks). Applying analysis of covariance to a more advanced experimental design simply involves making the appropriate adjustments to the corresponding multi-way analysis of variance.

In the one-way situation outlined in the section on Carrying Out an Analysis of Covariance, this is achieved by splitting the various sums of squares and cross-products into overall, between-groups, and within-groups (error) components.

SOURCE	df
Between groups	$k - 1$
Error	$N - k$
Total (between groups + error)	$N - 1$

These are used to make the appropriate overall and error adjustments. The adjusted between-groups sum of squares is then determined from the overall and error adjusted figures and tested for significance by comparing it (after appropriate division by the degrees of freedom) with the adjusted error sum of squares.

Exactly the same approach is used with more complex designs, with one subtle distinction. This is best illustrated by considering a randomized block experiment with two categorical sources of variation, between treatments and between blocks.

SOURCE	df
Between treatments	$t - 1$
Between blocks	$b - 1$
Error	$N - t - b + 1$
Total	$N - 1$

To remove variation resulting from a continuous source from this experiment requires adjusting the treatment, block, and error terms. The various sums of squares and cross-products can easily be split up into these three component parts using the usual multi-way analysis of variance approach for the sums of squares and its counterpart (see the section on Carrying Out an Analysis of Covariance) for the sum of cross-products. Where we must be slightly careful is how we then use these to make the adjustments for any covariate influence.

In the one-way situation we were able to calculate the error (within-groups) adjustment directly. This adjustment is also carried out in exactly the same way for more complex experimental designs.

However, remember that the between-groups adjustment is not usually calculated directly. The total (that is, between-groups + error) adjustment is calculated and the between-treatments adjustment then found by subtracting the error adjustment from the total adjustment. This works very nicely because there are no other influences involved in this simple situation. Basically,

$$(\text{Between groups} + \text{error}) - \text{error} = \text{Between groups}$$

In the more complex situation of multi-way analysis of variance, the total cannot be used in this way because it contains several sources of variation. In the randomized block situation:

$$\text{Total} = \text{Between treatments} + \text{between blocks} + \text{error}$$

Calculating the total adjustment and the error adjustment would not, for instance, enable us to determine the between-treatments adjustment because the between-blocks adjustment is also incorporated in the total − error difference.

$$(\text{Between treatments} + \text{between blocks} + \text{error}) - \text{error} = \text{Between treatments} + \text{between blocks}$$

If we can't use the overall total to help us make the between-treatments adjustment, and we can't make the between-treatments adjustment directly, then what can we do? The problem is that the overall total contains variation from a variety of sources. The solution is to create and use a "relevant total," that is, a subtotal that excludes all but the source that interests us. For, say, the between-treatments adjustment, the relevant total would be (between treatments + error). With this simple modification, the approach is exactly as before. The error adjustment is calculated, and then the adjustment for the appropriate "relevant total" is calculated. The adjustment for the particular "between" effect involved is then found by subtraction. This procedure is applied to all the effects included in the analysis of variance, with all adjustments involving the same error term, but each involving a different "relevant total."

In summary, the steps involved in a two-way analysis of covariance are:

1. Split up the sums of squares and cross-products:

SOURCE				df
Between treatments	SSY_t	SSX_t	SCP_t	$t - 1$
Between blocks	SSY_b	SSX_b	SCP_b	$b - 1$
Error	SSY_e	SSX_e	SCP_e	$N - t - b + 1$
Total	SSY	SSX	SCP	$N - 1$

where $SSY = SSY_t + SSY_b + SSY_e$, and so forth.

2. Calculate the error adjustment:
Error variation explained by covariate:

$$= \frac{SCP_e^{\,2}}{SSX_e} \text{ with 1 df}$$

Adjusted error variation:

$$= SSY_e - \frac{SCP_e^2}{SSX_e} \text{ with } N - t - b \text{ df}$$

3. To calculate, say, between-treatments adjustment:
   a. Form the appropriate "relevant totals":

$$SSY_t + SSY_e \quad SSX_t + SSX_e \quad SCP_t + SCP_e$$

   b. Calculate the "relevant total" adjustment:

$$\frac{(SCP_t + SCP_e)^2}{SSX_t + SSX_e}$$

that is, the adjusted "relevant total" variation

$$= (SSY_t + SSY_e) - \frac{(SCP_t + SCP_e)^2}{SSX_t + SSX_e}$$

   c. Obtain the adjusted between-treatments variation from the difference of these two.
   That is,

Adjusted "relevant total" SSY − adjusted error SSY

= Adjusted between-treatments SSY with $t - 1$ df

4. Use the same approach to calculate the adjusted between-blocks variation by using the appropriate "relevant totals,"
That is,

$$SSY_b + SSY_e \quad SSX_b + SSX_e \quad SCP_b + SCP_e$$

5. Complete the analysis of covariance by testing the adjusted between-treatments and between blocks-terms against the adjusted error term. (Although we have illustrated a two-way analysis without interaction, this crafty device of creating "relevant totals," that is, the error term + the effect to be adjusted, enables us to analyze three-way or even more complex experiments and to incorporate interaction adjustments in an analysis of covariance.)

ASSUMPTIONS AND FUNCTIONS

Like any statistical technique, the analysis of covariance depends on certain assumptions. Since it is effectively a composite of analysis of variance and regression analysis, it combines the assumptions of both these techniques. It assumes that the outcome (y) variable follows a Normal distribution and that it has effectively the same variability in each of the groups. It assumes that the outcome/covariate relationship is linear and that the values of the covariate can be measured without error. The analysis of covariance is reasonably robust to departures from these assumptions (provided that they are not grossly wrong), but they should never be taken for granted.

Analysis of covariance in its basic form also assumes that the outcome/covariate relationship (that is, the slope) is the same within each group. However, as we saw earlier, it is possible to adapt the technique to cope with the "different slopes" situation.

Analysis of covariance fulfills two functions. By considering the effect of another source of variation, it reduces the unexplained variation and makes the analysis more sensitive. By adjusting the groups to equivalent covariate values, it removes any imbalance in the structure of the groups and makes their comparison fairer.

For designed experiments in which the subjects have been randomly allocated to the groups, the reduction in error variation is likely to be much more important. Here there is no reason to believe that, say, the mean ages of the treatment groups should be very dissimilar. Any between-groups adjustment is, therefore, likely to be small.

However, when the groups are formed according to inherent properties of the individuals, and the individuals are not randomly allocated to the groups, imbalances in the covariate are very possible. For instance, in a comparison of blood pressures between smokers and nonsmokers, the average age of the smokers might well be substantially higher than that of the nonsmokers (a reflection of different generational attitudes). This would create an additional difference in blood pressure that has nothing to do with the effect under study (smoking). The between-groups comparison would, therefore, be unfairly distorted. In these uncontrolled situations, the ability of the analysis of covariance to eliminate these imbalances and distortions is one of the major reasons for its use.

—— **Key Terms** ——

Analysis of covariance
Covariate
Adjusted means

—— **PROBLEMS** ——

1. In a study of the influence of gender on susceptibility to anesthesia, the amount of theopentone (in mg) required to induce the abolition of the eyelash reflex was measured for 10 men and 10 women.
   a. Carry out an analysis of variance and determine if the sexes differ in their dose requirements.
   b. Variations in patients' weights are likely to have a substantial effect on dose requirements. Using weight (quoted below in kg) as a covariate, carry out an analysis of covariance and test for gender differences.

   Calculate the adjusted group means and determine why the results of the two analyses differ.

MALES		FEMALES	
Dose (mg)	Weight (kg)	Dose (mg)	Weight (kg)
266	72	238	64
285	84	265	74
250	82	220	62
212	55	242	60
280	74	254	72

MALES		FEMALES	
Dose (mg)	Weight (kg)	Dose (mg)	Weight (kg)
244	68	204	48
262	76	226	54
302	88	260	66
254	72	250	76
274	70	232	50

2. Initial weights of 24 patients were recorded, then weight loss was compared after the patients were assigned one of two diets. Use an analysis of covariance to remove the influence, if any, of initial weight from the observed weight losses and test for diet differences.

The weight losses and associated initial weights (in kg) are given below.

DIET A		DIET B	
Weight Loss (kg)	Initial Weight (kg)	Weight Loss (kg)	Initial Weight (kg)
4	60	5	65
8	75	7	70
8	88	8	78
11	80	13	90
3	60	5	62
9	85	8	80
6	70	10	67
7	70	11	72
12	84	18	80
12	88	22	95
7	72	10	80
6	64	11	78

3. As part of an investigation of the effects of smoking on respiration, vital lung capacity (in liters) was measured for eight smokers, eight ex-smokers and eight non-smokers. However, age also was suspected to have some influence on vital lung capacity. Using age as a covariate, test whether differences in vital lung capacity exist between the three smoking groups. Calculate the adjusted means for each group.

SMOKERS		EX-SMOKERS		NONSMOKERS	
Vital Lung Capacity (L)	Age	Vital Lung Capacity (L)	Age	Vital Lung Capacity (L)	Age
3.8	52	3.9	50	4.3	42
3.2	48	4.6	44	5.3	20
4.6	30	3.3	62	3.7	52
2.9	65	4.8	30	3.3	55
3.5	50	3.7	52	4.8	44
4.8	42	4.2	34	4.4	34
4.0	26	3.6	30	5.2	26
4.9	38	4.3	28	4.0	36

4. The following data represents the temperature increase results for the male patients mentioned in problem 2 of Chapter 11, together with their ages. Test for the existence of between-dose differences, after adjusting for age.

DOSE OF BUPIVOCAINE

5 ml		10 ml		15 ml	
Age	° C	Age	° C	Age	° C
34	0.5	26	1.6	33	1.9
20	1.4	29	0.8	25	1.1
48	0.3	44	0.7	49	1.3
37	1.7	37	1.4	41	1.8
58	0.2	52	1.0	63	0.6
66	0.1	56	0.3	54	1.3

5. The effects of alcohol abuse and gender differences on the time (in seconds) taken to complete a specific task were examined in problem 1 in Chapter 11. Age was also included in that data set as a blocking variable. If age had been recorded for each of the participating subjects, the results obtained would have been as given below.

Re-analyze this set of data using an analysis of covariance (that is, treating age as a covariate rather than a blocking variable).

(To simplify your calculations, you may ignore the gender × abuse interaction. You can also assume, without testing, that the use of a common slope is acceptable.)

ABUSE

	No		Yes	
	Time (sec)	Age	Time(sec)	Age
Male	23	24	28	26
	16	20	22	18

ABUSE

	No Time (sec)	Age	Yes Time(sec)	Age
Male	21	33	30	30
	19	44	25	36
	28	51	40	54
	23	55	43	60
Female	18	20	24	23
	26	22	29	19
	24	33	32	40
	21	38	22	32
	26	50	41	53
	31	54	46	62

CHAPTER 13

Multiple Regression

THE PROBLEM OF MULTIPLE EXPLANATIONS

Regression analysis (see Chapter 7) provides an elegant and informative way of testing the possible existence of, and describing, a relationship between a continuous outcome or dependent variable (such as diastolic blood pressure) and an explanatory or independent variable (such as a patient's age). This technique enables us to examine the proposition that a patient's age has some real influence on his or her diastolic blood pressure. If this is confirmed, regression also enables us to quantify the relationship and determine the extent to which changes in age are reflected in corresponding changes in diastolic blood pressure.

In many research situations, however, there is very unlikely to be only one explanatory variable that might potentially influence the outcome variable that happens to interest us. Blood pressure, for instance, is known to be influenced by a host of factors, some demographic (age, gender, race, socioeconomic status, and so on), some dietary (sodium, potassium, or alcohol consumption, and so on), some genetic, and others. Therefore, a health researcher frequently will have a number of potential explanatory variables that he or she suspects may have some influence on the criterion or outcome variable under study, particularly in epidemiological investigations. Some of these explanatory or independent variables may have a real effect on the outcome or dependent variable; others may have no influence at all on it. The objectives of the investigation are to determine which (if any) of the various independent variables have a genuine relationship with the dependent variable, to determine the strength and nature of the relationship and to make sense of the tangle of influences that may be present when there are a number of possible explanatory variables. This is the role of multiple regression analysis.

REGRESSION REVISITED

Multiple regression analysis is a very simple and direct extension of ordinary regression analysis (that is, with one explanatory variable). To help us appreciate just how straightforward and useful this extension is, let's briefly reacquaint ourselves with the key ideas underlying regression analysis.

Regression analysis tackles the problem of describing the relationship between an independent variable, x, and a dependent variable, y, by finding the straight line that best describes the relationship (see Chapter 7). Best, in this context, means the line that misses the various values of y by as little as possible overall. The "misses" or errors are squared to avoid complications of sign and summed over all results to give an error sum of squares. The best line, and hence the line used in regression, has the smallest possible error sum of squares and is known as the least squares line or least squares fit.

This error sum of squares represents the variation in the dependent variable, y, that the regression relationship cannot describe or explain. Subtracting this unexplained variation from the total variation (the overall sum of squares) in y leaves, of course, the variation in the dependent or outcome variable that can be successfully described by the regression relationship and that is, therefore, influenced by changes in the independent or explanatory variable. Once the variation in y is split up in this way (into explained and unexplained) it is possible to test the significance of the relationship using the standard analysis of variance. The degrees of freedom involved are a total of $n - 1$ (that is, sample size less one) since variation in y is measured about its mean, the "explained by regression" df is 1, since one independent variable is involved, and hence the "unexplained" df are $n - 2$. If the variance explained by the regression relationship is not substantially greater than the unexplained or error variance, then there would seem little reason to take the relationship at all seriously. On the other hand, if the explained variance is very clearly larger than the error (unexplained) variance, then it seems sensible to conclude that the relationship is real and that the independent variable has some genuine connection with the dependent variable.

The proportion of the total variation in y that has been successfully explained by regression, that is,

$$r^2 = \frac{\text{SS in } y \text{ explained by regression}}{\text{total SS in } y}$$

is often used as a measure of the strength or success of the regression relationship (clearly the larger this is, the stronger and more clear the regression relationship is). The square root of this proportion is, as we saw in Chapter 7, the correlation between y and x.

The regression relationship itself is described by two parameters, a (the intercept) and b (the slope). The intercept represents the value of the dependent variable when the independent variable is zero. The intercept will be meaningful only if the idea of the independent variable taking a zero value is sensible (0 cigarettes smoked/day is certainly meaningful; a patient weight of 0 kg is just as certainly meaningless). Either way, the intercept defines an appropriate starting point for the regression line.

The slope represents the change in the outcome or dependent variable that occurs (on average) for each unit change in the independent variable, for example, the change in blood pressure that occurs for each year increase in age. This will be potentially meaningful for every regression relationship and clearly tells us a lot about the connection between the dependent and independent variables.

It should always be borne in mind, however, that the slope, b, expresses the regression relationship in terms of the units in which the dependent and independent variables were actually measured. If, for some reason, age had been measured in months rather than years, then the regression slope (that is, change in blood pressure/month increase in age) would be only one twelfth the value of the slope if age had been measured in years. The existence and strength of the blood pressure/age relationship being described are, of course, totally unaffected by this arbitrary choice of measurement units. One way out of this ambiguity might be to describe the change in terms of standardized units, that is, the standardized change in y/unit standardized change in x. (By "standardized" we mean measurements expressed in terms of standard deviations; see Chapter 2). The relationship between this standardized slope or standardized regression coefficient (usually denoted as β) and the ordinary slope or regression coefficient, b, is as follows.

$$b = \text{change in } y/\text{unit change in } x$$

$$\beta = \text{standardized change in } y/\text{unit standardized change in } x$$

$$= b \frac{1}{SD_y} \Big/ \frac{1}{SD_x}$$

$$= b \frac{SD_x}{SD_y}$$

Actually there is very little point in calculating a standardized regression coefficient when we are dealing with an ordinary (that is, with only one independent variable and hence only one regression coefficient) regression analysis. It is more convenient, and more immediately meaningful, to see a relationship expressed in terms of the original units of measurement rather than in abstract, albeit universal, standardized units. However, when we face a number of independent variables, standardized regression coefficients can be very informative, as we shall see later in this chapter.

Now let's see these ideas set in a practical context. In a study of the influence of patient characteristics on susceptibility to anesthetic agents, the dose of thiopentone required to abolish the eyelash reflex was noted for 903 patients and various physical (such as age, gender, height and weight), social (use of tobacco, alcohol, and other drugs), and physiological (systolic blood pressure, heart rate) characteristics were recorded. The relationship between dose of thiopentone (the outcome or dependent variable) and a particular patient characteristic, height (the explanatory or independent variable), can be assessed using regression analysis, the results of which are as follows:

$$y = -250.442 + 2.929\, x$$

where

$$y = \text{Induction dose of thiopentone in mg}$$

$$x = \text{patient's height in cm}$$

SOURCE	SS	df	MS	F
Explained by height	1,186,852	1	1,186,852.00	272.13
Unexplained	3,929,522	901	4361.29	
Total	5,116,374	902		

There is therefore a very highly significant relationship between a patient's height and the dose of thiopentone required to induce anesthesia in the patient. The regression coefficient shows that, on average, the dose required increases by 2.929 mg for each 1 cm increase in height. We might infer from this that increase in height tends to correspond to increase in body mass and hence to an increased dose requirement.

The regression equation can also be used to make predictions. For instance, it predicts that a patient who is 180 cm tall will require a dose of

$$-250.442 + (2.929 \times 180) = -250.442 + 527.22$$

$$= 276.778 \text{ mg}$$

(As we commented in Chapter 7, prediction only makes numerical, and common, sense within the range of normal patient heights.)

The strength of the dose/height relationship is given by

$$\text{Proportion variation explained, } r^2 = \frac{\text{SS explained}}{\text{Total SS}}$$

$$= \frac{1,186,852}{5,116,374}$$

$$= .2320$$

$$\text{Height/dose correlation, } r = \sqrt{.2320}$$

$$= +.4816$$

(Note that the sign of the correlation corresponds to the sign of the regression coefficient.)

The standardized regression coefficient, β, is

$$+2.929 \times \frac{12.385}{75.314} = +.482$$

where 12.385 is the standard deviation of height in cm and 75.314 is the standard deviation of dose in mg.

Any changes in the unit of measurement of either variable would leave this standardized regression coefficient unaltered. Measuring height in meters would increase b (that is, dose change/meter change in height) 100-fold but would decrease the standard deviation of x 100-fold (since height would be measured by much smaller numerical values), leaving β unaffected.

STUDYING SEVERAL EXPLANATORY VARIABLES

When a number of potential explanatory or independent variables are available for study, we face a major new problem. The various explanatory variables may or may not be related to the outcome variable. In addition they may also be related to one another to varying degrees. For instance, age and routine recreational exercise levels (explanatory or independent variables) might have a real effect on a patient's heart rate after one minute of intense exercise (the outcome or dependent variable). However, to further complicate matters, it seems very likely that age and exercise levels will themselves be related, with older subjects perhaps tending to take less regular exercise. (This potential connection between the various explanatory variables illustrates just how inappropriate the term "independent variables" actually is.)

How might we analyze data like this? One solution might be to carry out separate regressions of age on heart rate and exercise level on heart rate. This approach has major limitations. It involves multiple analyses, which is always undesirable. More important, while it would give us information about the individual age/heart rate or exercise level/heart rate relationships, it would tell us absolutely nothing about the possible age/exercise level connection and just how this affects heart rate. If, for instance, there is a strong relationship between age and heart rate, how much of this is uniquely due to the aging process and how much is simply due to the fact that older subjects tend to exercise less frequently? If these older subjects had followed routine exercise regimes that were similar to those of their younger compatriots, then their performance might have been considerably better. In that situation any relationship between heart rate and age would be uniquely due to the aging process itself and not (even partially) due to associated exercise habits.

The capacity to filter out other associated (confounding) influences and effectively make the subjects truly similar to one another clearly would greatly increase our capacity to understand exactly how various possible explanatory variables affect an outcome variable and enable us to build up a very comprehensive and subtle picture of the relationships involved. (The idea of removing, or "controlling for" the influence of other variables is something we have, in fact, already met in Analysis of Covariance, Chapter 12.)

The ideal method of analysis would therefore describe the joint influence of the various explanatory variables on the outcome variable in one integrated package, rather than as a series of separate analyses. In addition, and even more important, for each explanatory variable the analysis would measure the effect on the outcome variable that is uniquely attributable to the influence of that explanatory variable alone and that is not the possible result of other explanatory variables that happen to have some connection with the first explanatory variable.

Multiple regression analysis meets both these requirements admirably.

___ THE MULTIPLE REGRESSION APPROACH ___

Multiple regression analysis meets our first requirement by yielding a regression equation that relates the outcome variable to all the explanatory variables in the study. This has exactly the same basic form as the familiar univariable (one explanatory variable) regression equation and describes the combined influence of the various explanatory variables on the outcome variable. The only innovation is the fact that, since several explanatory variables are now involved, each will have a "slope" or regression coefficient associated with it (this also requires some extra care in notation). For instance, a multiple regression analysis relating heart rate to age and exercise level would yield a regression equation of the form

$$y = a + b_1 x_1 + b_2 x_2$$

when

$$y = \text{Heart rate}$$

$$x_1 = \text{Age}$$

$$x_2 = \text{Exercise level}$$

$$b_1 = \text{Change in heart rate/unit change in age}$$

$$b_2 = \text{Change in heart rate/unit change in exercise level}$$

and

$$a = \text{Intercept (heart rate when } x_1 \text{ and } x_2 \text{ are zero)}$$

For a practical example of a multiple regression equation, let's reconsider the thiopentone induction dose study. We have already looked at the relationship between a patient's height and the dose of thiopentone required to induce anesthesia in the patient. We will now broaden our interest and explore how the induction dose depends on the patient's age, weight, and height. The resulting multiple regression equation is

$$y = -172.127 - .807\, x_1 + 1.814\, x_2 + 1.954\, x_3$$

when

$$y = \text{Dose of thiopentone in mg}$$

$$x_1 = \text{Age in years}$$

$$x_2 = \text{Weight in kg}$$

$$x_3 = \text{Height in cm}$$

The equation shows that increasing age is associated with increased susceptibility to thiopentone (more precisely, each year increase in age results in a .807 mg decrease in the dose required to induce anesthesia). Similarly it shows that increases in weight and in height are both associated with increased dose requirements.

Like a simple regression equation, a multiple regression equation can be used to predict the value of the outcome variable that one would expect to observe for any given combination of the explanatory variables. For instance, for a subject of age 45, weight 76 kg, and height 180 cm, the equation predicts an induction dose requirement of

$$-172.127 - (.807 \times 45) + (1.814 \times 76) + (1.954 \times 180)$$

$$= 281.142 \text{ mg}$$

In principle, the multiple regression intercept and regression coefficients are calculated in the same general way as for simple regression. They are the coefficients that enable us to predict the actual outcome values as accurately as possible. In other words, the error (that is, actual minus predicted outcome) sum of squares is as small as possible, and a multiple regression equation, just like a univariable regression equation, represents a least squares fit. However, the calculations, although simple in principle, can be very tedious in practice, especially if a number of independent variables are involved. Because easy-to-use statistical software packages are widely available, it is not necessary, and in fact not desirable, to attempt a multiple regression analysis by hand.

The significance of a multiple regression equation is tested by comparing the outcome sum of squares successfully explained by the regression equation with the unexplained or error sum of squares, just as is done in univariable regression. (This comparison will be provided as part of the standard output from a multiple regression software package.) There is one minor modification. If the regression equation involves k independent variables, the "explained" or "due to regression" sum of squares will have k degrees of freedom associated with it. The "unexplained" or "error" degrees of freedom will correspondingly be $n - k - 1$, when n is the sample size.

The analysis of variance table for the regression of age, weight, and height on thiopentone dose is as follows.

SOURCE	SS	df	MS	F
Explained by age, weight, and height	1,580,112	3	526,704.0	133.9
Unexplained	3,536,262	899	3933.6	
Total	5,116,374	902		

This confirms that a subject's age, weight, and height jointly have a very highly significant effect on his or her thiopentone requirements and that a real relationship does exist. The strength of that relationship can, as before, be expressed as the proportion of the variation in dose that has been successfully explained by the regression equation.

$$\text{Proportion of variation explained by age, weight, and height } (r^2) = \frac{1,580,112}{5,116,374}$$

$$= .3088$$

Note that the proportion of dose variation explained by the influence of age, weight, and height (.3088) is larger than the proportion explained by the influence of height on its own (.2320). This, of course, makes sense. The use of additional explanatory variables must result in the gain of some additional explanatory power. (Whether this gain is large enough to be important is another matter and one we will explore in the next couple of sections.)

$$\text{Multiple correlation coefficient, } r = \sqrt{.3088}$$

$$= .5557$$

This proportion of variation successfully explained (the most frequently quoted measure of the success of a regression equation) can, again, be expressed as a correlation coefficient. Since this represents the correlation between the dependent variable (dose) and a combination of several independent variables (age, weight, and height) it is called a **multiple correlation coefficient.** (It is, in fact, simply the correlation between the actual outcome values and the outcome values predicted by the regression equation.)

INTERPRETING THE REGRESSION COEFFICIENTS

To date we have concentrated on assessing the overall picture presented by a multiple regression equation. However, confirmation that a dependent variable is influenced by a combination of several independent variables raises as many questions as it answers. Do all the independent variables have a real influence on the dependent variable, or do some of the independent variables have no real connection with it? Just what influence does each explanatory variable have on the outcome variable, and is it uniquely due to that variable or can it be partially explained by the influence of other, related, independent variables? The answers to all these questions are contained in the various regression coefficients.

A regression coefficient clearly describes the influence of a particular independent variable on the outcome variable. Our thiopentone multiple regression equation, for instance, tells us that every 1 cm increase in a patient's height results in a 1.954 mg increase in the patient's induction dose requirements. Actually, a reconsideration of the multiple regression equation shows that it is telling us even more than this.

$$\text{Dose (mg)} = -172.07 - .807 \text{ age (yr)} + 1.814 \text{ weight (kg)} + 1.954 \text{ height (cm)}$$

The regression coefficient for height shows that, when weight and age remain constant, every 1 cm increase in height will result in an increased dose requirement of 1.954 mg. In other words, it is describing the influence of height that remains after any associated variation in (and influence of) weight and age have been eliminated. The regression coefficients therefore fulfill the second requirement we laid down in the section on Studying Several Explanatory Varaibles, the ability to untangle the influences of several, possibly related, explanatory variables and measure the unique effect of any particular independent variable (unique at least in the sense of being totally unaffected by any associated variation in any of the other independent variables in the regression equation). This can be exceptionally useful in enabling researchers to determine the extent to which various, usually related, factors each uniquely contributes to a health problem (such as the effect of alcohol consumption, cigarette consumption, and obesity on hypertension). In experimental studies, the effect of various background factors, such as gender and age, can be artificially separated by designing balanced experiments (see Chapter 11). When these factors are matters of social choice this approach is, of course, impossible and multiple regression represents the only way to untangle the inevitable web of interrelated explanatory factors.

Although the various regression coefficients do measure the unique influence of each explanatory variable they cannot be directly compared with one another, since each is expressed in terms of potentially quite different measurement units (mg/yr, mg/kg and so forth). However, standardizing the coefficients in the manner outlined earlier, eliminates complications of units and enables direct comparisons to be made between the various explanatory variables.
That is,

$$\beta_i = \frac{SD_i}{SD_y} b_i$$

where β_i is the standardized regression coefficient for explanatory variable number i
SD_i is the standard deviation of explanatory variable number i
SD_y is the standard deviation of the outcome variable
and b_i is the regression coefficient for explanatory variable number i.

For the three explanatory variables (1=age, 2=weight, 3=height) involved in the induction dose study, the respective standardized regression coefficients are as given in Table 13-1.

■ **TABLE 13-1** ■

Standardized Regression Coefficients

i	b_i	SD_i	SD_y	β_i
1	−0.807	17.518	75.314	−0.188
2	+1.814	13.059	75.314	0.315
3	+1.954	12.385	75.314	0.321

The standardized coefficients confirm that weight and height are almost equivalent in terms of the strength of their (unique) effects on induction dose and that the influence of age on dose is considerably weaker than either of these.

Regression coefficients or standardized regression coefficients describe the unique influence of an explanatory variable on an outcome variable in alternate ways. However, neither of them directly tests whether that influence is real or merely illusory. Testing the reality of, say, the unique influence of height (in effect, testing the hypothesis that the true value of b_3 is 0) requires us to find the amount of dose variation that is explained solely by the effect of height, that is, that cannot be explained by possibly related variations in age and weight. Once this is known, it can be tested against the unexplained variation in the usual analysis of variance way.

This uniquely explained variation can be found by carrying out and comparing two regression analyses. In the first, age and weight (the variables whose influence we wish to control or eliminate) are regressed on induction dose. The resulting analysis of variance table is as follows.

SOURCE	SS	df
Explained by age and weight	1,279,561	2
Unexplained	3,836,813	900
Total	5,116,374	902

A regression analysis involving the variable to be tested (height) together with the variables whose influence we wish to control is then carried out. This is the age, weight, and height regression analysis whose results we have already discussed.

SOURCE	SS	df
Explained by age, weight, and height	1,580,112	3
Still unexplained	3,536,262	899
Total	5,116,374	902

The additional variation in dose explained in the age, weight, and height regression (relative to the age, and weight regression); that is,

	SS	df
Explained by age, weight, and height	1,580,112	3
− Explained by age and weight	1,279,561	2
= Explained by height after allowing for influence of age and weight	300,551	1

must be uniquely due to the influence of height since the influence of age and weight has already been fully used to explain variation in dose. The gain in explanatory power that is obtained by bringing in height as an explanation must reflect some influence of height that is totally unrelated to variations in age and weight (that is, unique to height, at least in the sense of being distinct from age and weight influences). Explanatory power must always increase when fresh explanatory variables are employed. That gain might be trivial and accidental (consistent with the null hypothesis that b is 0) or it may be substantial and may show a real relationship. Its importance can be assessed by comparing it to the variation that is still unexplained after all the explanatory variables have been used.

SOURCE	SS	df	MS	F
Explained by age, weight	1,279,561	2		
Explained by height after controlling for age and weight	300,551	1	300,551.00	76.41
Unexplained	3,536,262	899	3933.55	
Total	5,116,374	902		

Height, even after possible associated influences of age and weight have been controlled, still has a highly significant ($p < .001$) influence on a patient's thiopentone dose requirement. Two patients of the same age and weight but different heights (and therefore of different builds) will require genuinely different induction doses. This idea of controlling for other influences enables us to develop an understanding of the influence of height that is quite different from the "taller is heavier, hence more thiopentone" conclusion inspired by a simple univariable regression of height on dose.

Most, if not all, multiple regression software packages will automatically carry out this test of the various regression coefficients (that is, of the unique influence of each explanatory variable). Therefore, it is usually not necessary to carry out a sequence of regressions to build up the test. The idea of building up the explanatory power sequentially by adding in fresh explanatory variables is, however, right at the heart of an important practical variant of multiple regression analysis, stepwise multiple regression (see the following section). Since the F test involved always has one numerator degree of freedom, it is often quoted in statistical packages, quite equivalently, as a t value.

$$F = 76.41 \text{ with } 1, 899 \text{ df}$$

or

$$t = \sqrt{76.41} = 8.74 \text{ with 899 df}$$

The unique influence of one variable on another (say height on dose after eliminating the influence of age and weight) can be quantified in another way. After the influence of age and weight have been taken into account, a dose sum of squares of 3,836,813 remains unexplained.
That is,

$$\text{SS unexplained} \mid \text{age and weight} = 3,836,813$$

(for ∣ read "after allowing or controlling for")

Thanks to the unique influence of height, 300,551 of this variation can be explained. That is,

$$\text{SS due to height} \mid \text{age and weight} = 300,551$$

The proportion of the variation (available after controlling for age and weight) explained by height (after controlling for age and weight) is therefore

$$r^2 = \frac{300,551}{3,836,813} = .0783$$

and hence

$$r = \sqrt{.0783} = +.2799$$

This represents the correlation between height and dose that is uniquely due to variation in height and that is not influenced by related variation in age and weight. Such a correlation (that is, one in which the influence of other variables has been controlled and removed) is known as a **partial correlation** (its sign corresponds to the sign of the equivalent regression coefficient).

This partial dose, height ∣ age and weight correlation (.2799 or 7.83% explained) is substantially less than the raw dose, height correlation (.4816 or 23.20% explained), showing that some (although, as we have seen, certainly not all) of the variation in height simply reflected parallel variation in age and/or weight.

It cannot be stressed too much that the terms "unique" and "unique influence" mean unique only in the context of the other explanatory variables involved in the regression analysis. A particular independent variable could, of course, still be connected with other explanatory variables not incorporated in the study. The apparently "unique" influence of the variable in question could therefore be largely explained by the influence of this unconsidered variable. When seeking to control for the effects of possible confounding variables it is essential to ensure that they are all incorporated in the regression analysis.

STEPWISE MULTIPLE REGRESSION

Many investigations, particularly in the field of epidemiology, are a search for relationships between an outcome variable and a considerable number of potential explanatory variables. Some of those explanatory variables may, in fact, turn out to have no connection whatsoever with the outcome variable. Others may indeed have a genuine relationship with the outcome variable but simply replicate essentially the same relationship described by one or more of the other explanatory variables (caffeine and cigarette consumption, for example, tend to be closely related and to reflect much the same pattern of indulgence). Many of the explanatory variables in a study of this type are therefore usually redundant. As a consequence it is frequently possible to find a multiple regression equation that uses only some of the explanatory variables but still explains almost as much variation in the outcome variable as it would by using all the explanatory variables. Such a "cost effective" multiple regression equation has several advantages. It is a way of obtaining the best possible prediction of outcome for the least amount of explanatory information. More important, it identifies the major influences on outcome and the explanatory variables that merely duplicate information already available in another guise.

This "compact" multiple regression equation is built up in a series of steps, hence the name **stepwise multiple regression.** Each individual explanatory variable is first separately regressed on the outcome variable. The explanatory variable that can explain the largest proportion of the outcome variation is selected as the first variable to enter the regression equation. Each remaining explanatory variable then is regressed on the outcome variable jointly with the first variable. The explanatory variable that provides the largest gain in explanatory power (on top of that explained by the first variable on its own) is then added in as the second variable in the multiple regression equation. In the third step, each remaining variable is tried in combination with these two selected variables to see which provides the maximum gain in variation explained (that is, the maximum explanatory power after controlling for the variables already selected) and so on.

At each step the maximum gain in variation explained is tested against the variation still unexplained at that stage, in the way described in the previous section. The stepwise selection process is terminated when this gain is not significantly greater than pure random variation, in other words when the "gain" is an illusion. The resulting regression equation is the most compact possible, consistent with incorporating all real explanatory relationships.

TABLE 13-2

Building the Stepwise Regression Equation

Explanatory variable	Additional variation explained after step (%)					
	0	1	2	3	4	5
Gender	6.5	0.4	0.4	0.1	0.2	0.1
Age	0.1	2.2	3.3	2.9	—	—
Weight	22.3	4.5	—	—	—	—
Height	23.2	—	—	—	—	—
Body type	7.2	3.6	0.4	0.4	0.3	0.2
Cigarette consumption	3.0	1.0	1.1	0.0	0.0	0.1
Alcohol consumption	10.7	4.0	4.0	—	—	—
Degree of anxiety	2.3	1.3	1.5	1.2	1.0	0.8
Systolic BP	1.7	0.1	0.0	0.0	0.7	0.6
Heart rate	0.1	2.1	2.4	2.7	2.2	—

The thiopentone induction dose study we discussed earlier actually involved the investigation of 10 possible explanatory variables (Table 13-2).

Inspection of each independent variable on its own shows that height is the most powerful single predictor of dose (23.2% explained). Height is therefore the first variable to be entered into the multiple regression equation. Once we account for variation due to height, most of the other explanatory variables lose a great deal of their explanatory power, reflecting the fact that most of them vary to some degree with height. However, weight can explain a further 4.5% of the variation in dose, and it is entered into the stepwise equation next. In succeeding steps alcohol consumption, age, and heart rate are entered into the equation. After this, no further significant gain in variation explained can be achieved by incorporating any other independent variables. The final stepwise regression equation therefore involves five independent variables and explains 36.8% of the variation in induction dose (Table 13-3).

The influences of the five variables not included in the final equation are described by proxy. Gender differences are explained by height and weight differences, anxiety levels are effectively reflected in heart rates, and so forth. Controlling for the influence of other variables tends to reduce the explanatory power of a particular independent variable. However, the reverse can happen. On its own, age appears to have little influence on dose (.1% explained). After controlling for height and weight, however, its influence is dramatically increased (3.3% explained). Although older people are indeed more susceptible to anesthesia they also tend to be plumper than younger patients. The increase in dose dictated by weight gain tends to mask the decrease in dose resulting from the aging process. Only after controlling for height and weight is the true effect of aging fully evident.

MULTIPLE REGRESSION AND CATEGORICAL EXPLANATIONS

Regression analysis assumes that the dependent variable is continuous and follows a Normal distribution. There is, however, no such restriction on the independent variables. They can, quite legitimately, be categorical as well as continuous, and it is very

TABLE 13-3

The Final Stepwise Regression Model

Step	Variable added	Additional variation explained (%)	Influence
1	Height	23.2	Body mass/
2	Weight	4.5	body shape
3	Alcohol consumption	4.0	Drug tolerance
4	Age	2.9	Aging
5	Heart rate	2.2	Anxiety
	Total	36.8	

$$y = -239.77 + 1.78\,x_1 + 1.83\,x_2 + 20.49\,x_3 - .71\,x_4 + .86\,x_5$$

easy to incorporate categorical explanatory variables into a multiple regression equation, provided we exercise a little care in the values we give them. Remember that the categories in a categorical variable are simply different from one another and have no inherent order. The values we use to represent the categories must, therefore, simply distinguish between them without imposing any order on them. For a two-category variable such as gender, this can be easily achieved by representing males by the value 0 and females by the value 1 (or vice versa). The 0 category effectively serves as a baseline or reference category and the choice of males or females as the "reference" gender is, of course, totally immaterial. If the categories were, say, "received placebo" and "received active treatment," then the placebo group would seem the most logical choice as the reference category.

Three (or more) category variables can be represented in exactly the same way. A three category situation cannot be represented by a single variable taking the values 0, 1, 2 since these impose a very clear order on the categories, but it can be represented by two variables, each taking the values 0 and 1. (Such variables are often called dummy variables or indicator variables.)

CATEGORY	DUMMY VARIABLES	
	x_1	x_2
A	0	0
B	1	0
C	0	1

If x_1 takes the value 1, then the individual is in category B. If x_2 takes the value 1, then the individual is in category C. If neither x_1 nor x_2 takes the value 1 (that is, both are 0), then the individual must be in category A (that is, the individual is not in either category B or C). Here, category A is serving as the reference category. This coding system distinguishes the three categories, while imposing no arbitrary order on them. (Note how the number of dummy or indicator variables we need to represent a categorical variable corresponds exactly to the between-categories degrees of freedom.)

To see categorical explanations in action, let's first examine the relationship between thiopentone dose in mg (y), and gender (x), coded as 0 for male and 1 for female. The regression equation is

$$y = 251.73 - 43.21 x$$

This relationship predicts a dose of 251.73 mg for males and 208.52 mg (251.73 − 43.21) for females. These are, in fact, the mean (and, hence, most likely) doses of thiopentone needed to induce anesthesia in the two sexes. The regression coefficient therefore expresses the difference in the mean outcome between the indicated category (that is, that designated as 1) and the reference (or 0) category. The dose/gender relationship is very highly significant ($p < .001$) with $F = 62.96$ with 1 and 901 df. This corresponds exactly to the t value of 7.93, that is, $\sqrt{62.96}$, which results from a male/female comparison using the conventional t test, and which confirms that the two sexes do have very different dose requirements.

When height and weight are included in the analysis, the dose/gender relationship becomes

$$y = -133.52 - 11.92\,x_1 + 1.64\,x_2 + 1.61\,x_3$$

where

$$y = \text{Dose of thiopentone in mg}$$

$$x_1 = \text{Gender} \qquad 0 = \text{Male} \qquad 1 = \text{Female}$$

$$x_2 = \text{Height in cm}$$

$$x_3 = \text{Weight in kg}$$

This relationship indicates that a male 160.7 cm tall and weighing 61.1 kg (the average height and weight for the study) would require a dose of 228.40 mg, that is,

$$-133.52 + (1.64 \times 160.7) + (1.61 \times 61.1)$$

The prediction for a female of the same height and weight is a dose requirement of 216.48 mg, that is,

$$-133.52 - 11.92 + (1.64 \times 160.7) + (1.61 \times 61.1)$$

These values are, in fact, the adjusted means for males and females, that is, the means we would see if all the males and females in the study had the overall average height and weight, and hence we effectively eliminated or controlled for height and weight. The regression coefficient for gender is, therefore, the difference between the adjusted means or the difference between the genders (expressed as female relative to male) remaining after we have controlled for the other variables in the regression equation. Note that the difference between the genders is very greatly reduced once the influence of height and weight have been eliminated (an adjusted difference of -11.92, as opposed to a raw difference of -43.21).

A test of the unique influence of gender (using the approach discussed in the previous section), shows that no significant difference between the genders exists ($F = 3.39$ with 1 and 899 df or $t = 1.84$ with 899 df), after we control for the influence of height and weight on dose requirements. In other words, the dose requirement differences we see between men and women are simply a reflection of their differing (average) body sizes, and would effectively disappear for men and women of the same body size.

Multiple regression is a very elegant and convenient way to carry out an analysis of covariance using multiple covariates. Because we can, of course, have two or more categorical explanations represented in a multiple regression equation, we can also use the technique to untangle the effects of interrelated categorical variables. This makes multiple regression an ideal tool for analyzing unbalanced experiments (see Chapter 11), something we noted was impossible with the classic analysis of variance approach presented in Chapter 11. The capacity of multiple regression to control for other influences means that the tangle of effects resulting from an unbalanced design can be separated, and the variation uniquely attributable to a particular factor can be isolated and tested.

Multiple regression can be employed to analyze virtually any possible data analysis situation involving a continuous outcome variable. It can describe and test the relationship between such an outcome and one (or many) continuous variables, carry out the equivalent of a simple t test, or analyze an unbalanced experiment with multiple covariates. It is sometimes called, in the more technical literature, the **general linear model,** a tribute to its unique ability as a general analysis technique.

Key Terms

Multiple regression analysis

Multiple correlation coefficient

Partial correlation

Stepwise multiple regression

General linear model

CHAPTER 14

Nonparametric Techniques

PROBLEMS WITH PARAMETRIC TESTS

All of the statistical tests that we have discussed to date except the Chi-squared test depend on two fundamental assumptions. The first is that we can meaningfully and sensibly calculate some basic descriptions of our data set, such as its mean and standard deviation. This implies that the data is measured on a ratio scale (or at the very least, on an interval scale; see Chapter 1). The second is that our data follows a Normal distribution. If both assumptions are reasonable, then we can, for instance, express the difference between two means in terms of standardized units (see Chapter 2) and argue that if the difference is bigger than approximately two such units, then the difference is unlikely (a probability of less than .05) to have arisen by chance. We would then conventionally reject a null hypothesis of no real difference between the means.

If either of these assumptions is not reasonable, then the whole logical argument on which all these techniques (the t test, analysis of variance, regression and correlation analysis, and the analysis of covariance) rest falls to pieces. Calculating a standardized result is pointless if the means and standard deviations involved are of dubious validity. The fact that less than 5% of such results will exceed a standardized value of approximately 2 is true only if the results follow a Normal pattern and cannot be assumed to be true for any other pattern. In this situation, these techniques (known as parametric methods because they use and test information about the parameters, that is, the mean and variance, of the Normal distribution) should not be used. We should also remember that the individual parametric tests may also have additional assumptions (such as equality of variance in the case of the independent t test) that must be met if the test is to be used.

What can we do if any of these key assumptions do not seem to be met in practice? There is a group of statistical techniques that parallel the more frequently used parametric tests, but that do not assume that the data is measured on at least an interval scale or that it follows a Normal distribution. They are referred to as **nonparametric tests** or distribution-free tests (since they assume nothing about the distribution of the data).

ANALYZING PAIRED DATA—THE SIGN TEST

Perhaps the best way to introduce the ideas underlying the nonparametric approach is to consider the analysis of a set of paired results (say, the difference in the assessments of two analgesics presented to patients in a crossover design). If we thought that calculating a mean difference was reasonable and that the differences followed a Normal distribution, then we would test for differences between the analgesics using the paired t test of Chapter 4.

PAIRED t TEST		SIGN TEST
-2		$-$
$+7$	H_0	$+$
$+1$	Difference $= 0$	$+$
$+4$		$+$
.		.
.		.

What would we do, however, if we felt unhappy with these assumptions? We might decide to disregard the actual numerical values (since we think we cannot use them to calculate a truly meaningful mean difference) and instead simply note the direction or sign of the difference (in other words, the analgesic for which the patient expressed a preference). We would, therefore, record the number of results that favored one analgesic (the $+$ results) and the number that favored the other (the $-$ results). These observations certainly would be valid and very meaningful.

If the null hypothesis of no real difference is actually true, then we expect no consistent evidence would emerge from the experiment. In fact, of the patients actually expressing a preference, we expect we would see half of them favoring one analgesic and half the other. If, however, one treatment genuinely yields more pain relief than the other, then we would almost certainly see this reflected in a substantial majority of preferences for this treatment. We can, therefore, test the null hypothesis of no difference between the analgesics by checking how well the observed preferences fit the expected (under H_0) situation of a 50/50 split between the preferences.

Let n_+ denote the number of $+$ preferences observed,
and n_- denote the number of $-$ preferences observed.
Let $n_p = n_+ + n_-$, that is, the total number of preferences expressed.

Remember that n_p is the total sample size less the number of tied results (0 differences). (Only individuals who express a preference can help us solve the riddle of which analgesic is superior.)

Under H_0, we would expect to see $n_p/2$ $+$ preferences and $n_p/2$ $-$ preferences.

	Preference		Total
	$+$	$-$	
OBSERVED	n_+	n_-	n_p
EXPECTED	$n_p/2$	$n_p/2$	n_p

The level of agreement between the observed and expected results can now be tested using the Chi squared goodness of fit test of Chapter 6.

$$\chi^2 = \Sigma \frac{(O - E)^2}{E} \qquad \text{with 1 df}$$

Since only 1 degree of freedom is involved, it is customary to utilize Yates' Correction (that is, reduce the absolute value of $O - E$ by .5).

An acceptable fit shows that the null hypothesis is a reasonable explanation and should be retained. Significant lack of fit, however, suggests that we should reject the null hypothesis and conclude that the treatment with more preferences is indeed superior.

Note that this test (known as the **Sign test**) makes no assumptions about the results, other than that the preferences are all independent decisions. By totally ignoring the size of the differences, we have cunningly avoided having to assume anything at all about their distribution or nature.

ANALYZING PAIRED DATA—THE WILCOXON SIGNED RANK SUM TEST

The Sign test represents a possible solution to the problem of dubious assumptions. It is, however, a rather extreme solution. By totally ignoring the size of the differences, we avoid having to make assumptions of any kind about them. However, we also throw away a great deal of potentially useful information. The Sign test, for instance, regards a difference of $+10$ as being effectively the same as a difference of $+1$. We might have grave doubts as to whether a difference of $+10$ is really 10 times more impressive than a difference of $+1$. However, it does seem reasonable and fair at least to acknowledge that $+10$ represents a bigger difference (a more clear-cut and dramatic preference) than $+1$. In addition, we would probably feel comfortable with the idea that a difference of 5 was, in a real sense, less clear cut than a difference of 10 but more clear cut than a difference of 1, distinctions that are completely lost in the sign test.

As a consequence, the Sign test is rarely used in practice and an alternate test, the **Wilcoxon signed rank sum test,** which retains this order information, is generally used in this situation.

The basic idea underlying practically all the nonparametric methods is that, while the magnitudes of the various results perhaps cannot be taken too seriously, they do at least allow us to rank the results. The first step in carrying out a Wilcoxon signed rank sum test is, therefore, to rank the differences in order from smallest to largest, disregarding the sign of the difference. (As in the Sign test, results expressing no preference or difference are ignored.)

This has the effect of ranking a difference of -5 between differences of, say, $+10$ and $+1$. This is, in fact, very sensible, since it gives us a ranking of the conviction level of the various study participants (-5 is a less decisive view of the perceived treatment differences than $+10$, but clearly more decisive than $+1$). However, the direction of difference is, of course, important evidence of possible treatment superiority and clearly must be taken into account. The next step in the Wilcoxon signed rank sum test is to tag each rank with its appropriate sign. Then the $+$ ranks are summed and the $-$ ranks are summed.

Under a null hypothesis of "no real difference" we would expect to see no consis-

■ **TABLE 14-1** ■
Paired Data—No Real Difference

Difference	Rank	Sign	+ Rank	− Rank
−12.1	6	−		6
+3.7	2	+	2	
+8.4	4	+	4	
−7.2	3	−		3
−1.6	1	−		1
+9.3	5	+	5	
			$T_+ = 11$	$T_- = 10$

tent pattern in the rankings (see Table 14-1). Positive signs should be just as common as negative signs. Additionally, the + and − signs should be scattered throughout the rankings, and the + rank sum should equal the − rank sum or be very close to it.

If there is a real difference between the two treatments under comparison (Table 14-2), then we expect we would see most (and in a very clear-cut situation, all) of the differences lying in one particular direction (that is, with one particular sign). We also certainly expect we would see the biggest differences (that is, those with the highest rankings) all correspond to this direction of real difference. All in all, therefore, we would expect one of the rank sums to be very large and the other to be correspondingly very small.

The smaller of the two rank sums is used as the test criterion. Although a very large rank sum suggests a real difference just as much as a very small rank sum, the smaller of the two values is the more convenient to assess. A perfect separation will always result in a smaller rank sum of 0, whereas the value of the larger rank sum will vary depending on the study size, and hence the number of ranks being totalled.

The smaller the minimum value is, the more evidence there is to suggest that real differences do indeed exist between the paired results.

How impressive the rank sum value is has to be assessed in the context of the total number of ranks (total sample size) involved. (A minimum rank sum of 10 would be very unimpressive if we had only 6 subjects (see Table 14-1) but very impressive if we had 60 subjects.) The key question is, "How likely is it that we could get a rank sum this small or smaller if there were really no genuine difference there?" Tables have been calculated for various sample sizes, giving the rank sum that would be equalled or exceeded (that is, a smaller rank sum observed) less than 5% of the time, and so forth, if there were no real difference. A rank sum less than or equal to the tabulated value, therefore, shows that we should reject the null hypothesis at the corresponding level of significance. An abbreviated table of Wilcoxon signed rank sum critical values is presented in Table 14-3. More detailed tables can be found on page 399 of the Handbook of Tables for Probability and Statistics published by CRC Press.*

There is one potential problem we should consider at this point. The basis of the Wilcoxon signed rank sum test (and of most nonparametric tests) is the ranking of the

---

*Beyer WH, editor: CRC Handbook of tables for probability and statistics, ed 2, Boca Raton, Florida, 1985, CRC Press, Inc.

TABLE 14-2

Paired Data—Dramatic Differences

Difference	Rank	Sign	+ Rank	− Rank
+2.7	3	+	3	
+3.6	5	+	5	
−0.1	1	−		1
+1.5	2	+	2	
+6.9	6	+	6	
+2.9	4	+	4	
			$T_+ = 20$	$T_- = 1$

TABLE 14-3

Critical Values of the Wilcoxon Signed Rank Sum Test

Sample size		6	8	10	12	14	16	18	20
Critical values	.05	1	4	8	14	21	30	40	52
	.01	−	0	3	7	13	19	28	37

results. What do we do if we observe two or more results that are exactly the same? In this situation the tied results are given the average of the ranks for which they are "eligible." For example, consider the following hypothetical results:

Result	1.6	0.7	2.1	2.1	3.5
Rank	2	1	3.5	3.5	5

3, 4

Here the tied results are clearly intermediate between the second and fifth ranked results and hence, are eligible for ranks 3 and 4. As a result of their equality, they both receive the mean ranking of 3.5.

This is the standard approach to the ranking of tied results in nonparametric analyses. The presence of ties tends to make the analyses somewhat less powerful since they have less information available to use. In certain situations, some of the nonparametric calculations have to be modified slightly to take account of the presence of ties. However, this becomes important only if the number of ties is relatively large, and in most situations, modifications are unnecessary.

Wilcoxon's signed rank sum test illustrates practically all the basic ideas of nonparametric analysis. It makes no assumptions about the data other than that the results are independent and that the relative size of the results at least reflects their relative stature. It ranks the results and measures an aspect of the ranking (the rank sums) that reflects potential deviations from the null hypothesis. The significance of the test value obtained is then assessed against a table of critical values, which indicates the chance of a value of that size or less resulting from chance alone.

ANALYZING TWO INDEPENDENT GROUPS—THE MANN-WHITNEY *U* TEST

What would we do if we wished to compare two independent sets of results rather than one paired set? The independent *t* test is, of course, the appropriate way to compare the means of two groups of results, provided we can reasonably assume that the results follow a Normal distribution, the variances of the two groups are comparable, and the means themselves are actually meaningful. Should we feel unhappy with any of these assumptions, we can utilize a nonparametric equivalent of the *t* test known as the **Mann-Whitney *U* test.** (We can think of this test as comparing the medians of the two groups, since the median is a "nonparametric" measure of the typical result, obtained by ranking the available evidence.)

The Mann-Whitney *U* test operates in the same general way as the Wilcoxon signed rank sum test. All the results are pooled and ranked from smallest to largest, ignoring the group to which they belong. The rankings are then tagged with the identity of the group from which they originated, and the sum of the ranks for each group is obtained.

Let n_a = Number of results in group A,

T_a = Rank sum for group A, and so forth.

The appropriate test statistic U is given by the smaller of:

$$T_a - \frac{n_a(n_a + 1)}{2}$$

or

$$T_b - \frac{n_b(n_b + 1)}{2}$$

We can illustrate the calculations involved with a simple example:

A	B
6.2	5.3
4.8	10.0
12.1	7.3
3.9	4.3

Result	3.9	4.3	4.8	5.3	6.2	7.3	10.0	12.1
Rank	1	2	3	4	5	6	7	8
Group	A	B	A	B	A	B	B	A

$$T_a = 1 + 3 + 5 + 8 = 17$$

$$T_b = 2 + 4 + 6 + 7 = 19$$

$$U = 17 - \frac{4 \times 5}{2} = 7$$

or

$$U = 19 - \frac{4 \times 5}{2} = 9$$

The test value is therefore 7.

The concepts underlying the Mann-Whitney U test are very similar to those underlying the Wilcoxon signed rank sum test. If the null hypothesis is true, then there should be no clear pattern. The two groups should be randomly scattered over the rankings. Hence, the rank sums (and values of U) should be very similar.

If, however, one group tends to yield consistently higher results than the other, a clear pattern will appear in the rankings. Results from the group with the lower median will tend to correspond to the smallest rankings, and the other group will dominate the higher rankings. As a result, one of the rank sums (and, hence, one U value) will be very small. The smaller the U value, therefore, the stronger the evidence in favor of a real difference in the group medians.

Why do we subtract a term from the rank sum, rather than simply use the rank sum itself? Even in the most clear-cut situation possible, one in which there is no overlap at all between the rankings of the two groups, the rank sum of the group with the smaller median cannot be zero. For example, in the following hypothetical situation in which the four results for treatment A are all smaller than any of the six results for treatment B, the rank sum for group A is 10.

Rank	1	2	3	4	5	6	7	8	9	10
Group	A	A	A	A	B	B	B	B	B	B

$$T_a = 1 + 2 + 3 + 4 \qquad\qquad = 10$$

$$T_b = 5 + 6 + 7 + 8 + 9 + 10 = 45$$

This minimum possible rank sum for a particular group is given by $n(n + 1)/2$ where n is the size of the particular group.

By subtracting this from the actual rank sum, we achieve a measure of how close the results actually came to achieving a perfect separation between the two groups. Therefore, this gives us a more immediately meaningful measure of the evidence in favor of group differences than is provided by the raw rank sum.

For the situation depicted above;

$$U = 10 - (4 \times 5)/2 = 0$$

confirming the perfect separation of the rankings.

Allowing either group A or group B to provide the test (that is, smallest) value of U is appropriate in the usual two-tailed test situation. After all, the alternative hypothesis is simply that the two medians differ, that is, either group A or group B could conceivably have the smaller median and thus produce the lower rankings and lower U value.

There may be situations, however, in which we have a specific alternative hypothesis in mind, that is, that one specific group should have the lower median. In this situation, a one-tailed test is required, and the group hypothesized to have the lower median must be used to calculate the test value of U. As in the parametric situation, a two-tailed .10 critical value corresponds to a one-tailed .05 critical value.

Again, the appropriate critical values for U are tabulated by sample size. (Remember that a U value equal to or smaller than a tabulated value indicates rejection of the null hypothesis at the corresponding level of significance.) Mann-Whitney U tables are rather cumbersome (as are most nonparametric tables), since they require you to specify the sizes of both the groups involved. For this reason we will not present them in this textbook. Detailed U tables are available on page 405 of Beyer (1985).* However, when the size of either of the two groups exceeds 20, the significance of U can be tested by calculating

$$\frac{U - n_1 n_2/2}{\sqrt{n_1 n_2 (n_1 + n_2 + 1)/12}}$$

where n_1, n_2 are the respective sample sizes.

This approximates a t statistic with infinite (∞) degrees of freedom (in other words, a standardized Normal deviate) and can be tested against the appropriate tabulated value.

The term $n_1 n_2/2$ is the U value we would expect to see if the null hypothesis were true and the two groups were perfectly intermingled. The top line therefore measures how dramatically the sample evidence differs from the null hypothesis situation. The term

$$n_1 n_2 (n_1 + n_2 + 1)/12$$

measures how variable this U value will tend to be. The larger the sample sizes, the more potential there is for the U value to fluctuate, and the larger this term will be. This ratio of difference to variability is very similar in principle to the approach taken by the t test, and its correspondence to a t value is not as surprising as it might seem.

COMPARING SEVERAL GROUPS—THE KRUSKAL-WALLIS TEST

The nonparametric equivalent of the one-way analysis of variance is the **Kruskal-Wallis test.** The Kruskal-Wallis test is, in many ways, a generalization of the Mann-Whitney test to enable it to cope with more than two groups. As in the Mann-Whitney test, the results are pooled and ranked from smallest to biggest. The rankings are then tagged with the identity of the groups from which they originated, and a rank sum is calculated for each group.

Let n_i denote the number of individuals in group number i, N denote the total number of individuals, S_i denote the rank sum for group number i, and k denote the number of groups involved.

The appropriate test statistic for the Kruskal-Wallis test is given by

$$H = \frac{12}{N(N+1)} \left(\frac{S_1^2}{n_1} + \ldots + \frac{S_k^2}{n_k} \right) - 3(N+1)$$

This formula does, in fact, greatly resemble the familiar calculations in the analysis of variance (Chapter 5).

---

*Beyer, WH, editor: CRC Handbook of tables for probability and statistics, ed 2, Boca Raton, Florida 1985, CRC Press, Inc.

The term:

$$\frac{S_1^{\,2}}{n_1} + \ldots + \frac{S_k^{\,2}}{n_k}$$

corresponds closely to the between-treatments sum of squares, which also involves squared totals divided by the number of results involved. The more clear cut the separation of the groups (that is, the more marked the differences in the rank sums), the larger this "between-groups" term will be, as we can see in the following example.

For three groups with four results/group and a great deal of group overlap,

Rank	1	2	3	4	5	6	7	8	9	10	11	12
Group	A	C	B	A	C	B	C	B	B	A	C	A

$$S_a = 27 \qquad S_b = 26 \qquad S_c = 25$$

$$\frac{27^2}{4} + \frac{26^2}{4} + \frac{25^2}{4} = 507.5$$

However, in the same situation, but with very clear-cut differences between the groups,

Rank	1	2	3	4	5	6	7	8	9	10	11	12
Group	A	A	A	B	A	B	B	C	B	C	C	C

$$S_a = 11 \qquad S_b = 26 \qquad S_c = 41$$

$$\frac{11^2}{4} + \frac{26^2}{4} + \frac{41^2}{4} = 619.5$$

The term $N(N + 1)/12$ measures how much the rank sums will tend to vary (the larger the numbers involved, the more potential there is for variation), and corresponds to the "error" or "within-groups" variation in analysis of variance. (Note the use of a very similar term when testing the Mann-Whitney U statistic.)

The use of both these terms in the formula for H mirrors the way we balance between-groups and within-group variation when assessing group differences using analysis of variance. In addition, we can think of the $3(N + 1)$ term as corresponding to a correction factor, removing the built-in rank sums that will be present even when there are no real differences between the groups.

In the Kruskal-Wallis test the larger the value of the test statistic, H, the stronger the evidence in favor of real group differences. The first data set, with substantial group overlap, results in a very small H value of .038. The presence of very clear-cut differences in the second data set is reflected in its much larger H value of 8.654.

Tables of the Kruskal-Wallis statistic, H, are given on page 430 of Beyer (1985).*

*Beyer, WH, editor: CRC Handbook of tables for probability and statistics, ed 2, Boca Raton, Florida, 1985, CRC Press, Inc.

It turns out (provided the group sizes are all greater than 5) that the Kruskal-Wallis statistic, H, follows a χ^2 distribution with $k - 1$ degrees of freedom. This means that it is often more convenient to test it by simply consulting a standard set of χ^2 tables. If the appropriate .05 critical value of Chi squared is exceeded, then we can reject the null hypothesis of no group differences (that is, equality of group medians) at the .05 level of significance.

──── TWO-WAY NONPARAMETRIC ANALYSIS OF VARIANCE — FRIEDMAN'S TEST

Flexibility is one of the major advantages of the parametric methods. The analysis of variance, for instance, can be used to test for differences between groups (one-way analysis of variance; see Chapter 5). However, it can easily be extended to control and eliminate variation from various sources and to test for the significance of a variety of influences (multi-way analysis of variance; see Chapter 11). This is much more difficult to do with the somewhat "rough and ready" methods of nonparametric statistics.

However, it is possible to extend the one-way nonparametric analysis of variance (the Kruskal-Wallis test) to allow it to control and eliminate one source of variation. This extension essentially allows us to carry out a nonparametric analysis of a randomized block experiment. In this, the between-treatments comparisons are carried out on a within-blocks basis. The between-blocks variation is thus eliminated completely from the analysis and, as a result, the between-treatments comparisons are rendered correspondingly more sensitive.

These within-blocks comparisons are achieved very directly by ranking the results separately within each block (that is, the k results in block 1 are ranked 1 . . . k, the k results in block 2 are ranked 1 . . . k, and so on). The ranks are then labelled with their corresponding treatment, and the rank sum is obtained for each treatment by summing over all the blocks.

The appropriate test statistic for this test, **Friedman's test,** is:

$$\chi^2 = \frac{12}{k(k + 1)} \left(\frac{S_1^{\,2}}{b} + \frac{S_2^{\,2}}{b} + \ldots + \frac{S_k^{\,2}}{b} \right) - 3b(k + 1)$$

where S_1 is the rank sum for treatment 1, and so on, b is the number of blocks, and k is the number of treatments.

(Note how the form of the Friedman statistic is very similar to that of the Kruskal-Wallis statistic, except that it emphasizes the block size, k, rather than the total sample size, N.)

Friedman's test is intended for use when the analysis of variance is unreplicated, that is, each treatment appears only once in each block. Each block, therefore, consists of k observations, each rank sum is based on b observations, and the total sample size is bk.

Like the Kruskal-Wallis statistic, the Friedman statistic follows a Chi squared distribution and can be tested against a set of Chi squared tables using $k - 1$ degrees of freedom. (Like the conventional Chi squared statistic that we met in Chapter 6, both the Kruskal-Wallis and Friedman statistics actually only approximate a Chi squared distribution. As a rough rule of thumb, the approximation is quite adequate when the rank sums are based on five or more rankings. For smaller sample sizes, detailed tables of the appropriate critical values are available.)

For example, the following results, comparing three treatments within five randomized blocks:

	T_1	T_2	T_3
Block 1	12.6	29.3	21.2
2	15.2	50.3	41.9
3	34.2	59.2	48.7
4	6.3	17.8	24.2
5	13.2	18.6	12.9

yield the rankings:

	T_1	T_2	T_3
Block 1	1	3	2
2	1	3	2
3	1	3	2
4	1	2	3
5	2	3	1

and the rank sums:

$$S_1 = 6 \qquad S_2 = 14 \qquad S_3 = 10$$

The Friedman's statistic is therefore

$$\frac{12}{3 \times 4} \left(\frac{6^2}{5} + \frac{14^2}{5} + \frac{10^2}{5} \right) - (3 \times 5 \times 4) = 6.40$$

which is significant [χ^2 (.05) = 5.99 for 2 df] and confirms the existence of significant differences among the three treatments.

CORRELATING RANKS—SPEARMAN'S RANK CORRELATION

The conventional correlation coefficient of Chapter 7 (more formally called the Pearson's product-moment correlation coefficient) requires the assumption that the measurements involved are on an interval (or ratio) scale and that the two variables being measured jointly follow Normal distributions. If these assumptions seem dubious, then a variety of nonparametric correlation coefficients are available to us. By far the most commonly employed, and the simplest to use, is the **Spearman's rank correlation coefficient.**

As we should expect by now, the Spearman's correlation coefficient is calculated from ranks rather than from the original measurements. In fact, the ranks are used in the most direct way possible. The x and y variables to be correlated are each ranked separately from smallest to biggest, and then the rankings are correlated with one another, exactly as if they were original observations (Table 14-4).

In fact, it turns out that, because the ranking situation is so much less complicated (Σx and Σy must both always equal $n(n + 1)/2$, and so on), it is possible to consider-

■ **TABLE 14-4** ■

Parametric and Nonparametric Correlation

Original Observations		Rankings		Difference
x	y	x	y	d
27.3	12.1	1	1	0
36.1	20.2	2	2	0
52.9	27.6	5	4	+1
50.3	25.3	4	3	+1
48.4	29.9	3	5	−2
72.9	40.2	6	6	0

$$\frac{\Sigma xy - (\Sigma x)(\Sigma y)/n}{\sqrt{[\Sigma x^2 - (\Sigma x)^2/n][\Sigma y^2 - (\Sigma y)^2/n)]}}$$

Pearson's
product-moment
correlation coefficient

$$1 - \frac{6\Sigma d^2}{n(n^2 - 1)}$$

Spearman's
rank
correlation coefficient

ably simplify the correlation formula in this special case.

Spearman's rank correlation coefficient is equivalently (and much more conveniently) given by

$$r_s = 1 - \frac{6\,\Sigma d^2}{n(n^2 - 1)}$$

where d is the difference between each pair of ranks, and n is the number of pairs involved.

For instance, for the data set quoted in Table 14-4,

$$r_s = 1 - \frac{6(1 + 1 + 4)}{6 \times 35} = .829$$

Like the product-moment correlation coefficient, the Spearman's rank correlation coefficient can take values between $+1$ and -1 corresponding to a perfect direct and a perfect reversed relationship between the ranks.

For sample sizes of 25 or more, Spearman's rank correlation can be tested against standard correlation coefficient tables with the usual $n - 2$ degrees of freedom. For sample sizes of less than 25, special tables of critical values for r_s should be used and are given on page 445 of the CRC Statistical Tables.*

In common with most nonparametric methods, the basic formula for the rank correlation coefficient should really be modified if tied ranks are present. However, unless a substantial number of ties are involved, the resulting modifications are minor and, for all practical purposes, can be ignored.

---

*Beyer, WH, editor: CRC Handbook of tables for probability and statistics, ed 2, Boca Raton, Florida 1985, CRC Press, Inc.

There is no really practical nonparametric equivalent of regression analysis (and hence of analysis of covariance). Although nonparametric techniques have been devised to deal with most hypothesis testing situations, it is very difficult to develop a way of describing relationships between measurements without making assumptions about the measurement scales. It has proven virtually impossible to extend the simplistic nonparametric philosophy to the sophisticated situations with which multi-way analysis of variance, analysis of covariance, and multiple regression can cope so easily.

TRANSFORMATIONS

There is often an alternative to carrying out a nonparametric analysis (an option that could be very attractive if we found ourselves requiring some relatively sophisticated analytic technique). When it is clear that the data violates one or more of the assumptions required to carry out the appropriate parametric test, it is often possible to remedy the situation by transforming the data (in other words, measuring it on another scale). The use of **transformations** often can effectively restore a data set to normality, equalize variation in groups, or help make relationships linear.

One of the key assumptions underlying all the parametric techniques is that the data follows a Normal distribution. Back in Chapter 6 we discussed how we could use the Chi squared goodness of fit test to determine if there was any evidence that a set of data was not following a Normal distribution. A simple visual inspection of the data will, however, usually reveal any marked or important departures from normality. One of the most frequently encountered departures from normality occurs when the results are skewed toward the right. (In other words, very high values occur more frequently than very low values.) Replacing the original measurements with their logarithms will pull high values down more than low values and effectively restore a skewed distribution to symmetry and normality (Figure 14-1). The appropriate parametric analysis can then be carried out on these transformed values.

There are some other relatively common research situations that can be troublesome and that can be easily corrected by a simple transformation. When we study measurements that are counts, it is important to remember that the variance of a Poisson distribution is equal to its mean (see Chapter 2). If, therefore, a *t* test or analysis of

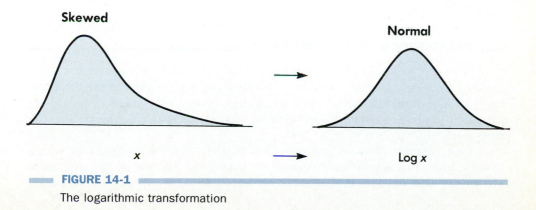

FIGURE 14-1

The logarithmic transformation

TABLE 14-5

Transformations and Their Objectives

Transformation	Measurement	Objective
Logarithmic Log x	Continuous	Alter skewed distribution to normality
Square root $\sqrt{x}$	Counts	Ensure equality of variance
Arcsine $\sin^{-1}\sqrt{p}$	Proportions	Ensure equality of variance

variance shows that the means of two or more groups of counts are unequal, it has also simultaneously shown that their variances are unequal. Since equality of variance is a basic assumption of both these tests, a significant result effectively proves that such a test should never have been carried out! It can be shown that replacing the counts by their square roots effectively stops the variance being controlled by the means and helps ensure equality of variance in the groups being compared. This "square root" transformation is, therefore, used extensively when working with counts data.

A very similar situation arises when dealing with measures that are proportions, since the variance of a proportion is dependent on the proportion itself (see Chapter 2), that is,

$$p = \text{Proportion} \qquad \text{Variance} = p(1 - p)/n$$

The variance can be effectively made independent of the proportion by replacing the original measurements with:

$$\sin^{-1}\sqrt{p}$$

that is, by the angle whose sine is $\sqrt{p}$. For example, we would replace a proportion of .25 by the value 30, since the sine of 30 degrees is .5, that is, the square root of .25.

This variance-equalizing transformation for proportions is known as the arcsine, angular, or inverse sine transformation.

The features of these most frequently encountered transformations are summarized in Table 14-5.

PARAMETRIC OR NONPARAMETRIC?

Since the nonparametric methods require far fewer assumptions than the corresponding parametric methods, wouldn't it be safer and more prudent to use them routinely for all statistical analysis? There are, in fact, several very good reasons why the parametric methods continue to be far more widely used than their distribution-free equivalents.

First, parametric tests are generally more powerful than the corresponding nonparametric tests. This is principally because the parametric tests use all the information available to them while the nonparametric methods discard information to some degree.

If the assumptions involved are met, it is much more sensible to use a parametric test, since this will reduce your risk of making a β error and missing a genuine effect.

Second, the parametric tests have been repeatedly shown to be robust to departures from their basic assumptions. In other words, even when there are moderate departures from normality or equality of variance, the parametric tests give reliable results with α and β errors very close to those expected under ideal conditions. This means that the parametric methods can be used with confidence in most practical research situations. This is not carte blanche to use a parametric method without asking yourself, "Are its assumptions reasonable?" You must establish, either by testing or inspection, that the relevant assumptions are not grossly violated. We have already mentioned the goodness of fit test for normality. It is also possible, and very simple, to test for the equality of two variances. Simply divide the larger variance by the smaller variance. The result is a variance ratio, in the most literal sense, and can be tested against a set of F tables with degrees of freedom defined by the sizes of the respective groups less one. A significant result confirms that the two variances are indeed different.

Third, the parametric tests are far more elegant than the nonparametric tests. In the simple two-group comparison situation, both parametric and nonparametric tests are available. However, when the comparisons are part of an even moderately complicated experimental design, it is usually impossible to find a nonparametric method that can cope with the analysis. In addition, the nonparametric methods can be cumbersome to use. When the numbers involved are small, there is no doubt that the nonparametric methods are much more convenient than the parametric tests. However, when sample sizes become reasonably large (say, over 30), the ranking of the results can become very tedious and time-consuming. (The larger the sample size, the more robust the parametric methods, since the more deviations from normality are effectively smoothed out. Therefore, with large sample sizes, the parametric methods are especially attractive.)

In general we can say that, if the assumptions required seem reasonable (either as the data stands or after it has been transformed), or at least are not seriously violated, then the appropriate parametric test should always be used. If, however, the assumptions are clearly inappropriate, then the nonparametric approach is the only justifiable method of analysis.

One situation in which nonparametric methods are clearly indicated arises when the results are measured on an ordinal scale. Here we cannot, or certainly should not, attempt to calculate means or claim that our data follows a Normal distribution, and the ordinal nature of the data lends itself ideally to the ranking philosophy of the nonparametric methods.

▬ Key Terms ▬▬▬▬▬▬▬▬▬▬▬▬▬▬▬▬▬▬▬▬▬▬▬▬▬▬▬▬

Nonparametric tests

Sign Test

Wilcoxon signed rank sum test

Mann-Whitney *U* test

Kruskal-Wallis test

Friedmann's test

Spearman's rank correlation coefficient

Transformation

PROBLEMS

1. In a study of the usefulness of an anticonvulsant (carbamazepine) in the management of acute schizophrenia, 20 patients meeting the admission criteria were randomly allocated to either a placebo or a carbamazepine treatment group. The patients were then assessed at weekly intervals (on a double-blind basis) on the In-
patient Multidimensional Rating Scale (IMRS). (A lower score is more favorable.)
The IMRS scores for the two groups after 5 weeks are as follows. Use the appropriate nonparametric technique to compare the treatment groups.

IMRS SCORE	
PLACEBO	**CARBAMAZEPINE**
23	8
12	3
6	13
15	16
11	7
14	5
6	11
15	4
9	5
16	10

2. The IMRS scores at 1 week of treatment and 5 weeks of treatment are given below
for the placebo patients mentioned in problem 1. Use the appropriate nonparametric technique to determine if there is any evidence of a change in these patients'
conditions.

PATIENT	1 WEEK	5 WEEKS
1	16	23
2	14	12
3	9	6
4	10	15
5	11	11
6	10	14
7	7	6
8	20	15
9	17	9
10	12	16

3. Ten arthritic patients suffering from chronic pain were asked to try three analgesics in random order, and rank them by preference. A rating of one indicates most preferred, and so on. The results are presented as follows. Is there any evidence of differences in perceived effectiveness between the analgesics?

PATIENT	ANALGESIC		
	A	B	C
1	1	3	2
2	1	3	2
3	2	3	1
4	3	2	1
5	1	3	2
6	1	2.5	2.5
7	1.5	3	1.5
8	1	3	2
9	2	3	1
10	1	2	3

4. A 1-day alcohol education program was administered to health workers from three different specialties. After completion of the program, the participants' knowledge of alcohol-related problems was measured on a 20-point scale, with higher scores showing more knowledge. Use the appropriate nonparametric technique to test for evidence of any difference between the various specialties.

GENERAL PRACTICE	ACCIDENT AND EMERGENCY	SOCIAL WORK
6	13	14
11	8	17
10	10	16
15	7	9
9	12	18
	4	11

5. As part of the education program outlined in problem 4, the 17 participants were also rated in their attitudes toward working with alcoholic patients. (This rating was based on a 16-point scale, with a higher rating corresponding to a more positive attitude.) Use the appropriate nonparametric technique to determine whether any relationship exists between knowledge and attitude.

KNOWLEDGE	ATTITUDE
6	10
11	7
10	9
15	13
9	4
13	11
8	3
10	10
7	6
12	14
4	3
14	11
17	16
16	15
9	5
18	14
11	12

CHAPTER 15

The Design of Observational Studies

THE ROLE OF OBSERVATIONAL STUDIES

The designed experiment (which we discussed in Chapter 8) is, in many ways, the ideal way to carry out a health investigation project. In it, randomly selected groups of patients can be subjected to a variety of well-defined treatments or interventions, and the relative merits of these interventions can be evaluated. Because of the random assignment, the progression from intervention to outcome, and the high degree of control inherent in an experiment or clinical trial, we can be fully confident in concluding that any differences between the patient groups must be due to the different interventions involved.

However, there are many situations in which we wish to study conditions that simply cannot be manipulated in this way. In a study of the effects of smoking on respiratory disease, it would clearly be totally unethical and totally impractical to randomly assign one group of subjects to a smoking regime for 20 or more years and another group to a comparable nonsmoking regime. All that can be done in this and similar situations is to passively observe the smoking habits and respiratory problems of an appropriate group of subjects, without any attempt to control their behavior or to impose an intervention. These two problems of ethics and practicality occur in many areas of health research. However, they are characteristically encountered in the social environment, where individuals make their own health/lifestyle decisions, or have their decisions imposed on them by economic or other circumstances.

Observational studies, or surveys, have strengths and weaknesses. Since they are much less controlled than designed experiments, we clearly must be much more cautious about the conclusions we draw from them. If patients who smoke heavily are observed to have considerably more respiratory problems than nonsmokers, could this be due, at least in part, to other lifestyle features, such as lack of exercise or heavy alcohol consumption, possibly associated with heavy smoking? Might the respiratory problems be due to the fact that individuals who choose to smoke are inherently different from those who choose not to and are, for whatever reason, inherently more prone to respiratory disease? Such ambiguities are present in most observational situations because we cannot control the interventions that the subjects impose on themselves. Without the protection of random allocation, we have no guarantee of inherent comparability. However, observational studies also have their advantages. Because surveys take place in the real world (the ultimate laboratory), their results and conclusions are usually highly relevant and widely generalizable, qualities that experimental studies often badly lack.

THE OBJECTIVES OF OBSERVATIONAL STUDIES

Observational studies usually have one of two primary aims. The first is to describe the health status of a particular population by measuring the prevalence of a disease (the proportion of affected individuals in a population at a fixed point in time) or by estimating the mean level of a health parameter. We might, for instance, wish to ascertain the prevalence of glaucoma in institutionalized over-60-year-olds, or to estimate the mean alcohol consumption of unemployed males between 18 and 29 years old.

The second aim is to improve our understanding of the etiology of a disease process by determining the factors or influences that are associated with the disease. We might, for instance, wish to establish whether prolonged heavy drinking is associated with the occurrence of coronary heart disease in young males.

The descriptive approach usually is the initial method of investigation. Once a general understanding of the disease has been established, the etiologic approach is then used to determine which, if any, of a variety of suspected risk factors play a role in the disease process.

Surveys can be categorized into two major types, depending on the involvement of a time element in the survey process.

Cross-sectional Surveys

A **cross-sectional survey** is primarily concerned with describing or measuring a population at one fixed point in time. In a sense, it represents a numerical "photograph" of some aspect of a population's health. For this reason, cross-sectional surveys are sometimes called static surveys.

Cross-sectional surveys are primarily concerned with the descriptive approach to health investigation. By their nature they can provide information on the prevalence of a particular condition and also describe other aspects (gender and age distributions, and so forth) of the population under study. These descriptive features can, however, be combined to provide clues to potential risk factors. A cross-sectional survey might, for example, be intended to establish the prevalence of mental illness, specifically clinical depression, in a certain urban community and how that prevalence varies among subgroupings within the community. The results might show that the proportion of unemployed individuals suffering from clinical depression is significantly higher than the proportion observed of individuals who are currently employed.

Such a result establishes a link between depression and employment status. It does not, however, pinpoint the direction of the relationship. Is depression a consequence of unemployment, or are depressed individuals less able to function effectively in the workplace and hence at greater risk of dismissal? Although the results of a cross-sectional survey can never be regarded as proof of the existence of a risk factor or causal connection, they can play an important part in establishing potential disease/exposure links that can be subsequently explored in more detail by the etiologically oriented longitudinal surveys to be discussed in the next section.

Longitudinal Surveys

A **longitudinal survey** is concerned with describing or measuring a population at several points in time. This might involve repeating the measurement process at successive points in time, carrying out the survey once but obtaining descriptions of events that occurred at different points in time, or following the study individuals over time until a certain event (such as death or a coronary episode) occurs.

In all these situations, a longitudinal survey offers a picture of the health of a population as it changes with time. (It can be likened to a "movie" rather than a "photograph.") For this reason, they are sometimes called "dynamic" surveys. Because the time element enables us to explore relationships between cause and effect in a more natural and logical way, longitudinal surveys are primarily concerned with etiological aspects of health investigation.

(With the existence of a time axis, we could now determine whether loss of employment preceded a depressive episode. This information would place us in a much stronger position to identify unemployment as a potential precipitator of, or a risk factor for, clinical depression.)

Longitudinal surveys (especially the prospective studies discussed in the following section) can, of course, also fulfill an important descriptive function.

Longitudinal surveys fall into two types, depending on the direction of the time process: prospective studies and retrospective studies.

Prospective studies. In a **prospective study**, the subjects are followed forward in time. In a sense, it aims to move from (potential) cause to effect. In a typical prospective study, a group or cohort of individuals is followed over a period of years, during which they select themselves (either by choice or circumstance) into various subgroups. Such a cohort is frequently subdivided on the basis of the risk factor under investigation into those individuals exposed to the factor in question and those unexposed. The rates of occurrence of the disease in question for the exposed and unexposed groups can be compared, and the question of whether exposure results in increased risk of disease can be tested.

For example, in a classic study that helped establish the link between smoking and lung cancer, family doctors in Britain were followed over a period of 20 years to compare the incidence of lung cancer in those doctors who smoked, those who never smoked, and those who had given up smoking.

Prospective studies are also variously called follow-up studies, cohort studies, or incidence studies. (Since they follow a cohort of individuals forward over time, they can yield information on the incidence of a disease (the proportion of initially unaffected individuals in a population who manifest the disease over a certain specified time period), something that a cross-sectional or prevalence study just cannot do.)

Cohorts are often drawn from groups of individuals assumed to be at high risk of the disease under investigation (thus ensuring that an appreciable number of disease occurrences will become available for study) or from groups of individuals whose personal and health histories are well documented and for whom long term follow-up and disease detection is likely to be practical. For both these reasons, various occupational groups have frequently served as the basis for cohort studies, because of the availability of company employment and medical records and varying quantifiable levels of exposure to the suspected risk factor.

When detailed records are already in existence, a "historical" cohort can sometimes be created. By learning that an individual joined the company in 1957 and was diagnosed with lung cancer at a company medical examination in 1971, after having worked in an asbestos-rich environment for 12 years, we can instantly create a follow-up profile for that individual. Cohort studies assembled in this way are sometimes called "retrospective cohort" studies, since the information was generated in the past. However, since the focus of such studies is still on the natural progression forward from exposure to disease, they are mentioned in this section. When this kind of detailed information is not available, then a cohort must be recruited in the present and tracked into the future.

Prospective studies are, in many ways, the ideal way to look at disease/risk factor relationships. They approach the cause/effect relationship in the most logical way, and the quality and nature of the information collected is under the complete control of the investigator. However, prospective studies are usually expensive and inevitably time-consuming. (In Manitoba, Canada, a cardiac-oriented cohort study of almost 4000 World War II airmen has run for almost 50 years, and follow-up periods longer than 20 years are not uncommon.) In addition, keeping track of the study individuals over a long time can present major problems, and patients lost from follow-up may differ in some systematic way from those retained, resulting in study bias.

In general, prospective studies are more suitable for common diseases. When a disease is relatively rare, a very large and unwieldy cohort would be needed to ensure that the eventual disease numbers are large enough to produce an acceptably powerful study. (About 50,000 doctors were involved in the British study of smoking and lung cancer.)

Retrospective studies. A **retrospective study** looks back in time. Therefore, it moves from effect back to potential cause. In the typical retrospective study, a group of individuals who already suffer from the disease in question are compared to a control group of individuals who do not suffer from the disease. It is then ascertained whether the individuals in the two groups have been exposed to a particular risk factor. The rates of exposure are then compared in the disease and control groups. (Groups of lung cancer and lung cancer–free patients, for instance, might be questioned about their smoking habits and history.)

Retrospective studies (also called case-control studies, for obvious reasons) have several attractive features. They are relatively inexpensive and can be completed very much more quickly than prospective studies. (The major risk factors associated with AIDS were discovered by noting evidence of differing sexual practices between a group of infected individuals and a non-infected group. The ability to pinpoint risk factors after weeks, rather than decades, of study meant that preventive measures could be implemented as soon as possible.)

However, retrospective studies also have some potentially serious drawbacks. The fact that the evidence must be obtained from historical sources can pose serious problems of incompleteness, inaccuracy, or bias. When the evidence has to be gleaned from medical records or other documentary sources, there is no guarantee that the relevant information will be recorded accurately or consistently, or even recorded at all. Even when the information can be obtained by patient interview, the risk of the patient unintentionally exaggerating, minimizing, or totally forgetting past exposure experiences ("biased recall") must be viewed as very real. Although this is true for both cases and controls, it is perhaps an even greater danger in the case or disease group (raising the specter of between-group bias). Guilt may cause diseased patients to perceive past experiences as having been more extensive than they truly were, or it may result in a denial process that underplays their historical behavior patterns.

Another major problem in case-control studies is the selection of the control group. In follow-up studies, the groups to be compared (say, smokers and nonsmokers) select themselves quite naturally. In case-control studies this is definitely not the situation. The cases will effectively define themselves. The controls, however, must somehow be selected from the population of individuals who do not have the disease in question. Is it really possible to do this in a way that ensures that the groups are truly comparable to one another and that the comparison is fair? If the cases and controls are inherently different from the beginning, other than in their levels of exposure to the potential risk factor, then the whole comparison process is meaningless.

Various approaches are taken to control selection. Because the cases are usually encountered in a hospital or other health institution setting, controls are often selected from patients at the institution who are suffering from diseases assumed to be totally unrelated to the one under study. This is usually a convenient and inexpensive way to recruit controls. It could, however, create problems if the diseases represented in the control group turn out to be also linked to the risk factor under investigation. (For this reason, it is preferable to recruit the controls from as many different disease sources as possible.)

A more appealing, but certainly more expensive and less convenient, alternative would be to recruit the controls from the general, non-diseased population from which the cases themselves ultimately derive. To ensure optimal comparability, the cases and controls are often matched with one another (age and gender are two very frequently used matching characteristics) on an individual basis, or else these characteristics are built into the study design as blocking factors. Comparing a group of predominantly young, male, AIDS cases to a group of age and gender matched controls is clearly far more meaningful than comparing them to a group of men and women of widely differing ages. Matching is sometimes also achieved by recruiting controls who live close to the respective cases ("neighbor controls") and thus generally have comparable lifestyles. The lack of a truly natural control group is, however, an unavoidable cause for concern in the case-control situation.

Retrospective studies are particularly well suited to the study of relatively rare diseases. To investigate such diseases prospectively would require very large cohorts, very long follow-up times, or indeed both. A prospective investigation of a possible link between contraceptive use and cervical cancer might require a cohort of 10,000 women to be followed for at least 5 years to ensure that an adequate number of cases develop. A case-control study of 100 cases and 100 matched controls might well result in a study of appreciably greater power at a very much lower cost.

SAMPLING FRAMES

An inescapable fact that underlies practically all health research is that the study of a relatively small, suitably selected sample of individuals usually offers the only practical means of obtaining information about the behavior and characteristics of a very much larger population of individuals that primarily interest us. Indeed, biostatistics as a discipline is concerned with telling us just what we can infer about the behavior of a population, given evidence from a sample. Cross-sectional, cohort, and case-control studies all generally depend on sample selection to make them feasible. (A case-control study being carried out at a particular research center will frequently employ all the available cases, but the corresponding controls will usually have to be selected from a population of potential controls. Historic cohort studies may draw on all the available records, or all employees recruited in a particular year may be followed into the future, but cohort selection from some population of interest is generally the only practical approach.) To successfully implement an observational study, therefore, requires us to come to grips with the principles and practice of selecting an appropriate sample.

Since the aim of a sample is get information on a larger group or population, the logical first step in any investigation is to define the population of interest (the "**target population**"). The next step is to list all the individuals in that population as a prelude to selecting a sample of individuals for detailed study. The individual entities that form the focus of the study are called **sampling units** (these might be individual people, individual family groups, individual hospitals, and so forth, depending on the objectives of the study), and the list of sampling units is known as a sampling list or a **sampling frame.**

Defining the target population is usually easy (for example "all males between the ages of 18 and 65 on June 1, 1990, and resident in the state of California on that date"). Creating the corresponding sampling frame is usually considerably more difficult and can be impossible. A surprising number of sampling frames do exist, however, or can be created with a little effort. For health studies targeted at the general population, electoral registers and similar lists can serve as sampling frames. For the study of specific target groups, hospital records, health clinic records, and so on, can be used as the basis for a list of potential subjects. Other specialized populations (such as schoolchildren) may well be comprehensively documented by various administrative systems.

Ideally, a sampling frame should be a perfect list of all the individuals in the target population. In practice, this is rarely the case. Inevitably, a sampling frame will be out of date by the time it is compiled. Individuals will have died or changed address, new individuals will have joined the target population, other individuals will never have been placed on the list at all, and so on. Probably not much can be done to remedy this situation completely. Investigators should at least be aware of the potential shortcomings, remedy them where possible, and have some idea of the reliability of the sampling frame they propose to use.

Particular attention should be paid to the common situation in which particular subgroups of individuals are systematically excluded from the sampling frame, either intentionally or, more often, inadvertently. The homeless would be excluded from sampling frames based on address; using a list of telephone numbers as a sampling frame may exclude a disproportionate number of the poor; individuals living in remote areas may be deliberately excluded from sampling frames to minimize study costs. In a situation like this, the population actually studied by the selected sample ("the study population") is not the same as the population originally intended for study (the target population). When the missing elements of the population are distinctive and important (such as particular ethnic groups or patients with particularly severe conditions), the survey results should be interpreted with an appropriate degree of caution.

Developing a sampling frame is usually a challenge. Indeed, there are situations in which it is impossible to prepare in advance a detailed list of individuals even when we can define precisely the population we wish to study (such as all patients using the hospital emergency rooms in a given metropolitan area over the next 6 months). In such circumstances a sample can still be selected by using an appropriate sample selection technique, even in the absence of a detailed sampling frame. When the population of interest can be clearly defined but a list of the individuals in the population cannot be made, we are said to be using a **notional sampling frame.**

SAMPLE SELECTION

Once the sampling frame has been specified (either a literal sampling frame that lists all the individuals in the population, at least in theory, or a notional sampling frame that merely defines the population), the next step is to select a sample of individuals from the frame for detailed study. The actual sample selection can be approached in three basic ways.

Purposive Selection

The objective of selecting a sample is, after all, to obtain results that, we hope, represent the population as a whole. Therefore, there can be a great temptation to deliberately or purposively select individuals who seem to represent the population under study. In a study of malnourishment in an urban school population, for instance, 20 "representative" children might be picked for detailed examination and diet assessment.

Purposive selection is very easy to carry out and does not require the preparation of a sampling frame. However, it is fraught with disaster and is never advisable. It has been repeatedly confirmed that human beings, no matter how experienced and highly skilled, are woefully incapable of deliberately selecting a truly representative sample. The final results obtained will inevitably be biased and will consistently overestimate or underestimate the true population value. In effect, results obtained this way merely (and inevitably) reflect and reinforce the preconceived opinions and prejudices of the person selecting the sample. The bias (or, at the very least, the potential bias) of the investigator is built into the results, and the study outcome cannot be viewed as an independent, objective, and unbiased assessment of the problem under investigation. The parallels with the allocation of patients to treatments in experimental studies (see Chapter 8) are very real. Even the possibility of bias is enough to completely undermine a study's credibility.

In addition, purposive selection is very likely to produce results that substantially underrepresent the variability of the population under study. If the individuals are all selected to represent the typical individual (whether, in fact, they do or not), they will, by definition, be similar to one another. As a result, the variability of the sample will be very much less than the variability of the population. In many studies, it is as valuable (and often more valuable) to obtain a picture of the range of variation in the population as it is to describe the "typical" individual. All in all, purposive selection has little, if anything, to recommend it.

Random Sampling

A good sample should be a fair reflection of the population from which it is drawn, in both its mean value and its variability. Any attempt to influence the sample selection by human intervention is very likely to result in a sample that is biased and no longer representative of the whole population. (The suspicion of bias will certainly be raised.) The only way to ensure that a sample is free from even the risk of bias is to remove the human element, and the resulting risk of prejudice, entirely. An "untainted" sample should, therefore, be selected on some purely random basis. Drawing names from a hat is a perfect example of random sample selection.

However, **random sampling** is most conveniently implemented through the use of random number tables (see Chapter 8). Typically, each individual in the sampling frame is identified by a sequence number. Random number tables are then used to select the sequence numbers and, hence, the individuals who will constitute the sample. The malnourishment study to which we alluded in our discussion of Purposive Selection should, therefore, be based on a random sample drawn from a list of all the children attending school in the defined urban area.

Random sampling ensures that sample selection is bias-free. It must, therefore, form the basis of any acceptable sampling system. However, by its very nature, it requires a sampling frame to detail the individuals who are eligible for selection. This can constitute a major practical problem.

Systematic Sampling

It is often simply impossible to draw up a sampling frame, even when we have a very clear idea of the population we wish to study. We clearly cannot specify in advance the particular individuals who will use a particular health resource in the next year, and hence we cannot hope to have a detailed sampling frame in place before the study starts. How, then, can we hope to select an acceptable sample?

Even in the absence of a formal sampling frame, it is still usually possible to select a sample that has the unbiased characteristics of a random sample by using an approach known as **systematic sampling.** This requires us to decide the proportion of the total population (say, 1 in 50) to be included in the sample. This is known as the **sampling fraction.** To do this we need to make an educated guess as to the size of our population of interest and, of course, calculate the required sample size. If we anticipate that about 20,000 individuals will use the health facility in the next year, and our sample size calculations show that we will need 400 individuals to achieve an acceptable level of power, then a sampling fraction of 1 in 50 (400/20,000) is indicated. A random number between 1 and 50 is selected to define the initial sample individual. The remaining individuals are obtained by systematically selecting every fiftieth subsequent individual.

Use of a hospital emergency room over the next year could be surveyed, for instance, by selecting, say, the thirty seventh admission (determined on a random basis) and then selecting the eighty seventh admission, the one hundred and thirty seventh admission, and so on, for detailed study. Since the composition of the final sample is defined solely by the initial random selection, there is clearly no risk of human bias. Since it does not require the preparation and use of a sampling frame, it is a highly flexible and practical sampling method and can be adapted to most sample selection situations.

Even when a detailed sampling frame can be constructed (for example, in a survey of previous use based on the health records of all the individuals who used the emergency room in the preceding year), it may be more convenient to sample the records on a systematic, rather than purely random, basis.

Systematic sampling does have one potentially serious, but admittedly uncommon, disadvantage over simple random sampling. If a particular phenomenon tends to occur in the population periodically, and if its period coincides with the period of sample selection (that is, every fiftieth unit in our example), then the final systematic sample could seriously misrepresent the population as a whole. If, for example, the emergency admissions selected tended to occur at weekends, they might substantially overrepresent the unit's alcohol-related workload. Periodic phenomena of this type are rare, but when they occur, their effects can be disastrous. The possibility of a periodic influence should, therefore, always be considered when deciding to use systematic sampling.

SURVEY DESIGN

Selecting the sample from the sampling frame, using either random or systematic sampling as appropriate, is the fundamental component of survey design. However, there may be very real practical problems to face when organizing the subsequent field work and sample collection, especially when the survey is large. Care at the initial design and organization stage of a survey can often reduce expenses, improve efficiency, and yield additional useful information.

Stratified Sampling

When the population of interest consists of a number of clearly defined subpopulations or strata, it often may be more convenient to take this into account and explicitly select separate samples in each stratum, a process called **stratification.** In effect, we treat each stratum as a separate population and carry out a series of parallel surveys. This approach has a number of practical advantages. Different staff members can be assigned responsibility for data collection in each stratum, thus simplifying administration. Using stratification on a regional basis will considerably reduce expenses and administrative problems since the field staff assigned to a particular stratum will have to function with only a restricted geographic area. A nationwide survey of alcohol abuse in Canada, for instance, might be stratified on a provincial basis; a similar survey in the U.S. could be stratified by regional groupings of states.

Stratification offers additional advantages. Results from the various strata can be compared very easily, and inter-strata differences can be detected. Stratified surveys are, therefore, inherently more informative than unstratified surveys. Another important advantage is that they are also more flexible. If, for example, a particular stratum is thought to require more intense examination (because it is more variable, less common, or simply of more interest) than the other strata, then a higher sampling fraction can be employed for that particular stratum.

In an alcohol abuse survey, for instance, special interest might be focused on teenage alcohol abuse. Stratification on age would facilitate between-ages comparison and would also facilitate a more intense sampling of the target teenage population. An unstratified sample would contain relatively few teenagers, since they constitute a relatively small proportion of the alcohol-consuming population. As a result, estimates of the behavior of this subgroup would have relatively wide confidence limits, and comparisons with other groups would have relatively low power, problems that could be solved by increasing the sampling fraction for this special-interest group.

Although stratified sampling is an extremely useful approach, it does nothing to help us overcome the often difficult problems of developing a detailed sampling frame for our target population. (In stratified sampling, we simply define distinct sampling frames for each stratum.) There are organization strategies, however, that can help reduce these problems.

Cluster Sampling

One such strategy uses the fact that the individuals who are the focus of a survey frequently occur in natural groups or clusters. Human beings tend to form family groups. They also tend to live clustered in buildings (on the small scale) or in villages and towns (on the large scale). It is often relatively easy to construct a sampling frame of those natural groupings (such as a list of addresses or a list of villages) even when it is very difficult or impossible to construct a sampling frame for the individuals. In these circumstances, one option is to randomly select some of the clusters or groupings (from a sampling frame of clusters) and then to include in the final sample all the individuals in the selected clusters. This is known as **cluster sampling.** In the developing world, for example, there are likely to be few sampling frames of individual people, whereas sampling frames of clusters such as villages can be constructed fairly easily. A health survey of a developing nation could, therefore, be carried out by selecting certain villages at random and using a health team to survey all the individuals in each community.

Cluster sampling has a number of practical advantages. Sample selection problems are very greatly reduced. There are also considerable reductions in field costs when data collection is concentrated in a relatively small number of locations. Provided the clusters are selected at random, the results will still be unbiased. However, the results will be inherently somewhat less reliable (have wider confidence limits) than results obtained by simple random sampling or stratified sampling because coverage of the overall population is inevitably more patchy.

Multi-stage Sampling

With a little imagination, the problems of constructing a detailed sampling frame can be defused (or at least delayed) in other ways. The individuals in a population can often be defined in a hierarchical manner (such as employees within a division, divisions within a factory, factories within a company). In such circumstances, the final sample can be selected in an approach analogous to these hierarchies. At each stage of the hierarchy, some of the basic units are selected at random. For each of these, some of the corresponding subunits are then selected at random, and so on; this technique is called hierarchical or **multi-stage sampling.**

For example, the school population of a particular city can be viewed as a hierarchy of children within classes within schools. A hierarchical approach to sampling such a population as part of a dental caries survey could be implemented as follows. First, a number of schools are selected at random from a list of all city schools. Next, for each selected school, a class is selected at random. From the selected class in each school, a number of children then are selected at random to yield the final sample.

Again, a sampling strategy of this type offers a number of practical advantages in that field work is concentrated in the units selected (such as schools) rather than spread randomly over the entire school population. Perhaps even more important, detailed individual sampling frames have to be prepared only for the final subunits selected. In the school situation, lists of individual students are required only for each selected class. If a simple random sampling approach had been used, a list of all school children in the entire city would have been required. This might be feasible, but the list would certainly be inconvenient to assemble. Developing such a list for a nationwide survey would, however, be totally impractical. Simply adding a couple of layers to the hierarchy (a list of regions, a list of schools for each selected region) would make a multi-stage nation-wide survey quite practical.

As with cluster sampling, these advantages must be weighed against the loss of precision and reliability resulting from the patchy coverage of the total population in the final sample. This loss is, however, less for hierarchical sampling than for cluster sampling, since the patchiness of coverage is less severe. In both cases, the practical advantages tend to outweigh the loss of precision. A feasible and unbiased survey with somewhat less precision is clearly preferable to a theoretically superior design that is just not practical.

SAMPLE SIZE CONSIDERATIONS

In Chapter 9 we discussed the important subject of how one can determine the sample size needed to estimate a parameter to a certain level of precision (very much an observational objective) or to compare two groups with a certain level of power (an objective that could arise in both surveys and experiments). There is just one problem that we still have to address.

The comparison of sample means, for example, results in the following sample size formula:

$$n = 2 \left(PI \frac{\sigma}{\mu_1 - \mu_2} \right)^2$$

where PI is the desired power index,

    $\mu_1 - \mu_2$ defines the mean difference we wish to detect,

    σ defines the anticipated standard deviation,

and n is the number of individuals in each group.

(This formula was introduced in Chapter 9; if you need a refresher on the ideas that underlie it, reread the chapter.)

We might recall that the value 2 appears in this formula because, if the groups are the same size, the variance of the difference between the means is given by

$$\frac{\sigma^2}{n} + \frac{\sigma^2}{n} = 2 \frac{\sigma^2}{n}$$

Equal size groups are a practical proposition in the experimental environment or in a case-control study. However, what do we do if we are planning the comparison as part of a cross-sectional or cohort study? If we wished to compare blood pressures in smokers and nonsmokers, it seems unlikely that a cohort would self-select itself into equal numbers of smokers and nonsmokers.

What implications does this have for our sample size calculations? Do we have to maintain a rather dubious belief in equal size groups when planning an observational study? Provided we can anticipate the likely relative sizes of the two groups, we can easily adapt our sample size formulae. Suppose we project that one group will be R times larger than the other (that is, with sample sizes of n and Rn respectively). The variance of the mean difference will now be

$$\frac{\sigma^2}{n} + \frac{\sigma^2}{Rn} = \frac{R\sigma^2}{Rn} + \frac{\sigma^2}{Rn} = \frac{(R+1)}{R}\frac{\sigma^2}{n}$$

Modifying the sample size formula simply involves replacing the value 2 with the value $(R+1)/R$ (which, of course, takes the value 2 when the groups are of equal size).

The sample size required to compare two means with a desired level of power, when the group sizes are different, is therefore given by

$$n = \frac{(R+1)}{R}\left(PI\ \frac{\sigma}{\mu_1 - \mu_2}\right)^2$$

where PI is the appropriate power index,

$\mu_1 - \mu_2$ is the anticipated mean difference,

σ is the anticipated standard deviation,

R is the relative size of the two groups,

and n is the size of the smaller group.

The total sample size required is, of course, $(1+R)n$.

Exactly the same approach can be used to generalize the comparison of two proportions (reread Chapter 9, if necessary) to the unequal group situation.

$$n = \frac{(R+1)}{R}PI^2\ \frac{\bar{p}(1-\bar{p})}{(p_1 - p_2)^2}$$

where n is the size of group 1 (the smaller group),

R is the relative size of the two groups,

PI is the appropriate power index,

p_1 is the anticipated proportion in group 1,

p_2 is the anticipated proportion in group 2,

and $\bar{p}$ is the overall mean proportion,

      calculated as

$$\bar{p} = \frac{p_1 + Rp_2}{1+R}$$

that is, taking the relative size of the two groups into account.

Key Terms

Observational study	Purposive selection
Cross-sectional survey	Random sampling
Longitudinal survey	Systematic sampling
Prospective study	Sampling fraction
Target population	Stratification
Sampling units	Cluster sampling
Sampling frame	Multi-stage sampling
Notional sampling frame	

Appendix: Statistical Tables

TABLE T/1

Normal Distribution (Proportion of Population Between 0 and z)

z	.00	.01	.02	.03	.04	.05	.06	.07	.08	.09
0.0	.0000	.0040	.0080	.0120	.0160	.0199	.0239	.0279	.0319	.0359
0.1	.0398	.0438	.0478	.0517	.0557	.0596	.0636	.0675	.0714	.0753
0.2	.0793	.0832	.0871	.0910	.0948	.0987	.1026	.1064	.1103	.1141
0.3	.1179	.1217	.1255	.1293	.1331	.1368	.1406	.1443	.1480	.1517
0.4	.1554	.1591	.1628	.1664	.1700	.1736	.1772	.1808	.1844	.1879
0.5	.1915	.1950	.1985	.2019	.2054	.2088	.2123	.2157	.2190	.2224
0.6	.2257	.2291	.2324	.2357	.2389	.2422	.2454	.2486	.2517	.2549
0.7	.2580	.2611	.2642	.2673	.2704	.2734	.2764	.2794	.2823	.2852
0.8	.2881	.2910	.2939	.2967	.2995	.3023	.3051	.3078	.3106	.3133
0.9	.3159	.3186	.3212	.3238	.3264	.3289	.3315	.3340	.3365	.3389
1.0	.3413	.3438	.3461	.3485	.3508	.3531	.3554	.3577	.3599	.3621
1.1	.3643	.3665	.3686	.3708	.3729	.3749	.3770	.3790	.3810	.3830
1.2	.3849	.3869	.3888	.3907	.3925	.3944	.3962	.3980	.3997	.4015
1.3	.4032	.4049	.4066	.4082	.4099	.4115	.4131	.4147	.4162	.4177
1.4	.4192	.4207	.4222	.4236	.4251	.4265	.4279	.4292	.4306	.4319
1.5	.4332	.4345	.4357	.4370	.4382	.4394	.4406	.4418	.4429	.4441
1.6	.4452	.4463	.4474	.4484	.4495	.4505	.4515	.4525	.4535	.4545
1.7	.4554	.4564	.4573	.4582	.4591	.4599	.4608	.4616	.4625	.4633
1.8	.4641	.4649	.4656	.4664	.4671	.4678	.4686	.4693	.4699	.4706
1.9	.4713	.4719	.4726	.4732	.4738	.4744	.4750	.4756	.4761	.4767
2.0	.4772	.4778	.4783	.4788	.4793	.4798	.4803	.4808	.4812	.4817
2.1	.4821	.4826	.4830	.4834	.4838	.4842	.4846	.4850	.4854	.4857
2.2	.4861	.4864	.4868	.4871	.4875	.4878	.4881	.4884	.4887	.4890
2.3	.4893	.4896	.4898	.4901	.4904	.4906	.4909	.4911	.4913	.4916
2.4	.4918	.4920	.4922	.4925	.4927	.4929	.4931	.4932	.4934	.4936
2.5	.4938	.4940	.4941	.4943	.4945	.4946	.4948	.4949	.4951	.4952
2.6	.4953	.4955	.4956	.4957	.4959	.4960	.4961	.4962	.4963	.4964
2.7	.4965	.4966	.4967	.4968	.4969	.4970	.4971	.4972	.4973	.4974
2.8	.4974	.4975	.4976	.4977	.4977	.4978	.4979	.4979	.4980	.4981
2.9	.4981	.4982	.4982	.4983	.4984	.4984	.4985	.4985	.4986	.4986
3.0	.4987	.4987	.4987	.4988	.4988	.4989	.4989	.4989	.4990	.4990

Reprinted with permission from Mosteller F, Rourke REK, and Thomas GB Jr: Probability and statistical applications, ed 2, Reading, Mass, 1970, Addison-Wesley.

TABLE T/2
Critical Values of Students' t

df	p = .1	.05	.02	.01	.005	.002	.001
1	6.314	12.706	31.821	63.657	127.320	318.31	636.62
2	2.920	4.303	6.965	9.925	14.089	22.327	31.598
3	2.353	3.182	4.541	5.841	7.453	10.214	12.924
4	2.132	2.776	3.747	4.604	5.598	7.173	8.610
5	2.015	2.571	3.365	4.032	4.773	5.893	6.869
6	1.943	2.447	3.143	3.707	4.317	5.208	5.959
7	1.895	2.365	2.998	3.499	4.029	4.785	5.408
8	1.860	2.306	2.896	3.355	3.833	4.501	5.041
9	1.833	2.262	2.821	3.250	3.690	4.297	4.781
10	1.812	2.228	2.764	3.169	3.581	4.144	4.587
11	1.796	2.201	2.718	3.106	3.497	4.025	4.437
12	1.782	2.179	2.681	3.055	3.428	3.930	4.318
13	1.771	2.160	2.650	3.012	3.372	3.852	4.221
14	1.761	2.145	2.624	2.977	3.326	3.787	4.140
15	1.753	2.131	2.602	2.947	3.286	3.733	4.073
16	1.746	2.120	2.583	2.921	3.252	3.686	4.015
17	1.740	2.110	2.567	2.898	3.222	3.646	3.965
18	1.734	2.101	2.552	2.878	3.197	3.610	3.922
19	1.729	2.093	2.539	2.861	3.174	3.579	3.883
20	1.725	2.086	2.528	2.845	3.153	3.552	3.850
21	1.721	2.080	2.518	2.831	3.135	3.527	3.819
22	1.717	2.074	2.508	2.819	3.119	3.505	3.792
23	1.714	2.069	2.500	2.807	3.104	3.485	3.767
24	1.711	2.064	2.492	2.797	3.091	3.467	3.745
25	1.708	2.060	2.485	2.787	3.078	3.450	3.725
26	1.706	2.056	2.479	2.779	3.067	3.435	3.707
27	1.703	2.052	2.473	2.771	3.057	3.421	3.690
28	1.701	2.048	2.467	2.763	3.047	3.408	3.674
29	1.699	2.045	2.462	2.756	3.038	3.396	3.659
30	1.697	2.042	2.457	2.750	3.030	3.385	3.646
40	1.684	2.021	2.423	2.704	2.971	3.307	3.551
60	1.671	2.000	2.390	2.660	2.915	3.232	3.460
120	1.658	1.980	2.358	2.617	2.860	3.160	3.373
∞	1.645	1.960	2.326	2.576	2.807	3.090	3.291

Reprinted with permission from Pearson ES and Hartley HO: Biometrika tables for statisticians, ed 3, 1966.

TABLE T/3

Critical Values of the Variance Ratio, F

a) $p = .05$

v_2 \ v_1	1	2	3	4	5	6	7	8	9	10	12	15	20	24	30	40	60	120	∞
1	161.4	199.5	215.7	224.6	230.2	234.0	236.8	238.9	240.5	241.9	243.9	245.9	248.0	249.1	250.1	251.1	252.2	253.3	254.3
2	18.51	19.00	19.16	19.25	19.30	19.33	19.35	19.37	19.38	19.40	19.41	19.43	19.45	19.45	19.46	19.47	19.48	19.49	19.50
3	10.13	9.55	9.28	9.12	9.01	8.94	8.89	8.85	8.81	8.79	8.74	8.70	8.66	8.64	8.62	8.59	8.57	8.55	8.53
4	7.71	6.94	6.59	6.39	6.26	6.16	6.09	6.04	6.00	5.96	5.91	5.86	5.80	5.77	5.75	5.72	5.69	5.66	5.63
5	6.61	5.79	5.41	5.19	5.05	4.95	4.88	4.82	4.77	4.74	4.68	4.62	4.56	4.53	4.50	4.46	4.43	4.40	4.36
6	5.99	5.14	4.76	4.53	4.39	4.28	4.21	4.15	4.10	4.06	4.00	3.94	3.87	3.84	3.81	3.77	3.74	3.70	3.67
7	5.59	4.74	4.35	4.12	3.97	3.87	3.79	3.73	3.68	3.64	3.57	3.51	3.44	3.41	3.38	3.34	3.30	3.27	3.23
8	5.32	4.46	4.07	3.84	3.69	3.58	3.50	3.44	3.39	3.35	3.28	3.22	3.15	3.12	3.08	3.04	3.01	2.97	2.93
9	5.12	4.26	3.86	3.63	3.48	3.37	3.29	3.23	3.18	3.14	3.07	3.01	2.94	2.90	2.86	2.83	2.79	2.75	2.71
10	4.96	4.10	3.71	3.48	3.33	3.22	3.14	3.07	3.02	2.98	2.91	2.85	2.77	2.74	2.70	2.66	2.62	2.58	2.54
11	4.84	3.98	3.59	3.36	3.20	3.09	3.01	2.95	2.90	2.85	2.79	2.72	2.65	2.61	2.57	2.53	2.49	2.45	2.40
12	4.75	3.89	3.49	3.26	3.11	3.00	2.91	2.85	2.80	2.75	2.69	2.62	2.54	2.51	2.47	2.43	2.38	2.34	2.30
13	4.67	3.81	3.41	3.18	3.03	2.92	2.83	2.77	2.71	2.67	2.60	2.53	2.46	2.42	2.38	2.34	2.30	2.25	2.21
14	4.60	3.74	3.34	3.11	2.96	2.85	2.76	2.70	2.65	2.60	2.53	2.46	2.39	2.35	2.31	2.27	2.22	2.18	2.13
15	4.54	3.68	3.29	3.06	2.90	2.79	2.71	2.64	2.59	2.54	2.48	2.40	2.33	2.29	2.25	2.20	2.16	2.11	2.07
16	4.49	3.63	3.24	3.01	2.85	2.74	2.66	2.59	2.54	2.49	2.42	2.35	2.28	2.24	2.19	2.15	2.11	2.06	2.01
17	4.45	3.59	3.20	2.96	2.81	2.70	2.61	2.55	2.49	2.45	2.38	2.31	2.23	2.19	2.15	2.10	2.06	2.01	1.96
18	4.41	3.55	3.16	2.93	2.77	2.66	2.58	2.51	2.46	2.41	2.34	2.27	2.19	2.15	2.11	2.06	2.02	1.97	1.92
19	4.38	3.52	3.13	2.90	2.74	2.63	2.54	2.48	2.42	2.38	2.31	2.23	2.16	2.11	2.07	2.03	1.98	1.93	1.88
20	4.35	3.49	3.10	2.87	2.71	2.60	2.51	2.45	2.39	2.35	2.28	2.20	2.12	2.08	2.04	1.99	1.95	1.90	1.84
21	4.32	3.47	3.07	2.84	2.68	2.57	2.49	2.42	2.37	2.32	2.25	2.18	2.10	2.05	2.01	1.96	1.92	1.87	1.81
22	4.30	3.44	3.05	2.82	2.66	2.55	2.46	2.40	2.34	2.30	2.23	2.15	2.07	2.03	1.98	1.94	1.89	1.84	1.78
23	4.28	3.42	3.03	2.80	2.64	2.53	2.44	2.37	2.32	2.27	2.20	2.13	2.05	2.01	1.96	1.91	1.86	1.81	1.76
24	4.26	3.40	3.01	2.78	2.62	2.51	2.42	2.36	2.30	2.25	2.18	2.11	2.03	1.98	1.94	1.89	1.84	1.79	1.73
25	4.24	3.39	2.99	2.76	2.60	2.49	2.40	2.34	2.28	2.24	2.16	2.09	2.01	1.96	1.92	1.87	1.82	1.77	1.71
26	4.23	3.37	2.98	2.74	2.59	2.47	2.39	2.32	2.27	2.22	2.15	2.07	1.99	1.95	1.90	1.85	1.80	1.75	1.69
27	4.21	3.35	2.96	2.73	2.57	2.46	2.37	2.31	2.25	2.20	2.13	2.06	1.97	1.93	1.88	1.84	1.79	1.73	1.67
28	4.20	3.34	2.95	2.71	2.56	2.45	2.36	2.29	2.24	2.19	2.12	2.04	1.96	1.91	1.87	1.82	1.77	1.71	1.65
29	4.18	3.33	2.93	2.70	2.55	2.43	2.35	2.28	2.22	2.18	2.10	2.03	1.94	1.90	1.85	1.81	1.75	1.70	1.64
30	4.17	3.32	2.92	2.69	2.53	2.42	2.33	2.27	2.21	2.16	2.09	2.01	1.93	1.89	1.84	1.79	1.74	1.68	1.62
40	4.08	3.23	2.84	2.61	2.45	2.34	2.25	2.18	2.12	2.08	2.00	1.92	1.84	1.79	1.74	1.69	1.64	1.58	1.51
60	4.00	3.15	2.76	2.53	2.37	2.25	2.17	2.10	2.04	1.99	1.92	1.84	1.75	1.70	1.65	1.59	1.53	1.47	1.39
120	3.92	3.07	2.68	2.45	2.29	2.17	2.09	2.02	1.96	1.91	1.83	1.75	1.66	1.61	1.55	1.50	1.43	1.35	1.25
∞	3.84	3.00	2.60	2.37	2.21	2.10	2.01	1.94	1.88	1.83	1.75	1.67	1.57	1.52	1.46	1.39	1.32	1.22	1.00

v_1, v_2 are upper, lower df respectively.

b) $p = .01$

v_2 \ v_1	1	2	3	4	5	6	7	8	9	10	12	15	20	24	30	40	60	∞
1	4052	4999.5	5403	5625	5764	5859	5928	5982	6022	6056	6106	6157	6209	6235	6261	6287	6313	6366
2	98.50	99.00	99.17	99.25	99.30	99.33	99.36	99.37	99.39	99.40	99.42	99.43	99.45	99.46	99.47	99.47	99.48	99.50
3	34.12	30.82	29.46	28.71	28.24	27.91	27.67	27.49	27.35	27.23	27.05	26.87	26.69	26.60	26.50	26.41	26.32	26.13
4	21.20	18.00	16.69	15.98	15.52	15.21	14.98	14.80	14.66	14.55	14.37	14.20	14.02	13.93	13.84	13.75	13.65	13.46
5	16.26	13.27	12.06	11.39	10.97	10.67	10.46	10.29	10.16	10.05	9.89	9.72	9.55	9.47	9.38	9.29	9.20	9.02
6	13.75	10.92	9.78	9.15	8.75	8.47	8.26	8.10	7.98	7.87	7.72	7.56	7.40	7.31	7.23	7.14	7.06	6.88
7	12.25	9.55	8.45	7.85	7.46	7.19	6.99	6.84	6.72	6.62	6.47	6.31	6.16	6.07	5.99	5.91	5.82	5.65
8	11.26	8.65	7.59	7.01	6.63	6.37	6.18	6.03	5.91	5.81	5.67	5.52	5.36	5.28	5.20	5.12	5.03	4.86
9	10.56	8.02	6.99	6.42	6.06	5.80	5.61	5.47	5.35	5.26	5.11	4.96	4.81	4.73	4.65	4.57	4.48	4.31
10	10.04	7.56	6.55	5.99	5.64	5.39	5.20	5.06	4.94	4.85	4.71	4.56	4.41	4.33	4.25	4.17	4.08	3.91
11	9.65	7.21	6.22	5.67	5.32	5.07	4.89	4.74	4.63	4.54	4.40	4.25	4.10	4.02	3.94	3.86	3.78	3.60
12	9.33	6.93	5.95	5.41	5.06	4.82	4.64	4.50	4.39	4.30	4.16	4.01	3.86	3.78	3.70	3.62	3.54	3.36
13	9.07	6.70	5.74	5.21	4.86	4.62	4.44	4.30	4.19	4.10	3.96	3.82	3.66	3.59	3.51	3.43	3.34	3.17
14	8.86	6.51	5.56	5.04	4.69	4.46	4.28	4.14	4.03	3.94	3.80	3.66	3.51	3.43	3.35	3.27	3.18	3.00
15	8.68	6.36	5.42	4.89	4.56	4.32	4.14	4.00	3.89	3.80	3.67	3.52	3.37	3.29	3.21	3.13	3.05	2.87
16	8.53	6.23	5.29	4.77	4.44	4.20	4.03	3.89	3.78	3.69	3.55	3.41	3.26	3.18	3.10	3.02	2.93	2.75
17	8.40	6.11	5.18	4.67	4.34	4.10	3.93	3.79	3.68	3.59	3.46	3.31	3.16	3.08	3.00	2.92	2.83	2.65
18	8.29	6.01	5.09	4.58	4.25	4.01	3.84	3.71	3.60	3.51	3.37	3.23	3.08	3.00	2.92	2.84	2.75	2.57
19	8.18	5.93	5.01	4.50	4.17	3.94	3.77	3.63	3.52	3.43	3.30	3.15	3.00	2.92	2.84	2.76	2.67	2.49
20	8.10	5.85	4.94	4.43	4.10	3.87	3.70	3.56	3.46	3.37	3.23	3.09	2.94	2.86	2.78	2.69	2.61	2.42
21	8.02	5.78	4.87	4.37	4.04	3.81	3.64	3.51	3.40	3.31	3.17	3.03	2.88	2.80	2.72	2.64	2.55	2.36
22	7.95	5.72	4.82	4.31	3.99	3.76	3.59	3.45	3.35	3.26	3.12	2.98	2.83	2.75	2.67	2.58	2.50	2.31
23	7.88	5.66	4.76	4.26	3.94	3.71	3.54	3.41	3.30	3.21	3.07	2.93	2.78	2.70	2.62	2.54	2.45	2.26
24	7.82	5.61	4.72	4.22	3.90	3.67	3.50	3.36	3.26	3.17	3.03	2.89	2.74	2.66	2.58	2.49	2.40	2.21
25	7.77	5.57	4.68	4.18	3.85	3.63	3.46	3.32	3.22	3.13	2.99	2.85	2.70	2.62	2.54	2.45	2.36	2.17
26	7.72	5.53	4.64	4.14	3.82	3.59	3.42	3.29	3.18	3.09	2.96	2.81	2.66	2.58	2.50	2.42	2.33	2.13
27	7.68	5.49	4.60	4.11	3.78	3.56	3.39	3.26	3.15	3.06	2.93	2.78	2.63	2.55	2.47	2.38	2.29	2.10
28	7.64	5.45	4.57	4.07	3.75	3.53	3.36	3.23	3.12	3.03	2.90	2.75	2.60	2.52	2.44	2.35	2.26	2.06
29	7.60	5.42	4.54	4.04	3.73	3.50	3.33	3.20	3.09	3.00	2.87	2.73	2.57	2.49	2.41	2.33	2.23	2.03
30	7.56	5.39	4.51	4.02	3.70	3.47	3.30	3.17	3.07	2.98	2.84	2.70	2.55	2.47	2.39	2.30	2.21	2.01
40	7.31	5.18	4.31	3.83	3.51	3.29	3.12	2.99	2.89	2.80	2.66	2.52	2.37	2.29	2.20	2.11	2.02	1.80
60	7.08	4.98	4.13	3.65	3.34	3.12	2.95	2.82	2.72	2.63	2.50	2.35	2.20	2.12	2.08	1.94	1.84	1.60
120	6.85	4.79	3.95	3.48	3.17	2.96	2.79	2.66	2.56	2.47	2.34	2.19	2.03	1.95	1.86	1.76	1.66	1.38
∞	6.63	4.61	3.78	3.32	3.02	2.80	2.64	2.51	2.41	2.32	2.18	2.04	1.88	1.79	1.70	1.59	1.47	1.00

Reprinted with permission from Pearson ES and Hartley HO: Biomerika tables for statisticians, ed 3, 1966.

c) $p = .005$

v_2 \ v_1	1	2	3	4	5	6	7	8	9	10	12	15	20	24	30	40	60	∞
1	16211	20000	21615	22500	23056	23437	23715	23925	24091	24224	24426	24630	24836	24940	25044	25148	25253	25465
2	198.5	199.0	199.2	199.2	199.3	199.3	199.4	199.4	199.4	199.4	199.4	199.4	199.4	199.5	199.5	199.5	199.5	199.5
3	55.55	49.80	47.47	46.19	45.39	44.84	44.43	44.13	43.88	43.69	43.39	43.08	42.78	42.62	42.47	42.31	42.15	41.83
4	31.33	26.28	24.26	23.15	22.46	21.97	21.62	21.35	21.14	20.97	20.70	20.44	20.17	20.03	19.89	19.75	19.61	19.32
5	22.78	18.31	16.53	15.56	14.94	14.51	14.20	13.96	13.77	13.62	13.38	13.15	12.90	12.78	12.66	12.53	12.40	12.14
6	18.63	14.54	12.92	12.03	11.46	11.07	10.79	10.57	10.39	10.25	10.03	9.81	9.59	9.47	9.36	9.24	9.12	8.88
7	16.24	12.40	10.88	10.05	9.52	9.16	8.89	8.68	8.51	8.38	8.18	7.97	7.75	7.65	7.53	7.42	7.31	7.08
8	14.69	11.04	9.60	8.81	8.30	7.95	7.69	7.50	7.34	7.21	7.01	6.81	6.61	6.50	6.40	6.29	6.18	5.95
9	13.61	10.11	8.72	7.96	7.47	7.13	6.88	6.69	6.54	6.42	6.23	6.03	5.83	5.73	5.62	5.52	5.41	5.19
10	12.83	9.43	8.08	7.34	6.87	6.54	6.30	6.12	5.97	5.85	5.66	5.47	5.27	5.17	5.07	4.97	4.86	4.64
11	12.23	8.91	7.60	6.88	6.42	6.10	5.86	5.68	5.54	5.42	5.24	5.05	4.86	4.76	4.65	4.55	4.44	4.23
12	11.75	8.51	7.23	6.52	6.07	5.76	5.52	5.35	5.20	5.09	4.91	4.72	4.53	4.43	4.33	4.23	4.12	3.90
13	11.37	8.19	6.93	6.23	5.79	5.48	5.25	5.08	4.94	4.82	4.64	4.46	4.27	4.17	4.07	3.97	3.87	3.65
14	11.06	7.92	6.68	6.00	5.56	5.26	5.03	4.86	4.72	4.60	4.43	4.25	4.06	3.96	3.86	3.76	3.66	3.44
15	10.80	7.70	6.48	5.80	5.37	5.07	4.85	4.67	4.54	4.42	4.25	4.07	3.88	3.79	3.69	3.58	3.48	3.26
16	10.58	7.51	6.30	5.64	5.21	4.91	4.69	4.52	4.38	4.27	4.10	3.92	3.73	3.64	3.54	3.44	3.33	3.11
17	10.38	7.35	6.16	5.50	5.07	4.78	4.56	4.39	4.25	4.14	3.97	3.79	3.61	3.51	3.41	3.31	3.21	2.98
18	10.22	7.21	6.03	5.37	4.96	4.66	4.44	4.28	4.14	4.03	3.86	3.68	3.50	3.40	3.30	3.20	3.10	2.87
19	10.07	7.09	5.92	5.27	4.85	4.56	4.34	4.18	4.04	3.93	3.76	3.59	3.40	3.31	3.21	3.11	3.00	2.78
20	9.94	6.99	5.82	5.17	4.76	4.47	4.26	4.09	3.96	3.85	3.68	3.50	3.32	3.22	3.12	3.02	2.92	2.69
21	9.83	6.89	5.73	5.09	4.68	4.39	4.18	4.01	3.88	3.77	3.60	3.43	3.24	3.15	3.05	2.95	2.84	2.61
22	9.73	6.81	5.65	5.02	4.61	4.32	4.11	3.94	3.81	3.70	3.54	3.36	3.18	3.08	2.98	2.88	2.77	2.55
23	9.63	6.73	5.58	4.95	4.54	4.26	4.05	3.88	3.75	3.64	3.47	3.30	3.12	3.02	2.92	2.82	2.71	2.48
24	9.55	6.66	5.52	4.89	4.49	4.20	3.99	3.83	3.69	3.59	3.42	3.25	3.06	2.97	2.87	2.77	2.66	2.43
25	9.48	6.60	5.46	4.84	4.43	4.15	3.94	3.78	3.64	3.54	3.37	3.20	3.01	2.92	2.82	2.72	2.61	2.38
26	9.41	6.54	5.41	4.79	4.38	4.10	3.89	3.73	3.60	3.49	3.33	3.15	2.97	2.87	2.77	2.67	2.56	2.33
27	9.34	6.49	5.36	4.74	4.34	4.06	3.85	3.69	3.56	3.45	3.28	3.11	2.93	2.83	2.73	2.63	2.52	2.29
28	9.28	6.44	5.32	4.70	4.30	4.02	3.81	3.65	3.52	3.41	3.25	3.07	2.89	2.79	2.69	2.59	2.48	2.25
29	9.23	6.40	5.28	4.66	4.26	3.98	3.77	3.61	3.48	3.38	3.21	3.04	2.86	2.76	2.66	2.56	2.45	2.21
30	9.18	6.35	5.24	4.62	4.23	3.95	3.74	3.58	3.45	3.34	3.18	3.01	2.82	2.73	2.63	2.52	2.42	2.18
40	8.83	6.07	4.98	4.37	3.99	3.71	3.51	3.35	3.22	3.12	2.95	2.78	2.60	2.50	2.40	2.30	2.18	1.93
60	8.49	5.79	4.73	4.14	3.76	3.49	3.29	3.13	3.01	2.90	2.74	2.57	2.39	2.29	2.19	2.08	1.96	1.69
120	8.18	5.54	4.50	3.92	3.55	3.28	3.09	2.93	2.81	2.71	2.54	2.37	2.19	2.09	1.98	1.87	1.75	1.43
∞	7.88	5.30	4.28	3.72	3.35	3.09	2.90	2.74	2.62	2.52	2.36	2.19	2.00	1.90	1.79	1.67	1.53	1.00

TABLE T/4

Critical Values of q

a) $p = 0.05$

Error df	Number of groups									
	2	**3**	**4**	**5**	**6**	**7**	**8**	**9**	**10**	**11**
5	3.64	4.60	5.22	5.67	6.03	6.33	6.58	6.80	6.99	7.17
6	3.46	4.34	4.90	5.30	5.63	5.90	6.12	6.32	6.49	6.65
7	3.34	4.16	4.68	5.06	5.36	5.61	5.82	6.00	6.16	6.30
8	3.26	4.04	4.53	4.89	5.17	5.40	5.60	5.77	5.92	6.05
9	3.20	3.95	4.41	4.76	5.02	5.24	5.43	5.59	5.74	5.87
10	3.15	3.88	4.33	4.65	4.91	5.12	5.30	5.46	5.60	5.72
11	3.11	3.82	4.26	4.57	4.82	5.03	5.20	5.35	5.49	5.61
12	3.08	3.77	4.20	4.51	4.75	4.95	5.12	5.27	5.39	5.51
13	3.06	3.73	4.15	4.45	4.69	4.88	5.05	5.19	5.32	5.43
14	3.03	3.70	4.11	4.41	4.64	4.83	4.99	5.13	5.25	5.36
15	3.01	3.67	4.08	4.37	4.59	4.78	4.94	5.08	5.20	5.31
16	3.00	3.65	4.05	4.33	4.56	4.74	4.90	5.03	5.15	5.26
17	2.98	3.63	4.02	4.30	4.52	4.71	4.86	4.99	5.11	5.21
18	2.97	3.61	4.00	4.28	4.49	4.67	4.82	4.96	5.07	5.17
19	2.96	3.59	3.98	4.25	4.47	4.65	4.79	4.92	5.04	5.14
20	2.95	3.58	3.96	4.23	4.45	4.62	4.77	4.90	5.01	5.11
24	2.92	3.53	3.90	4.17	4.37	4.54	4.68	4.81	4.92	5.01
30	2.89	3.49	3.85	4.10	4.30	4.46	4.60	4.72	4.82	4.92
40	2.86	3.44	3.79	4.04	4.23	4.39	4.52	4.63	4.73	4.82
60	2.83	3.40	3.74	3.98	4.16	4.31	4.44	4.55	4.65	4.73
120	2.80	3.36	3.68	3.92	4.10	4.24	4.36	4.47	4.56	4.64
∞	2.77	3.31	3.63	3.86	4.03	4.17	4.29	4.39	4.47	4.55

b) $p = .01$

Error df	Number of groups									
	2	3	4	5	6	7	8	9	10	11
5	5.70	6.98	7.80	8.42	8.91	9.32	9.67	9.97	10.24	10.48
6	5.24	6.33	7.03	7.56	7.97	8.32	8.61	8.87	9.10	9.30
7	4.95	5.92	6.54	7.01	7.37	7.68	7.94	8.17	8.37	8.55
8	4.75	5.64	6.20	6.62	6.96	7.24	7.47	7.68	7.86	8.03
9	4.60	5.43	5.96	6.35	6.66	6.91	7.13	7.33	7.49	7.65
10	4.48	5.27	5.77	6.14	6.43	6.67	6.87	7.05	7.21	7.36
11	4.39	5.15	5.62	5.97	6.25	6.48	6.67	6.84	6.99	7.13
12	4.32	5.05	5.50	5.84	6.10	6.32	6.51	6.67	6.81	6.94
13	4.26	4.96	5.40	5.73	5.98	6.19	6.37	6.53	6.67	6.79
14	4.21	4.89	5.32	5.63	5.88	6.08	6.26	6.41	6.54	6.66
15	4.17	4.84	5.25	5.56	5.80	5.99	6.16	6.31	6.44	6.55
16	4.13	4.79	5.19	5.49	5.72	5.92	6.08	6.22	6.35	6.46
17	4.10	4.74	5.14	5.43	5.66	5.85	6.01	6.15	6.27	6.38
18	4.07	4.70	5.09	5.38	5.60	5.79	5.94	6.08	6.20	6.31
19	4.05	4.67	5.05	5.33	5.55	5.73	5.89	6.02	6.14	6.25
20	4.02	4.64	5.02	5.29	5.51	5.69	5.84	5.97	6.09	6.19
24	3.96	4.55	4.91	5.17	5.37	5.54	5.69	5.81	5.92	6.02
30	3.89	4.45	4.80	5.05	5.24	5.40	5.54	5.65	5.76	5.85
40	3.82	4.37	4.70	4.93	5.11	5.26	5.39	5.50	5.60	5.69
60	3.76	4.28	4.59	4.82	4.99	5.13	5.25	5.36	5.45	5.53
120	3.70	4.20	4.50	4.71	4.87	5.01	5.12	5.21	5.30	5.38
∞	3.64	4.12	4.40	4.60	4.76	4.88	4.99	5.08	5.16	5.23

Reprinted with permission from Pearson ES and Hartley HO: Biometrika tables for statisticians, ed 3, 1966.

TABLE T/5

Critical Values of χ^2

df	p = .100	.050	.025	.010	.005	.001
1	2.71	3.84	5.02	6.63	7.88	10.83
2	4.61	5.99	7.38	9.21	10.60	13.82
3	6.25	7.81	9.35	11.34	12.84	16.27
4	7.78	9.49	11.14	13.28	14.86	18.47
5	9.24	11.07	12.83	15.09	16.75	20.52
6	10.64	12.59	14.45	16.81	18.55	22.46
7	12.02	14.07	16.01	18.48	20.28	24.32
8	13.36	15.51	17.53	20.09	21.96	26.13
9	14.68	16.92	19.02	21.67	23.59	27.88
10	15.99	18.31	20.48	23.21	25.19	29.59
11	17.28	19.68	21.92	24.73	26.76	31.26
12	18.55	21.03	23.34	26.22	28.30	32.91
13	19.81	22.36	24.74	27.69	29.82	34.53
14	21.06	23.68	26.12	29.14	31.32	36.12
15	22.31	25.00	27.49	30.58	32.80	37.70
16	23.54	26.30	28.85	32.00	34.27	39.25
17	24.77	27.59	30.19	33.41	35.72	40.79
18	25.99	28.87	31.53	34.81	37.16	42.31
19	27.20	30.14	32.85	36.19	38.58	43.82
20	28.41	31.41	34.17	37.57	40.00	45.32
21	29.62	32.67	35.48	38.93	41.40	46.80
22	30.81	33.92	36.78	40.29	42.80	48.27
23	32.01	35.17	38.08	41.64	44.18	49.73
24	33.20	36.42	39.36	42.98	45.56	51.18
25	34.38	37.65	40.65	44.31	46.93	52.62
26	35.56	38.89	41.92	45.64	48.29	54.05
27	36.74	40.11	43.19	46.96	49.64	55.48
28	37.92	41.34	44.46	48.28	50.99	56.89
29	39.09	42.56	45.72	49.59	52.34	58.30
30	40.26	43.77	46.98	50.89	53.67	59.70
40	51.81	55.76	59.34	63.69	66.77	73.40
50	63.17	67.50	71.42	76.15	79.49	86.66
60	74.40	79.08	83.30	88.38	91.95	99.61
70	85.53	90.53	95.02	100.43	104.21	112.32
80	96.58	101.88	106.63	112.33	116.32	124.84
90	107.57	113.15	118.14	124.12	128.30	137.21
100	118.50	123.34	129.56	135.81	140.17	149.45

Reprinted with permission from Pearson ES and Hartley HO: Biometrika tables for statisticians, ed 3, 1966.

TABLE T/6
Critical Values of the Correlation Coefficient, r

df	p = .1	.05	.02	.01	.005	.001
1	0.9877	0.9^2692	0.9^3507	0.9^3877	0.9^4692	0.9^5877
2	0.9000	0.9500	0.9800	0.9^2000	0.9^2500	0.9^3000
3	0.805	0.878	0.9343	0.9587	0.9740	0.9^2114
4	0.729	0.811	0.882	0.9172	0.9417	0.9741
5	0.669	0.754	0.833	0.875	0.9056	0.9509
6	0.621	0.707	0.789	0.834	0.870	0.9249
7	0.582	0.666	0.750	0.798	0.836	0.898
8	0.549	0.632	0.715	0.765	0.805	0.872
9	0.521	0.602	0.685	0.735	0.776	0.847
10	0.497	0.576	0.658	0.708	0.750	0.823
11	0.476	0.553	0.634	0.684	0.726	0.801
12	0.457	0.532	0.612	0.661	0.703	0.780
13	0.441	0.514	0.592	0.641	0.683	0.760
14	0.426	0.497	0.574	0.623	0.664	0.742
15	0.412	0.482	0.558	0.606	0.647	0.725
16	0.400	0.468	0.543	0.590	0.631	0.708
17	0.389	0.456	0.529	0.575	0.616	0.693
18	0.378	0.444	0.516	0.561	0.602	0.679
19	0.369	0.433	0.503	0.549	0.589	0.665
20	0.360	0.423	0.492	0.537	0.576	0.652
25	0.323	0.381	0.445	0.487	0.524	0.597
30	0.296	0.349	0.409	0.449	0.484	0.554
35	0.275	0.325	0.381	0.418	0.452	0.519
40	0.257	0.304	0.358	0.393	0.425	0.490
45	0.243	0.288	0.338	0.372	0.403	0.465
50	0.231	0.273	0.322	0.354	0.384	0.443
60	0.211	0.250	0.295	0.325	0.352	0.408
70	0.195	0.232	0.274	0.302	0.327	0.380
80	0.183	0.217	0.257	0.283	0.307	0.357
90	0.173	0.205	0.242	0.267	0.290	0.338
100	0.164	0.195	0.230	0.254	0.276	0.321

Reprinted with permission from Pearson ES and Hartley HO: Biometrika tables for statisticians, ed 3, 1966.

TABLE T/7
Random Numbers

12	67	73	29	44	54	12	73	97	48	79	91	20	20	17	31	83	20	85	66
06	24	89	57	11	27	43	03	14	29	84	52	86	13	51	70	65	88	60	88
29	15	84	77	17	86	64	87	06	55	36	44	92	58	64	91	94	48	64	65
49	56	97	93	91	59	41	21	98	03	70	95	31	99	74	45	67	94	47	79
50	77	60	28	58	75	70	96	70	07	60	66	05	95	58	39	20	25	96	89
00	31	32	48	23	12	31	08	51	06	23	44	26	43	56	34	78	65	50	80
01	67	45	57	55	98	93	69	07	81	62	35	22	03	89	22	54	94	83	31
24	00	48	34	15	45	34	50	02	37	43	57	36	13	76	71	95	40	34	10
77	52	60	27	64	16	06	83	38	73	51	32	62	85	24	58	54	29	64	56
36	29	93	93	10	00	51	34	81	26	13	53	26	29	16	94	19	01	40	45
94	82	03	96	49	78	32	61	17	78	70	12	91	69	99	62	75	16	50	69
23	12	21	19	67	27	86	47	43	25	25	05	76	17	50	55	70	32	83	36
77	58	90	38	66	53	45	85	13	93	00	65	30	59	39	44	86	75	90	73
92	37	51	97	83	78	12	70	41	42	01	72	10	48	88	95	05	24	44	21
28	93	48	44	13	02	49	32	07	95	26	47	67	70	72	71	08	47	16	18
09	68	01	98	80	27	49	78	56	67	49	22	13	66	61	33	53	18	36	03
61	73	92	33	89	48	20	42	32	33	79	37	68	88	44	59	35	17	97	61
82	35	37	33	53	42	52	04	16	54	08	25	48	89	57	87	59	89	96	76
39	20	77	72	55	19	66	58	57	91	38	43	67	97	52	66	45	29	74	67
51	90	71	05	82	38	37	40	94	52	24	09	35	44	37	33	35	20	65	89
97	49	53	79	17	25	02	65	77	70	88	45	53	51	63	30	89	66	42	03
73	18	91	38	25	82	29	71	56	89	86	74	68	58	75	36	93	13	33	31
17	79	34	97	25	89	01	17	67	92	62	25	54	70	52	88	28	05	61	17
97	27	26	86	17	67	59	56	95	07	49	05	70	06	70	35	21	35	26	18
56	06	63	00	07	40	65	87	09	49	70	34	67	02	33	39	04	40	01	51

Reprinted with permission from Kendall MG and Babington Smith B: Tables of random sampling numbers: tracts for computers, no XXIV, London, 1939, Cambridge University Press.

Glossary

A brief definition can give only a limited insight into a concept. Why not follow up your use of the glossary by reading the portion of the text where the term first appears? (The appropriate page number is given in parentheses.) This will set the ideas in a broader context.

adjusted mean (224) The *mean* outcome value that would result if every subject in a group had the same *covariate** value. Results are usually adjusted so that all subjects have the mean covariate value of the overall study.

alternative hypothesis (61) The converse of the *null hypothesis*. This can be either a general alternative (for example, that two drugs differ in their effectiveness) or a specific alternative (for example, that an active drug is superior to a placebo).

analysis of covariance (218) An *analysis of variance* that isolates and tests variation from a source that is continuous (for example, age) as well as variation from one or more *categorical* sources (for example, treatments or genders).

analysis of variance (82) A widely used statistical technique that determines the presence of an effect by estimating the *variance* associated with it and comparing this with a benchmark variance known to be purely random.

balance (196) An experiment is said to be balanced when the experimental material is organized in such a way that the various *main effects* are truly distinct and have no built-in connection with one another.

between-treatments variation (78) Variation between different treatment groups (as characterized by their respective mean performances). Part of this variation is random, and part may be due to genuine differences in treatment effectiveness.

bias (153) The occurrence of some systematic influence in a study that consistently tends to give an advantage or disadvantage to one or more subgroups in the study.

binomial distribution (26) The pattern usually followed by a set of *binomial measurements*, provided that the individuals involved do not influence one another's behavior.

binomial measurement (1) A measurement in which the individual is placed in one of two mutually exclusive categories.

categorical measurement (1) A measurement in which the individual is located in one of a number of mutually exclusive categories that have no inherent order to them (for example, blood types).

chi squared statistic (101) A measure of the level of agreement between a set of *observed* and *expected* values.

chi squared test (111) A test for possible association between two categorical variables, based on a *Chi squared statistic*.

clinical trial (183) A study that evaluates the effectiveness of one or more interventions in human subjects by comparing the intervention group(s) with a suitable control group. The participants are followed forward in time from initial intervention to final outcome.

cluster sampling (290) A sampling procedure often used when *sampling units* naturally form themselves into groups or clusters. A number of such clusters are randomly selected from a list of clusters, and all the sampling units in each selected cluster are entered into the final sample.

comparison-oriented study (167) Study in which the focus of the investigation is the comparison of two or more study subgroups and the testing of hypotheses concerning possible differences between these subgroups.

completely randomized experiment (158) An experiment in which the available subjects are *randomly allocated* to the various treatments under investigation, with no attempt to

---

*All terms in italic are defined elsewhere in this glossary.

impose any additional grouping or structure on the subjects.

confidence limits (44)   A range of values within which we are confident (often 95% confident) that the true but unknown *population* value lies.

confounding (153)   A situation in which there is more than one possible explanation for a particular study outcome.

contingency table (109)   A table that describes the relationship between two *categorical* variables by specifying the number of individuals falling into each possible combination of categories.

correlation coefficient, *r* (128)   An index of the tendency of two continuous measurements to vary together. It can range in absolute value from a minimum of 0 (no relationship) to a maximum of 1 (a perfect relationship). The sign of the coefficient shows the direction of the relationship. It is, in fact, a *standardized covariance*.

covariance (126)   A measure of the tendency of two continuous measurements to vary together. It is obtained by dividing the sum of cross products (see *Co-variation*) by the *degrees of freedom* and hence is measured in composite units.

covariate (218)   A continuous variable (also known as a control or concomitant variable) that is believed to have some influence on the study outcome and that therefore might result in additional variation in the outcome variable.

co-variation (124)   The tendency of two continuous measurements to vary together. The total amount of co-variation present in a data set is obtained by measuring each individual's deviation from the sample mean on both variables, multiplying these deviations, and summing the products over all the individuals in the data set. This quantity (known as the sum of cross products) is measured in composite units reflecting the two variables involved.

cross-over design (160)   An experimental design in which each subject receives all the treatments under comparison, in random order. Each subject therefore corresponds to a block in a *randomized block design*.

cross-sectional survey (282)   A survey that describes or measures a population at one fixed point in time.

Declaration of Helsinki (192)   A statement by the World Health Organization that provides ethical guidelines to researchers involved in experiments on human subjects.

degrees of freedom (9)   The number of truly independent or informative items of information in a set of data.

dependent variable (134)   A variable that depends on or is influenced (directly or indirectly) by another variable (known as the *independent variable*). For example, blood pressure (dependent variable) is influenced by age (independent variable).

discrete variable (23)   A *variable* that can take only a limited range of values.

double-blind experiment (157)   An experiment in which both the subjects and the investigators who interact with the subjects are unable to distinguish the treatments being compared.

error variation (194)   Variation that has no known cause and therefore must be assumed to be purely random.

estimation (38)   The act of using the results from a *sample* to learn about the characteristics of a *population*.

estimation-oriented studies (179)   Studies in which the primary focus of the investigation is the *estimation* of the value of some *population* characteristic.

expected values (100)   The number of individuals that some theory or hypothesis predicts should occur in the various categories of a *categorical* variable.

F tables (82)   Tables (also known as variance ratio tables) that document the critical values used in *analysis of variance* testing.

factor (162)   An overall treatment concept, such as diet or drugs. The various treatment categories within the concept (different diets, different dosages of a drug, and so on) are known as the levels of the factor.

factorial experiment (162)   An experiment that involves the simultaneous study of more than one *factor*.

first-order interaction (203)   An *interaction* between two *main effects*. (Interactions between more than two main effects are possible. When three main effects are involved, such an interaction is called a second-order interaction).

Fisher's exact test (112)   A test for the presence of an association between two categorical variables, used when the numbers involved are too small to permit the use of a *Chi squared test*.

Friedman's test (272)   A *nonparametric* equivalent of the *analysis of variance*, which permits the analysis of an unreplicated *randomized block design*.

general linear model (262)   A more technical name for *multiple regression analysis* that recognizes its ability to relate a continuous out-come (or *dependent*) variable to a linear combi-nation of continuous and/or categorical explan-atory (or *independent*) variables, thus providing a very general and flexible method of analysis.

goodness of fit test (104)   A test, based on a *Chi squared statistic*, that evaluates whether the real world evidence fits some particular pattern or hypothesis.

histogram (3)   A graphical representation of the pattern of variation present in a set of data. It is obtained by splitting a measurement scale into intervals and displaying the number of individ-uals falling into each interval.

hypothesis testing (38)   The act of using the re-sults from a sample-based investigation to an-swer research questions (for example, "Do men and women differ in their times of reaction to a stimulus?")

independent variable (134)   A variable (also known as an explanatory variable) that explains or influences (at least to some degree) the be-havior of another variable known as the *depen-dent variable*. For example, age (independent variable) influences blood pressure (dependent variable).

interaction (163)   A situation in which the rela-tive performance of the levels of one *factor* de-pends on the level of another factor. For in-stance, one drug might be markedly superior to another in young people but actually inferior to its competitor in elderly patients.

interval measurement (2)   A measurement in which the individuals are located on a scale, the intervals of which have a consistent interpreta-tion (for example, temperature in degrees Farenheit).

Kruskal-Wallis test (270)   A *nonparametric test* used to compare the *medians* of several independent samples. It is the nonparametric equivalent of the *one-way analysis of vari-ance*.

lambda, λ (33)   The single value needed to de-fine a *Poisson distribution*. It is both the *mean* and the *variance* of the distribution.

Latin square (161)   An experimental design that allows two *categorical* sources of variation to be controlled, while requiring the use of rel-atively few experimental subjects. It does, however, require that the number of categories in each source of variation be the same as the number of treatments being compared.

least significant difference (87)   The smallest difference between two group means that would yield a significant result if the two groups were compared using a *t test*. Using this technique to compare multiple groups will result in an in-creased risk of a *type I error*.

least squares fit (136)   The procedure by which we obtain a *regression line*. In such a line, the sum of the squared discrepancies between the actual *dependent* values and the corresponding values predicted by the line are as small as pos-sible, hence the name "least squares."

linear contrast (90)   A weighted combination of group *means* that reflects some specific ques-tion about the nature of the group differences.

longitudinal survey (283)   A survey that de-scribes or measures a *population* at several points in time.

main effects (200)   The various *categorical* sources of variation (such as differences be-tween drugs or differences between genders) that are built into a designed experiment.

Mann-Whitney *U* test (268)   A *nonparametric test* used to compare the *medians* of two inde-pendent samples. It is the nonparametric equiv-alent of the *t test*.

McNemar's test (116)   A variant of a *Chi squared test*, used when the data is paired.

mean (4)   A measure of the "typical" individual obtained by averaging (totalling the various in-dividual results and dividing by the number of results involved) a data set.

mean deviation (7)   A measure of the extent of the variation in a set of data, obtained by taking the average of the absolute deviation between each individual result and the *mean*. It is easier to calculate than the *standard deviation* but much less useful.

median (5)   A measure of the "typical" individ-ual, obtained by ranking the individual results in a data set from smallest to largest and select-ing the middle value.

mode (5)   The most frequently occurring result in a set of data. (It is meaningful only if the measurement involved takes a limited range of values).

mu, μ (39)   The *mean* of a *population*. In prac-tice this is usually unknown and must be es-timated by the mean of a *sample* (indicated by $\bar{x}$).

multiple correlation coefficient (253)   A *cor-relation coefficient* that measures the strength of the association between a *dependent* variable and a linear combination of *independent* vari-ables. It represents the correlation between the actual dependent values and the corresponding values predicted by a *multiple regression* equa-tion.

multiple regression analysis (251)   An exten-sion of regression analysis (see *regression line*) that describes and tests the relationship between a *dependent* variable and a linear combination of several *independent* variables.

multi-stage sampling (290) A sampling procedure often used when the *sampling units* can be defined in a hierarchical manner (for example, children within classes within schools). The final sample is selected in a corresponding series of steps (a number of schools are selected at random, for each selected school a number of classes are selected, and so on).

multi-way analysis of variance (195) An *analysis of variance* in which several *main effects* are tested simultaneously.

nonparametric tests (263) A family of statistical tests (also called distribution-free tests) that do not require any assumptions about the distribution the data set follows and that do not require the testing of distribution parameters such as *means* or *variances*.

normal distribution (15) The pattern followed by very many sets of continuous measurements. It is characterized by a symmetric, bell-shaped curve.

notional sampling frame (287) A *sampling frame* in which the *target population* can be defined but the individuals that constitute this population cannot (for example, all individuals who will use a particular hospital emergency room in the next month).

null hypothesis (61) The initial belief that no effect is actually present in the study (for example, that two drugs being compared are equally effective).

observational studies (281) Studies (often called surveys) in which the behavior of the subjects is merely observed and no attempt is made to impose interventions on them.

observed values (100) The number of individuals actually observed in practice in the various categories of a *categorical* variable.

one-tailed testing (63) A test procedure that evaluates the possibility that a specific *alternative hypothesis* is true.

one-way analysis of variance (194) An *analysis of variance* in which only one *main effect* is tested.

ordinal measurement (2) A measurement in which the individual is located in one of a number of mutually exclusive categories that have an inherent logical order.

paired *t* test (66) A variant of the *t test* used when the test results form logical pairs.

partial correlation (257) The correlation (see *correlation coefficient*) between two continuous variables that remains after the influence of one

or more other variables has been controlled or eliminated.

placebo (156) A sham or inert treatment that is superficially similar to the active treatment(s) being investigated.

Poisson distribution (33) The pattern usually followed by a set of results in which the measurements are counts. It is a special case of the *binomial distribution* in which the number of individuals involved is very large and the chance of one of the two possible outcomes occurring is very small.

population (38) The group of all individuals (usually a very large number) who are the focus of the investigation (for example, all children born to mothers over 35 years of age).

power (165) The chance of successfully rejecting the *null hypothesis* when it is indeed false. The power is given by 1 minus the probability of committing a *type II error*.

power index (174) A numerical value used in the calculation of sample size that incorporates a measure of the *type I* and *type II error* risks that the study planner believes are acceptable. In a comparison of two group means it is the value of the test statistic that results if the study data manifests the *true treatment effect*.

primary question (186) The research question that lies at the heart of an investigation. Answering this question is crucial to the success of an investigation, and the study sample size is calculated to ensure that it can be tested with an acceptable level of *power*.

probability (21) The chance or likelihood of a particular event happening expressed as a proportion of 1, with 0 denoting impossible and 1 absolutely certain.

prospective study (283) A study (also called a cohort or follow-up study) that follows its subjects forward in time from initial exposure (or nonexposure) to some suspected risk factor, to the eventual disease outcome.

protocol (184) A document that rigorously details the objectives of an investigation and how those objectives will be achieved in the investigation.

purposive selection (287) The deliberate selection of a *sample* because it is believed to be representative of the *target population*.

random allocation (154) A procedure for allocating experimental subjects to treatments (or groups) that removes the allocation decision entirely from the control of the investigator, hence eliminating the possibility of investigator-induced *bias*.

random sampling (288) A procedure for selecting a *sample* from a *population* that removes the selection decision entirely from the control of the investigator, hence eliminating the possibility of investigator-induced *bias*.

randomized block design (159) An experimental design in which the subjects are first formed into groups or blocks on the basis of similarity of subjects within a block. The subjects within each block are then *randomly allocated* to the treatments under investigation.

randomized clinical trial (187) A *clinical trial* in which the allocation of subjects to intervention groups or a control group is done randomly (see *random allocation*). A randomized clinical trial (RCT) is the only truly acceptable form of clinical trial.

range (6) The difference between the smallest and largest results in a set of data.

ratio measurement (3) A measurement in which the individual is located on a scale with a true zero point and intervals that have a consistent interpretation (for example, weight in kilograms).

regression line (135) The straight line that best describes the relationship between a *dependent variable* and an *independent variable*. It is obtained by calculating a *least squares fit* of the line to the data.

retrospective study (284) A study (also called a case-control study) that looks backward in time from final disease outcome to potential cause. In it, groups of affected individuals (cases) and unaffected individuals (controls) are compared in the extent of their exposure to some suspected risk factor.

sample (38) The group of individuals (usually a relatively small number) who are actually available for investigation. The sample is selected from the *population* of interest.

sampling fraction (288) The proportion of the *population* that will be included in the study *sample*.

sampling frame (286) An (ideally comprehensive) list of the *sampling units* that constitute the *target population*.

sampling units (286) The individual entities that form the focus of a survey. These are often individual people but might be other entities such as individual hospitals, depending on the objectives of the survey.

secondary question (186) A research question that, while interesting and potentially informative, does not form the central focus of an investigation. It is usually intended to supplement the issues addressed in the *primary question*.

sigma squared, σ^2 (139) The *variance* of a

population. In practice this is usually unknown and must be estimated as the variance of a *sample* (shown by s^2).

sign test (265) A *nonparametric test* used to analyze paired data. The test is based on counting the direction (or sign) of each within-pair difference.

single-blind experiment (156) An experiment in which the experimental subjects cannot distinguish the experimental conditions.

Spearman's rank correlation (273) A *nonparametric* equivalent of the *correlation coefficient* that measures the strength of the relationship between two variables using their rankings rather than the original measurements.

split-unit experiment (205) An experiment (also known as a split-plot, repeated-measures, or nested experiment) in which one of the *main effects* involves between-subjects comparisons while another involves comparisons within the same subject.

standard deviation (7) A measure of the magnitude of the variation present in a set of data. It is obtained by finding the square root of the *variance* and therefore is expressed in conventional measurement units.

standard error (43) A measure of the variability of the *mean* of a *sample* (in other words, the variation in mean values we would see if we collected a number of samples). It is obtained by dividing the *standard deviation* of the sample values by the square root of the sample size.

standardizing (21) Expressing the difference between two values in terms of *standard deviations* rather than the original units of measurement. Standardized values are often denoted by the letter z.

stepwise multiple regression (258) A *multiple regression analysis* in which the *independent* variables are entered into the analysis in a series of steps. At each step the independent variable that explains the largest amount of the so-far-unexplained variation in the *dependent* variable is selected to enter the analysis. This process terminates when the amount of variation explained is not significant.

stratification (289) The splitting of a *population* into a number of subpopulations (or strata).

sum of squares (6) A measure of the total amount of variation present in a set of data. It is obtained by summing the squared deviations between each individual result and the *mean*, and hence is measured in squared units.

systematic sampling (288) A *sample* selection procedure in which every kth member of the *population* under study is selected. The value of *k* is determined by the size of the desired *sampling fraction*.

t **distribution** (45)   A variation of the *Normal distribution* that allows for the fact that, in practice, we must use a *standard deviation* that is usually just a *sample estimate* of the true *population* value and hence is somewhat unreliable.

t **test** (59)   A statistical procedure used to test the equality of the *means* of two *samples*. It assumes that the results follow a *Normal distribution* and that the *variances* of the two samples are equal.

target population (286)   The *population* that a survey is intended to describe or measure.

transformation (275)   The act of replacing a data set with the equivalent values in another measurement system (for example, replacing the original values with their logarithms). Appropriately transforming the values can help counteract violations of basic statistical assumptions. For example, taking logarithms often will make a skewed data set normal.

treatment effect (77)   The presence of differences in mean performance between treatment groups resulting from genuine differences in treatment effectiveness.

true treatment effect (170)   The difference in the mean performance of two treatments that would be observed if the treatments could be administered to every individual in the *population* under study, expressed in *standardized units* (in other words, in terms of the population *standard deviation*).

Tukey's multiple comparison test (88)   Test used as a sequel to a significant *analysis of variance* test, to determine which of several groups are actually significantly different from one another. It has built-in protection against an increased risk of a *type I error*.

two-tailed testing (62)   A test procedure that evaluates the possibility that the general *alternative hypothesis* is true.

type I error (64)   Being misled by the sample evidence into rejecting the *null hypothesis* when it is in fact true.

type II error (65)   Being misled by the sample evidence into failing to reject the *null hypothesis* when it is in fact false.

variable (1)   Any measurement that can take a range of possible values.

variance (6)   A measure of the extent of the variation present in a set of data. It is obtained by taking the average of the sum of squares (dividing the *sum of squares* by the *degrees of freedom*) and hence is measured in squared units.

Wilcoxon signed rank sum test (265)   A *nonparametric test* used to analyze paired data. It is the nonparametric equivalent of the *paired t test*.

within-treatments   variation (78)   Variation between individuals who have been treated identically. This treatment therefore must be solely random.

Yates' correction (103)   An amendment to a *Chi squared statistic* that allows for the fact that some inherent disagreement between (discrete) observed values and (continuous) expected values is inevitable.

Solutions to Problems

___ **CHAPTER 1**

1. a) 0.875, 0.825 b) 1.10, 0.0807, 0.284
2. a) 213.5, 215.0 b)110, 1034.47, 32.16
3. 0.927, 0.193
4. 3.536, 0.3711, 0.609, 3.53

___ **CHAPTER 2**

1. a) yes b) 18 or 31.03% c) 0.549 to 1.305, 5.17%
2. Normal, 93.33%, yes
3. a) 0.0708 b) 0.6688 c) 0.3016
4. a) 0.2282 b) 0.0764 c) 0.6141 d) 0.0367 e) 0.3492
5. $p(0) = 0.0001$, $p(1) = 0.0118$, $p(2) = 0.0527$, etc., $p(6+) = 0.5265$
6. $p(0) = 0.0535$, $p(1) = 0.1567$, $p(2) = 0.2293$, etc.
7. a) $p(0) = 0.430$, $p(1) = 0.383$, $p(2) = 0.149$, etc. b) 0.038
8. $p(0) = 0.031$, $p(1) = 0.108$, $p(2) = 0.187$, etc.

___ **CHAPTER 3**

1. 0.927 ± 0.050, yes, no
2. 213.5 ± 15.03, ± 20.56, ± 27.92
3. 0.875 ± 0.151
4. 3.536 ± 0.158
5. 0.1667 ± 0.0667

___ **CHAPTER 4**

1. unpaired, two tailed, $t = 1.42$, 24 df, NS
2. paired, one tailed, $t = 1.51$, 8 df, NS

3. paired, two tailed, $t = 2.40$, 11 df, $p < .05$
4. unpaired, one tailed, $t = 1.94$, 22 df, $p < .05$
5. paired, two tailed, $t = 1.83$, 9 df, NS
6. unpaired, two tailed, $t = 0.84$, 18 df, NS, 20.6 ± 51.34
7. paired, two tailed, $t = 3.37$, 19 df, $p < .005$, 5.95 ± 3.69
8. unpaired, one tailed, $t = -0.86$, 72 df, NS

CHAPTER 5

1. $F = 3.65$ with 2, 15 df, NS
2. $F = 5.94$ with 3, 36 df, $p < .005$, Tukey's $= 0.598$
3. $F = 4.77$ with 2, 33 df, $p < .05$, Tukey's $= 0.997$
4. Linear renin trend, $F = 6.16$ with 1, 27 df, $p < .05$
5. Between drugs, $F = 24.33$ with 1, 36 df, $p < .005$
   Between routes, $F = 1.14$ with 1, 36 df, NS, yes
6. Linear smoking trend, $F = 16.65$ with 1, 20 df, $p < .005$
7. Linear diet trend, $F = 7.42$ with 1, 44 df, $p < .01$

CHAPTER 6

1. $\chi^2 = 127.06$, 5 df, $p < .001$, no
2. Binomial, $\chi^2 = 6.05$, 3 df (categories combined), NS, yes
3. $\chi^2 = 2.18$, 3 df (categories combined), NS, yes
4. $\chi^2 = 2.83$, 2 df (categories combined), NS, no
5. $\chi^2 = 4.26$, 1 df, $p < .05$, yes
6. $\chi^2 = 1.95$, 1 df, NS, no
7. $\chi^2 = 10.06$, 2 df, $p < .01$, yes
8. a) $\chi^2 = 6.26$, 1 df, $p < .025$, yes
   b) $\chi^2 = 61.83$, 4 df, $p < .001$, yes
9. paired, $\chi^2 = 4.27$, 1 df, $p < .05$, yes

CHAPTER 7

1. $r = -.312$, 12 df, NS
2. $r = -.683$, 11 df, $p < .02$, $r^2 = 0.467$
3. time $= 39.14 - 13.76$ dose, $F = 74.43$ with 1, 8 df, $p < .005$
4. $r = 0.857$, 13 df, $p < .001$, $r^2 = .735$
5. follow up $= -12.16 + 1.24$ initial, $F = 21.64$ with 1, 18 df, $p < .005$, i) 179 ± 73.32
   ii) outside range
6. rate $= 621.28 - 4.25$ age, $F = 18.35$ with 1, 8 df, $p < .005$; at age 25 rate $= 514.95 \pm 153.21$; at age 55 rate $= 387.36 \pm 142.03$

CHAPTER 9

1. two tailed, two means i) 176 in total ii) 106 in total iii) 950 in total
2. one tailed, two proportions i) 138 in total ii) 100 in total

3. one tailed, one mean, 16

4. one tailed, one mean, i) 16 ii) 9

5. two tailed, two means i) 208 in total ii) 126 in total

6. two tailed, two proportions, 672 in total

7. two tailed, two means i) 152 in total ii) 124 in total iii) 92 in total iv) 768 in total v) 3066 in total

8. two tailed, two proportions i) 144 in total ii) 118 in total iii) 88 in total

CHAPTER 11

1. abuse $\quad F = 28.37$ with 1, 12 df, $p < .005$

age $\quad F = 19.68$ with 2, 12 df, $p < .005$

sex $\quad F = 1.22$ with 1, 12 df, NS

abuse × age $\quad F = 4.07$ with 2, 12 df, $p < .05$

abuse × sex $\quad F = .25$ with 1, 12 df, NS

age × sex $\quad F = .07$ with 2, 12 df, NS

abuse × age × sex $\quad F = .04$ with 2, 12 df, NS

2. dose (linear trend) $\quad F = 6.85$ with 1, 18 df, $p < .05$

dose (other) $\quad F = .14$ with 1, 18 df, NS

age $\quad F = 5.59$ with 2, 18 df, $p < .05$

sex $\quad F = .14$ with 1, 18 df, NS

dose × age $\quad F = .20$ with 4, 18 df, NS

dose × sex $\quad F = .41$ with 2, 18 df, NS

age × sex $\quad F = .09$ with 2, 18 df, NS

dose × age × sex $\quad F = .23$ with 4, 18 df, NS

3. sex $\quad F = 1.60$ with 1, 8 df, NS

drug $\quad F = .18$ with 1, 8 df, NS

sex × drug $\quad F = 0.05$ with 1, 8 df, NS (split-unit)

4. missing values $\quad 0,2 = 13.31; 150,4 = 40.98$

dose (linear trend) $\quad F = 148.56$ with 1, 14 df, $p < .005$

dose (other) $\quad F = 9.21$ with 3, 14 df, $p < .005$

5. drug $\quad F = 12.25$ with 1, 22 df, $p < .005$

time $\quad F = 120.60$ with 1, 22 df, $p < .005$

drug × time $\quad F = 32.63$ with 1, 22 df, $p < .005$

(split-unit)

6. drain $\quad F = 15.44$ with 2, 12 df, $p < .005$

age $\quad F = 24.68$ with 1, 12 df, $p < .005$

sex $\quad F = .07$ with 1, 12 df, NS

drain × age $\quad F = 4.67$ with 2, 12 df, $p < .05$

drain × sex $\quad F = .41$ with 2, 12 df, NS

age × sex $\quad F = .84$ with 1, 12 df, NS

drain × age × sex $\quad F = .26$ with 2, 12 df, NS

7. missing values $\quad W2 = .828; B4 = .866$

diet $\quad F = 18.9$ with 2, 4 df, $p < .01$

CHAPTER 12

1. a) $F = 5.72$ with 1, 18 df, $p < .05$

b) $F = .10$ with 1, 17 df, NS

    (deviations from common slope $F = .78$ with 1, 16 df, NS)
    males, adjusted mean $= 252.16$ (unadjusted 262.9)
    females, adjusted mean $= 249.84$ (unadjusted 239.1)

2. Diet $F = 6.42$ with 1, 21 df, $p < .05$
    (deviations from common slope $F = 1.69$ with 1, 20 df, NS)

3. groups $F = .55$ with 2, 20 df, NS
    (deviations from common slope $F = .83$ with 2, 18 df, NS)
    adjusted means smokers $= 4.06$, ex-smokers $= 4.05$, nonsmokers $= 4.27$

4. doses $F = 3.69$ with 2, 14 df, NS
    (deviations from common slope $F = .51$ with 1, 13 df, NS)

5. sex $F = 1.31$ with 1, 20 df, NS
    abuse $F = 24.33$ with 1, 20 df, $p < .005$

CHAPTER 14

1. Mann-Whitney $U = 24$, $p < .05$

2. Wilcoxon signed rank sum $= 21.5$, NS

3. Friedman's test, $\chi^2 = 9.05$, 2 df, $p < .025$

4. Kruskal-Wallis test, $\chi^2 = 5.07$, 2 df, NS

5. Spearman's $r = .863$, $p < .01$

Index